T0305550

Multiaxial Notch Fracture and Fatigue

This book presents the unified fatigue life prediction equation for low/medium/high cycle fatigue of metallic materials relevant to plain materials and notched components. The unified fatigue life prediction equation is the Wöhler equation, in which the "stress-based intensity parameter" is calculated based on the linear-elastic analysis.

A local approach for the static fracture analysis for notched components is presented based on the notch linear-elastic stress field. In the local approach, a stress intensity parameter is taken as a stress-based intensity parameter. Experimental verifications show that the local approach is also suited for the static fracture analysis for notched components made of ductile materials.

The book is also concerned with a material failure problem under the multiaxial stress states. A concept of the material intensity parameter is introduced in this book. It is a material property parameter that depends on both Mode-I fracture toughness and Mode-II (or Mode-III) fracture toughness and the multiaxial parameter to characterize the variation of the material failure resistance (notch fracture toughness) with the multiaxial stresses states. The failure condition to assess mixed-mode fracture of notched (or cracked) components is stated as the stress-based intensity parameter being equal to the material intensity parameter.

With respect to the traditional S-N equation, a similar S-N equation is presented and verified to have high accuracy.

This book will be of interest to professionals in the field of fatigue and fracture for both brittle and ductile materials.

Multiaxial Notch Fracture and Fatigue

Xiangqiao Yan

CRC Press is an imprint of the
Taylor & Francis Group, an **informa** business

First edition published 2023
by CRC Press
6000 Broken Sound Parkway NW, Suite 300, Boca Raton, FL 33487-2742

and by CRC Press
4 Park Square, Milton Park, Abingdon, Oxon, OX14 4RN

CRC Press is an imprint of Taylor & Francis Group, LLC

ISBN: 978-1-032-41188-0 (hbk)
ISBN: 978-1-032-41194-1 (pbk)
ISBN: 978-1-003-35672-1 (ebk)

DOI: 10.1201/9781003356721

Typeset in Times
by Typesetter Apex CoVantage, LLC

Contents

Preface

"The unified prediction equation for a low/medium/high cycle fatigue of metallic materials, from plain materials to notched components" was originally to be the title of this book. The publisher suggested a tighter title: *Multiaxial Notch Fracture and Fatigue*.

This is a book on notch fatigue and fracture. However, why was the title of this book originally written by the author to be *"the unified prediction equation for a low/medium/high cycle fatigue of metallic materials, from plain materials to notched components"*? Obviously, the author wanted to use such a "book title" to attract one's attention. In fact, the work of this "book title" indeed attracted one's attention. Editor-in-Chief of Engineering Fracture Mechanics Meinhard Kuna said: "I do believe these results could be better considered by another journal, which is devoted to FATIGUE". A reviewer of this work said: "This work is very valuable and interesting in the fatigue life during low/medium/high cycle fatigue, which should be published."

The author considers that the work of this "book title" is outstanding. However, the work of this "book title" is so difficult to publish in relevant journals that the outstanding work has not been published in relevant journals up to March 12, 2022.

The work of this "book title" was submitted to a relevant journal in July 2019. The coeditor of the journal rejected this work in the unruly form ("不守规矩" in Chinese). For example, the author submitted the following two papers to the journal:

A unified lifetime estimation equation for a low/medium/high cycle fatigue of metallic materials under uniaxial and multiaxial loading—Part 1: plain materials

A unified lifetime estimation equation for a low/medium/high cycle fatigue of metallic materials under uniaxial and multiaxial loading—Part II: notched materials

The co-editor rejected the two papers with the following comment: "The journal does not publish the separate papers numbered as part I, part II and so on. One paper should be self-consistent and complete."

Further details how the co-editor rejected the work of this "book title" are given in the author's Blog (http://blog.sciencenet.cn/u/yanxq19591015) 杂谈："学术期刊不是个人菜园子"、规矩、评议 . . .

A reviewer who reviewed the work of this "book title" rejected this work using his concocted errors, including his concoct English word errors. The details are listed in the author's Blog (http://blog.sciencenet.cn/u/yanxq19591015) 审稿人臆造错误信息评议科技论文，你见过否？

A main contribution in this book is:

A Methodology of Fatigue and Fracture Failure Analysis for Notched Components Made of Brittle and Ductile Materials by Stress-based Intensity Parameter Calculated by Linear-Elastic Analysis.

Obviously, here, many field people have a suspicion that fatigue and fracture failure analysis for notched components made of ductile materials is performed by a stress-based intensity parameter calculated by linear-elastic analysis. It is natural. Susmel and Taylor [Susmel, L. and Taylor, D., The Theory of Critical Distances [TCD] estimate the static strength of notched samples of Al6082 loaded in combined tension and torsion. Part II: Multiaxial static assessment, *Engineering Fracture Mechanics* 77 (2010) 470–478.] said "it is much more difficult to understand why the *linear-elastic TCD* is so accurate also in estimating static strength in those situations in which notched materials undergo large-scale plastic deformations before final breakage occurs". An explanation by the author is that, "in those situations in which notched materials undergo large-scale plastic deformations before final breakage occurs", if their failure analysis can be well performed by a Stress-based Intensity Parameter (S-BIP), the Stress-based Intensity Parameter can be calculated by Linear-Elastic Analysis without needing time-consuming Elastic-Plastic Analysis. The author considers this an outstanding idea, which originated from a quick thought occurring on one day of the beginning of the year of 2019. [On one day of the beginning of the year of 2019, the author fixed his eyes on the well-known Rambrg–Osgood relationship (see Eqs (3) and (4)) on pages 215, 218 in a book (Luca Susmel, *Multiaxial Notch Fatigue*, Woodhead & CRC, Cambridge (2009)) and, after thinking for a moment, felt deeply that, for the LCF of metallic materials, such a way that a 'stress quantity', which is calculated based on the Linear-Elastic Analysis of the studied component, instead of a 'strain quantity', which is calculated based on the Elastic-Plastic Analysis of the studied component, is taken to be a *mechanical quantity, S,* to establish a relation of the *mechanical quantity, S,* to the fatigue life, *N*, is practicable], occurred in a word file (+stress-2019-3.doc))(2019/7/1:39) (title: A unified lifetime estimation equation for a low/medium/high cycle fatigue of metallic materials under uniaxial and multiaxial loading—Part 1:plain materials)]. Through this book, the author's idea appeared to have been proven, including *low cycle fatigue (LCF) of metallic materials [see Chapters 2 and 3] and extremely low cycle fatigue (ELCF) of metallic materials* [to be published in next paper], including also static fracture analysis for notched components made of ductile materials by the local approach proposed by the author (see Chapters 4 and 5).

The local approach for V-notches [pointed V-notches] proposed in this study [see Chapter 4] is called a *notch shape function* (NSF) approach because notch fracture toughness of different V-notches, denoted by K_{IVc}, are correlated through the notch shape function.

The local approach for sharp V-notches [rounded V-notches] proposed in this study [see Chapter 5] is called a *geometric material characteristic parameter* (GMP) approach because notch fracture toughness of different sharp V-notches, denoted by $K_{IV\rho c}$, are correlated through the geometric material characteristic parameter. In the local approach, the concepts of the critical stress and the critical distance by Taylor [Taylor D., The theory of critical distances: a new perspective in fracture mechanics. Oxford, UK: Elsevier; 2007] are employed. But in the author's local approach, the size of the critical distance is not taken to be a material constant, whereas it is assumed that the critical distances of different notches are different and are connected closely with the variation of notched geometry.

Recently, the author was also concerned with *material failure equation* for *multiaxial stresses state*, including cracked and notched members. This originated from the fact that the multiaxial fatigue limit equation proposed by Liu and Yan [Liu, B.W. and Yan, X.Q., A multiaxial fatigue limit prediction equation for metallic materials, ASME *Journal of Pressure Vessel Technology,* Vol. 142/034501 (2020)] employing the well-known Wöhler equation were both simple in computation form and highly accurate falling within ±10% interval: By using 53 sets of experimental data from the literature (see Figure 1.1 in the Introduction), the experimental verifications showed that the multiaxial fatigue limits predicted by the empirical failure equation are in excellent agreement with those measured experimentally. This accuracy was so exciting that the *empirical failure equation* was extended to the static failure model for the multiaxial stresses state, including cracked materials [Yan, X. Q., An empirical fracture equation of mixed mode cracks, *Theoretical and Applied Fracture Mechanics* 116 (2021) 103146] and notched members (see Chapter 7). By the way, a new concept of the *material intensity parameter* [MIP] is introduced in this study. It is a *material property parameter* that depends on both Mode-I fracture toughness, Mode-II (or Mode-III) fracture toughness and the multiaxial parameter to characterize the variation of the material failure resistance [notch fracture toughness] with the multiaxial stresses state. Thus the empirical failure equation to assess mixed-mode fracture of notched components is in fact a *notch failure condition*, which states that the *stress intensity parameter* [SIP] is equal to the *material intensity parameter* [MIP]. The notch failure equation, which is expressed mathematically to be $G(K_{\mathrm{IV}},K_{\mathrm{IIIV}},K_{\mathrm{Ic}},K_{\mathrm{IIIc}},\sigma_t,\tau_t)=0$ for sharp V-notches under Mode I/III loading, is obviously different from that by the TCD approach, by the volume-based SED approach [Strain Energy Density] and by failure criteria based on the maximum tangential stress [MTS] and the mean-stress (MS), which is expressed mathematically to be $h(K_{\mathrm{IV}},K_{\mathrm{IIIV}},K_{\mathrm{Ic}},\sigma_t)=0$ because of Mode I dominance failure assumption.

Based on the normalized values of the generalized stress intensity factor to failure by Berto and Lazzarin, [Recent developments in brittle and quasi-brittle failure assessment of engineering materials by means of local approaches, Materials Science and Engineering R75 (2014) 1–48], *an empirical equation of predicting the fracture toughness* K_{Ic} *(or* K_{Ivc}*)* was proposed in this study. And its applications given in this work appear to illustrate that the empirical equation has a wide application in the failure assessment analysis for notch components under static (or fatigue) loading.

As mentioned above, many field people have a suspicion that fatigue and fracture failure analysis for notched components made of ductile materials is performed by a stress-based intensity parameter calculated by linear-elastic analysis. In the author's book proposal, the suspicion point was written clearly. For example, *the Wöhler Curve Method is well suitable for the low-cycle fatigue life analysis of metallic materials.* And further the Wöhler Curve Method is well suitable for low/medium/high cycle fatigue of metallic materials, i.e., *a unified prediction equation for low/medium/high cycle fatigue of metallic materials is the well-known Wöhler equation.* The author hereby wishes to thank the reviewers of the book proposal for their suggesting the publisher publishing the book. And the author sincerely wishes to thank the publisher, Joseph Clements, for his accepting to publish the book.

This work is supported from the scientific contributions of many field experts, for example, Prof. L. Susmel, Prof. D. Taylor, Prof. P. Lazzarin, Prof. F. Berto, Prof. M. Zappalorto, Prof. S. Filippi, Prof. F.J. Gómez, Prof. M. Elices, Prof. A.R. Torabi, Prof. M.R. Ayatollahi, Prof. A. Seweryn, Prof. M. Strandberg; the author very much appreciates their contributions.

Finally, the author is very grateful to Ms. Furong Liu of Harbin Institute of Technology for her providing many literature to me.

Xiangqiao Yan (闫相桥)
Professor
Harbin Institute of Technology,
Harbin, P.R. China

Author

Xiangqiao Yan is a professor at Harbin Institute of Technology. His research interests include the fatigue and fracture of engineering materials and structures, and he has published more than 100 papers in international journals.

Abbreviations

NOMENCLATURE

a Notch depth
a_o El Haddad's short crack constant
$2a$ Crack length
b Fatigue strength exponent
c Fatigue ductility exponent
d Hole diameter
k Stress concentration factor
$k_{t,net}$ Stress concentration factor referred to the net area
k_e Stress intensity parameter (SIP)
k_R Material intensity parameter (MIP)
k^* Stress concentration factor eigenvalue
k^* Geometric material characteristic parameter
n Strain hardening exponent
n' Cyclic strain hardening exponent
r_n Notch root radius
t Time
t Plate thickness
A_0 Constant in S-N equation under torsional loading
A_1 Constant in S-N equation under tension loading
C_0 Constant in S-N equation under torsional loading
C_1 Constant in S-N equation under tension loading
D Notch depth
E Young's modulus
F LEFM geometrical factor
K Strength coefficient
K' Cyclic strength coefficient
K_I, K_{II}, K_{III} Stress intensity factors for Mode I, Mode II and Mode III cracks
K_{Ic}, K_{IIc}, K_{IIIc} Fracture toughness for Mode I, Mode II and Mode III cracks
K_{Iv}, K_{IIv}, K_{IIIv} Notch stress intensity factors for Mode I, Mode II and Mode III V-notches
K_{Ivc}, K_{IIvc}, K_{IIIvc} Notch fracture toughness for Mode I, Mode II and Mode III V-notches
K_{IU}, K_{IIU}, K_{IIIU} Notch stress intensity factors for Mode I, Mode II and Mode III U-notches
K_{IUc}, K_{IIUc}, K_{IIIUc} Notch fracture toughness for Mode I, Mode II and Mode III U-notches
K_c Fracture toughness
L Material characteristic length
L_c Critical distance

L_c^*	Geometric material characteristic parameter
L_T	Material characteristic length under Mode III loading
M_t	Torque
N_f	Number of cycles to failure
$N_{f,e}$	Estimated number of cycles to failure
R	Load ratio ($R = \sigma_{\text{min}} / \sigma_{\text{max}}$)
2α	Opening angle
$\lambda_1, \lambda_2, \lambda_3$	Mode I, II, III Williams' eigenvalues for stress distribution at V-notches
ξ	Multiaxial parameter
σ_e	von Mises equivalent stress
σ_{eo}	Stress intensity parameter for multiaxial fatigue limit analysis
$\sigma_{e,a}$	Amplitude of the von Mises equivalent stress
$\sigma_{kk,a}$	Amplitude of the first invariant of stress tensor
σ_o, τ_o	Material tension and torsion fatigue limits
σ_t, τ_t	Material tension and torsion critical stresses
σ_a, τ_a	Amplitude of bending (or tension) stress and torsion stress
σ_R	Material intensity parameter for multiaxial fatigue limit analysis
σ_{UTS}	Ultimate tensile stress
ρ	Notch radius
ρ^*	Geometric material characteristic parameter

ACRONYMS

BEM	Boundary Element Method
E.R. (%)	Error range
FEM	Finite Element Method
LEFM	Linear Elastic Fracture Mechanics
GMP	Geometric material characteristic parameter
M.A.E. (%)	Mean absolute error
M.E. (%)	Mean error
MIP	Material intensity parameter
NSF	Notch shape factor
N-SIF	Notch-stress intensity factor
PM	Point Method
R.E. (%)	Relative error
SED	Strain energy density
SIF	Stress intensity factor
SIP	Stress intensity parameter
TCD	Theory of Critical Distances

1 Introduction

Such an idea described by a sentence: "On one day of the beginning of the year of 2019, the author fixed his eyes on the well-known Rambrg–Osgood relationship (see eqns (3) and (4)) on Pages 215, 218 in a book (Luca Susmel, Multiaxial notch fatigue: From Nominal to Local Stress/Strain Quantities, Woodhead & CRC, Cambridge (2009) and, after thinking for a moment, felt deeply that, for the LCF of metallic materials, such a way that a 'stress quantity', which is calculated based on the linear elastic analysis of the studied component, instead of a 'strain quantity', which is calculated based on the elastic-plastic analysis of the studied component, is taken to be a *mechanical quantity, S*, to establish a relation of the *mechanical quantity, S*, to the fatigue life, *N*, is practicable," occurred in a word file (+stress-2019-3.doc)) (2019/7/1:39)(title: A unified lifetime estimation equation for a low/medium/high cycle fatigue of metallic materials under uniaxial and multiaxial loading—Part 1:plain materials). This, in fact, is a key of the research work: "the unified fatigue life prediction equation for a low/medium/high cycle fatigue of metallic materials, from plain materials to notched materials" reported in this book. The further explanation of the key is seen in Appendix A.

In an attempt to show that the unified fatigue life prediction equation for a low/medium/high cycle fatigue of metallic materials is used in notched metallic materials, a finding by Yang et al. (Xiaoguang Yang, Jingke Wang and Jinlong Liu, High temperature LCF life prediction of notched DS Ni-based super-alloy using critical distance concept, International Journal of Fatigue 33 (2011) 1470–1476) on the critical distance in the local approach dealing with notch failure problem deeply affected the author's research on notch fatigue and further notch fracture by using the local approach. The finding by Yang et al. is that the critical distances of different notches are different and are connected closely with the variation of notched geometry. Yang's idea on the critical distance is different from that in the usually used local approach in which the critical distance is taken as a material constant.

The book is also concerned with material failure equation at multiaxial stresses state, including cracked and notched materials. This originated from the fact that the multiaxial fatigue limit equation proposed by Liu and Yan [1] by using the well-known Wöhler equation and the multiaxial fatigue life equation proposed by Liu and Yan [2] was both simple in computation and high in accuracy falling within the ±10% interval: By using 53 sets of experimental data from the literature (see Figure 1.1), experimental verifications given in Ref. [1] showed that the multiaxial fatigue limits predicted by the empirical equation are in excellent agreement with those measured experimentally. This accuracy was so exciting that the empirical equation was extended to static failure model at the multiaxial stresses state, including cracked and notched materials Refs [3,4].

In order to let the reader fully understand the main contents of this book, a general description on the main contents is given in this introduction.

DOI: 10.1201/9781003356721-1

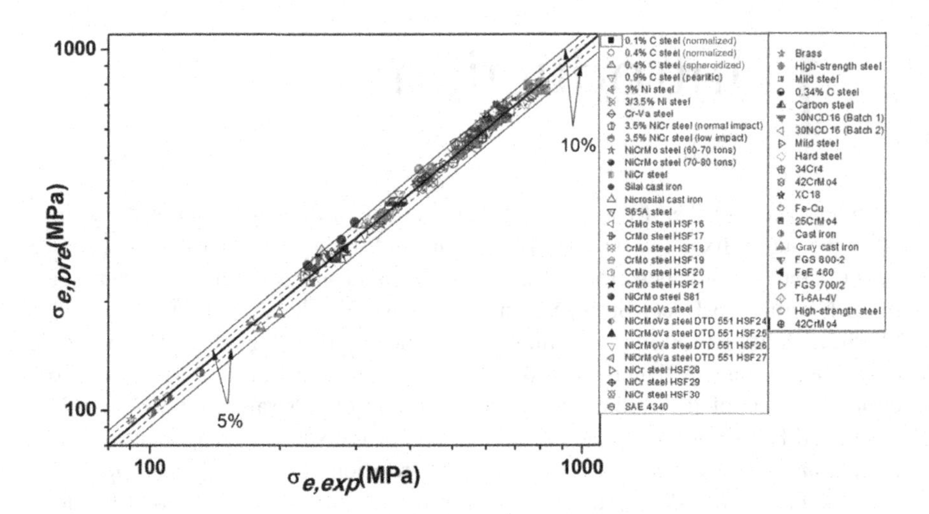

FIGURE 1.1 Comparison of experimental and predicted results of multiaxial fatigue limits.

Source: (Data from liu and Yan [1])

1.1 A BRIEF DESCRIPTION OF THE MULTIAXIAL FATIGUE LIMIT EQUATION

The multiaxial fatigue life equation proposed by Liu and Yan [1] was

$$\sigma_{eo} = \sigma_o^{\xi} (\sqrt{3}\tau_o)^{(1-\xi)} \tag{1}$$

where the **stress intensity parameter (SIP)** at the multiaxial stresses state is taken to be the well-known von Mises effective stress σ_{eo}, which is defined at tension-torsion or bending-torsion loading to be

$$\sigma_{eo} = \sqrt{(\sigma_a)^2 + 3(\tau_a)^2} \tag{2}$$

where σ_a and τ_a are the amplitudes of bending and torsion stresses, respectively. σ_o and τ_o in Eq. (1) are the tension and torsion fatigue limit, respectively, and the multiaxial parameter ξ is defined to be

$$\xi = \frac{\sigma_{kk}}{\sigma_{eo}} \tag{3}$$

σ_{kk} is the amplitude of the first invariant of stress tensor. It is evident that for the axial and pure torsion fatigue conditions, the values of the multiaxial parameter ξ are equal to 1 and 0, respectively.

From Eqs.(1) to (3), obviously, Eq.(1) is simplified, at the axial and pure torsion fatigue conditions, to be

$$\sigma_a = \sigma_o \tag{4}$$

and

$$\tau_a = \tau_o \tag{5}$$

From Eq. (1), here, the **material intensity parameter (MIP)** at the multiaxial stresses state, denoted by σ_R, is introduced as

$$\sigma_R = \sigma_o^{\xi}(\sqrt{3}\tau_o)^{(1-\xi)} \tag{6}$$

Thus, the multiaxial fatigue limit equation (1) is rewritten to be

$$\sigma_{eo} = \sigma_R \tag{7}$$

1.2 AN EXTENSION OF THE MULTIAXIAL FATIGUE LIMIT EQUATION TO MIXED MODE CRACKS

In this section, an extension of the multiaxial fatigue limit equation proposed in Ref. [1] to mixed mode cracks is briefly described.

For Mode I/II cracks, the failure condition is written as

$$K_e = K_R \tag{8}$$

where the **stress intensity parameter,** K_e, is defined as

$$K_e = K_e(K_I, K_{II}) = \sqrt{(K_I)^2 + 3(K_{II})^2} \tag{9}$$

which is the same as that of the von Mises effective stress at the multiaxial stresses state, the **material intensity parameter,** K_R, is expressed as

$$K_R = K_R(K_{Ic}, K_{IIc}, \xi) = (K_{Ic})^{\xi}(\sqrt{3}K_{IIc})^{1-\xi} \tag{10}$$

which is similar to formula (6), and multiaxial parameter ξ in (10) is defined as

$$\xi = K_I / K_e \tag{11}$$

which is similar to formula (3).

In fracture mechanics, K_e and K_R are usually called the crack extension force and the crack extension resistance, respectively.

For Mode I cracks, $\xi = 1$, the empirical fracture equation (8) is simplified as

$$K_I = K_{Ic} \tag{12}$$

while for Mode II cracks, $\xi = 0$, the empirical fracture equation (8) is simplified as

$$K_{II} = K_{IIc} \tag{13}$$

For Mode I/III cracks, obviously, the serious equations similar to (8) to (13) are obtained easily and not listed here.

Experimental verifications reported in Ref. [3] showed that the empirical failure equation (8), which is on the basis of the **stress intensity factors** in the linear elastic fracture mechanics, can be used to well perform the fracture analysis for mixed mode cracks made of both brittle materials (for example, soda-lime glass by Shetty et al. by the Disk Test [7] (see Table 1.1) and plastic materials (for example, I/II fracture behavior at room temperature of HY 130 steel tempered at 350°C reported by Maccagno and Knott [8] and Ductile fracture in HY100 steel under mixed I/II reported by Bhattachrjee and Knott [9] (see Tables 1.2 and 1.3).

TABLE 1.1
Comparison of the Stress Intensity Parameter with the Material Intensity Parameter (Soda-Lime Glass)

No.	K_I [7] (MPam$^{0.5}$)	K_{II} [7] (MPam$^{0.5}$)	ξ	K_e (MPam$^{0.5}$)	K_R (MPam$^{0.5}$)	R.E. (%)
1	0.73	0	1	0.73	0.73	0
2	0.715	0.13	0.9538	0.75	0.76	0.9
3	0.67	0.255	0.8349	0.8	0.83	3.1
4	0.63	0.371	0.7001	0.9	0.92	1.8
5	0.575	0.48	0.5688	1.01	1.01	0.2
6	0.5	0.56	0.4582	1.09	1.1	0.9
7	0.425	0.638	0.359	1.18	1.19	0.3
8	0.348	0.71	0.2723	1.28	1.27	−0.8
9	0.349	0.72	0.2695	1.3	1.27	−1.9
10	0.25	0.77	0.1842	1.36	1.36	−0.1
11	0.252	0.78	0.1834	1.37	1.36	−1.3
12	0.174	0.8	0.1246	1.4	1.42	1.6
13	0.1	0.823	0.07	1.43	1.48	3.4
14	0.11	0.87	0.0728	1.51	1.48	−2.4
15	0	0.9	0	1.56	1.56	0

Note: (*a*) $K_{Ic} = 0.73$ [7]. (*b*) $K_{IIc} = 0.9$ [7]. (*c*) M.E. = 0.4%, M.A.E. = 1.2% and E.R. = [−2.4,3.4]%.

TABLE 1.2
Comparison of the Stress Intensity Parameter with the Material Intensity Parameter (HY 130 Steel Tempered at 350°C)

No.	K_I [8] (MPam$^{0.5}$)	K_{II} [8] (MPam$^{0.5}$)	ξ	K_e (MPam$^{0.5}$)	K_R (MPam$^{0.5}$)	R.E. (%)
1	150.06	40.05	0.9077	165.32	167.92	1.6
2	117.72	67.81	0.7079	166.3	165.59	−0.4
3	82.21	82.36	0.4993	164.64	163.2	−0.9
4	52.58	90.93	0.3167	166.03	161.13	−3
5	23.34	90.97	0.1465	159.28	159.23	0
6	0	88.81	0	153.82	157.61	2.5

Note: $K_{Ic} = 169$ MPam$^{0.5}$, $K_{II} = 91$ MPam$^{0.5}$ [8]

TABLE 1.3
Comparison of the Stress Intensity Parameter with the Material Intensity Parameter (HY100 Steel)

No.	K_I [9] (MPam$^{0.5}$)	K_{II} [9] (MPam$^{0.5}$)	ξ	K_e (MPam$^{0.5}$)	K_R (MPam$^{0.5}$)	R.E. (%)
1	180.19	59.57	0.8678	207.64	195.66	−5.8
2	140.81	66.5	0.774	181.92	187.49	3.1
3	113.36	74.95	0.6578	172.34	177.82	3.2
4	117.16	77.46	0.6578	178.12	177.82	−0.2
5	73.32	80.8	0.4641	157.99	162.81	3.1
6	72.34	79.71	0.4641	155.87	162.81	4.5
7	50.4	83.31	0.3298	152.84	153.15	0.2
8	26.41	87.29	0.172	153.49	142.54	−7.1

Note: K_{Ic} = 207.8 MPam$^{0.5}$, K_{IIc} = 76.1 MPam$^{0.5}$

1.3 AN EXTENSION OF THE MULTIAXIAL FATIGUE LIMIT EQUATION TO MODE I/III V-NOTCHES

On the basis of the linear elastic stress field ahead of V-notches, by using the concepts of the critical stress and the critical distance by Taylor [18], and by using the assumption that the critical distances of different notches are different and are connected closely with the variation of notched geometry, the local stress field failure model proposed recently in Ref [5] has been proven to be both simple in computation and high in accuracy (see Table 1.4 and Table 1.5) for fracture analysis of V-notched components made of both *brittle* materials (for example, Plexiglas reported by Seweryn [10] by V-notch DENT specimens (see Figure 1.2)) and ***plastic*** materials (for example, duraluminum by Seweryn [10] by V-notch DENT specimens).

From the local stress field failure model proposed recently in Ref [5], the V-notch fracture toughness, denoted by K_{IVc}, is expressed mathematically to be

$$K_{IVc} = H_I(K_{Ic}, \sigma_o, \alpha) \tag{14}$$

where σ_o and α are the critical stress and notch angle, respectively. For Mode III V-notches, similarly, the fracture toughness, denoted by K_{IIIVc}, is expressed mathematically to be

$$K_{IIIVc} = H_{III}(K_{IIIc}, \tau_o, \alpha) \tag{15}$$

Because the notch stress intensity factors K_{Iv} and K_{IIIv} have different units, the **stress intensity parameter** employed here for Mode I/III V-notches is a nondimensional von Mises definition quantity:

$$k_e = \sqrt{(\bar{k}_{\mathrm{Iv}})^2 + 3(\bar{k}_{\mathrm{IIIv}})^2} \tag{16}$$

TABLE 1.4
Fracture Test Data and the Evaluation Calculation Results (Plexiglas) (V-Notch DENT Specimens)

β (degrees)	K_{vc}(exp) ($MPam^{1-\lambda}$)	λ	K_{vc}(cal) ($MPam^{1-\lambda}$)	R.E. (%)
10	1.866	0.5004	1.866	0
20	1.851	0.5039	1.9263	4.1
30	2.167	0.5117	2.0634	−4.8
40	2.436	0.5292	2.375	−2.5
50	3.059	0.5642	3.059	0
60	4.347	0.6161	4.36	0.3
70	8.861	0.6978	7.235	−18.4

Note: σ_o = 13.0 MPa, K_{Ic} = 1.857 MPam$^{0.5}$

TABLE 1.5
Fracture Test Data and the Evaluation Calculation Results (Duraluminum) (V-Notch DENT Specimens)

β (degrees)	K_{vc} (exp)	λ	K_{vc}(cal)	R.E.(%)
10	53.51	0.5004	53.51	0
20	57.1	0.5039	54.97	3.7
30	60.53	0.5117	58.23	3.8
40	66.34	0.5292	65.38	1.5
50	80.15	0.5642	80.15	0
60	102	0.6161	106.16	−4.1
70	150.44	0.6978	156.91	−4.3

Note: σ_o = 183.7 MPa, K_{Ic} = 53.28 MPam$^{0.5}$

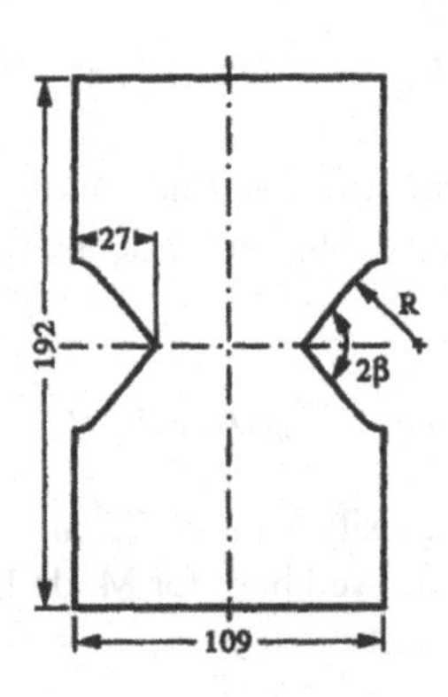

FIGURE 1.2 The geometric size of V-notch DENT specimens.

Source: (Data from Seweryn [14])

where

$$\overline{k}_{\text{Iv}} = \frac{K_{\text{Iv}}}{K_{\text{Ivc}}}, \qquad \overline{k}_{\text{IIIv}} = \frac{K_{\text{IIIv}}}{K_{\text{IIIvc}}} \tag{17}$$

and the corresponding **material intensity parameter** becomes

$$k_R = k_R(\overline{k}_{IVc}, \overline{k}_{IIIVc}, \xi) = (\sqrt{3})^{1-\xi} \tag{18}$$

which is obtained by replacing K_{Ic} and K_{IIIc} in Eq. (10) by $\overline{k}_{IVc} = 1$ and $\overline{k}_{IIIVc} = 1$, respectively. For Mode I/III V-notches, thus, the fracture condition can be written as

$$k_e = k_R \tag{19}$$

Multiaxial parameter ξ is here defined as

$$\xi = \overline{k}_{\text{Iv}} / k_e \tag{20}$$

For Mode I V-notches, $\xi = 1$, the fracture condition (19) becomes

$$K_{\text{Iv}} = K_{\text{Ivc}} \tag{21}$$

while for Mode III V-notches, $\xi = 0$, the fracture condition (19) becomes

$$K_{\text{IIIv}} = K_{\text{IIIvc}} \tag{22}$$

The fracture conditions (21) and (22), obviously, are the same as K criterions for pure Mode I cracks and pure Mode III cracks, respectively.

From (14) and (21), for Mode I V-notches, the fracture condition can be expressed mathematically to be

$$G_I(K_{IV}, K_{Ic}, \sigma_o, \alpha) = 0 \tag{23}$$

From (15) and (22), similarly, for Mode III V-notches, the fracture condition can be expressed mathematically to be

$$G_{III}(K_{IIIV}, K_{IIIc}, \tau_o, \alpha) = 0 \tag{24}$$

From (16) to (20), for Mode I/III V-notches, the failure condition can be expressed mathematically to be

$$S(K_{IV}, K_{IIIV}, K_{IVc}, K_{IIIVc}) = 0 \tag{25}$$

From (14), (15) and (25), the failure condition for Mode I/III V-notches can be also expressed mathematically to be

$$G(K_{IV}, K_{IIIV}, K_{Ic}, K_{IIIc}, \sigma_o, \tau_o, \alpha) = 0 \tag{26}$$

From (23), it is seen that the failure condition for Mode I V-notches is controlled by the fracture toughness K_{Ic} and the critical stress σ_o. From (24), similarly, the failure condition for Mode III V-notches is controlled by the fracture toughness K_{IIIc} and the critical stress τ_o. While from (26), the failure condition for Mode I/III V-notches is controlled by the fracture toughness, K_{Ic} and K_{IIIc}, and the critical stresses, σ_o and τ_o.

1.4 AN EXTENSION OF THE MULTIAXIAL FATIGUE LIMIT EQUATION TO MODE I/III ROUNDED V-NOTCHES

On the basis of the linear elastic stress field ahead of rounded V-notches with notch angle α, notch radius ρ_o, by using the concepts of the critical stress and the critical distance by Taylor [18] and by using the assumption that the critical distances of different notches are different and are connected closely with the variation of notched geometry, the local stress field failure model proposed recently in Ref [6] has been proven to be both simple in computation and high in accuracy for fracture analysis of rounded V-notched components made of both *brittle* materials (for example, ceramics (SI3N4 and Y-TZP) reported by Gogotsi [11]) and *plastic* materials (for example, center U-notch specimens made of Al 7075-T6 and Al 6061-T6 reported in Ref.[12], center circular hole specimens made of pure copper-T2 [13], and a center circular hole specimens made of structure steel (Q345) [14]).

From the local stress field failure model proposed recently in Ref [6], the rounded V-notch fracture toughness, denoted by $K_{IV\rho c}$, is expressed mathematically to be

$$K_{IV\rho c} = E_I(K_{Ic}, \rho_o, \sigma_o, \alpha) \tag{27}$$

where σ_o and α are the critical stress and notch angle, respectively. For Mode III rounded V-notches, similarly, the fracture toughness, denoted by $K_{IIIV\rho c}$, is expressed mathematically to be

$$K_{IIIV\rho c} = E_{III}(K_{IIIc}, \rho_o, \tau_o, \alpha) \tag{28}$$

As done for V-notches in Section 3, serious equations parallel to Eqs. (16) to (22) for rounded V-notches can be obtained and are not listed here.

For Mode I rounded V-notches, the fracture condition can be expressed mathematically to be

$$J_I(K_{IV\rho}, K_{Ic}, \rho_o, \sigma_o, \alpha) = 0 \tag{29}$$

which is similar to Eq.(23).

For Mode III rounded V-notches, similarly, the fracture condition can be expressed mathematically to be

$$J_{III}(K_{IIIV\rho}, K_{IIIc}, \rho_o, \tau_o, \alpha) = 0 \tag{30}$$

which is similar to Eq. (24).

For Mode I/III rounded V-notches, the fracture condition can be expressed mathematically to be

$$J(K_{IV\rho}, K_{IIIV\rho}, K_{Ic}, K_{IIIc}, \rho_o, \sigma_o, \tau_o, \alpha) = 0 \quad (31)$$

which is similar to Eq. (26).

1.5 THREE COMMENTS ON THE EMPIRICAL FAILURE EQUATION

Three comments given here are as follows:

(*a*) From Refs [1,3,4], it was found that the empirical failure equation at the multiaxial stresses state is not only simple in computation but also high in accuracy. What is the reason for it? It is due to the following respects:

(1) The von Mises definition of the **stress intensity parameter** (for example, the definition of the SIP by formula (2) for multiaxial fatigue limit analysis) is a proper quantity to characterize the failure force.

(2) A definition of the **material intensity parameter** (for example, the definition of the MIP by formula (6) for multiaxial fatigue limit analysis) is a proper material quantity to characterize the material failure resistance.

(3) The empirical failure equation at multiaxial stresses state (for example, Eq. (7) for multiaxial fatigue limit analysis) originated from the multiaxial fatigue limit equation (1), which was both simple in computation and high in accuracy falling within ±10% interval [1].

(4) The empirical failure equation at the multiaxial stresses state is largely different from the failure equation obtained from, for example for fracture analysis of Mode I/II cracks, the maximum $\sigma_{\theta\theta}$ theory originally proposed by Erdogan and Sih [15], and the strain energy density factor S theory by Sih [16]. By using the two well-known fracture criteria [15,16] of mixed cracks, the failure condition can be expressed as

$$f(K_I, K_{II}, K_{Ic}) = 0 \quad (32)$$

that is to say that fracture of the Mode I/II cracks is controlled by the fracture toughness K_{Ic} of Mode I cracks. For the materials (for example, high- or medium-strength steels as well as of nodular cast iron reported by Gao et al.[17]) whose crack extension resistance varies largely with the increase of K_{II}/K_I ratio, the failure load calculated by using the failure equation (32) is obviously different from that measured experimentally near to the region of pure II cracks.

For rounded V-notches, according to the two well-known local approaches (TCD, Theory of Critical Distance [18,19], the volume-based SED approach [20–22]), the failure condition can be expressed as

$$h(K_{IV\rho}, K_{IIV\rho}, K_{Ic}, \sigma_o) = 0 \quad (33)$$

For brittle materials, the critical stress σ_o is taken to be the ultimate tensile stress, σ_t. Obviously, the failure condition (33) is different from Eq. (31).

(*b*) The material failure analysis, by using the empirical failure equation, needs the material shear behavior to be known prior. But from previous studies [29–31] on the material *shear* behavior (for example, K_{IIc}), it is found that characterization of *tensile* properties is straightforward but finding *shear* properties is not as easy as tensile ones. The main difficulty in finding *shear* properties is to reach the "true" pure-shear in the test specimen or nearly pure-shear stress state at the "shear plane". By using the failure experimental data at the multiaxial stress state near to the "shear plane", thus, an approach to determine the material *shear* behavior was proposed in Ref. [3] on the basis of the multiaxial fatigue limit equation. As an example, here, an approach to determine the material torsion fatigue limit is given by using test data of tension/torsion fatigue and by using the multiaxial fatigue limit equation.

It is assumed here that there are M sets of tension/torsion fatigue test data near to pure torsion region:

$$\sigma_{ai}, \tau_{ai} \qquad (i = 1, 2, \cdots, M)$$

From formulas (2) and (3), one can obtain

$$\sigma_{eoi}, \xi_i \qquad (i = 1, 2, \cdots, M)$$

From equations (6) and (7), further, one can obtain

$$\tau_{oi} = \frac{1}{\sqrt{3}} \left\{ \frac{\sigma_{eoi}}{(\sigma_o)^{\xi_i}} \right\}^{\frac{1}{1-\xi_i}} \qquad (i = 1, 2, \cdots, M) \tag{34}$$

Then the mean value of τ_{oi} calculated by using formula (34) can be taken an approximate estimation of τ_o.

Experimental verifications given here are as follows:

By using 53 sets of tension/torsion fatigue limit test data of metallic materials collected in Ref.[19] (see Figure 1.1), as mentioned above, experimental verifications given in Ref.[1] showed that the multiaxial fatigue limits predicted by the multiaxial fatigue limit equation are in excellent agreement with those measured experimentally. Here, a part of tension/torsion fatigue limit test data collected in Ref.[19] is used to check the accuracy of the prediction equation of the material torsion fatigue limit (34) proposed here (see Table 1.6), in which HSF1, HSF2, all the way to HSF7, are the number of materials used in Ref.[19]. In Table 1.6, $R.E.(1)$ and $R.E.(2)$ are defined as: $R.E.(1) = \frac{(\sigma_R - \sigma_{eo}) \times 100}{\sigma_{eo}}$ and $R.E.(2) = \frac{(\tau'_o - \tau_o) \times 100}{\tau_o}$, respectively.

From the calculation results of $R.E.(1)$ and $R.E.(2)$ shown in Table 1.6, it is seen that

TABLE 1.6
Comparison of the Stress Intensity Parameter with the Material Intensity Parameter and Comparison of Experimental and Predicted Material Torsion Fatigue Limit

Materials	σ_a (MPa)	τ_a (MPa)	σ_{eo} (MPa)	ξ	σ_R (MPa)	R.E. (1) (%)	τ_o' (cal) (MPa)	R.E. (2) (%)
HSF1	80.9	147.1	267.3	0.30	264.0	−1.2	154.0	1.8
	Note: $\sigma_o = 268.6$ MPa, $\tau_o = 151.3$ MPa							
HSF2	101.7	189.7	343.9	0.30	350.3	1.9	201.6	−2.6
	Note: $\sigma_o = 331.9$ MPa, $\tau_o = 206.9$ MPa							
HSF3	76.9	145.6	263.7	0.30	271.4	2.9	149.6	−4.0
	Note: $\sigma_o = 274.8$ MPa, $\tau_o = 155.9$ MPa							
HSF4	113	208.1	377.7	0.30	396.4	5.0	224.8	−6.7
	Note: $\sigma_o = 352$ MPa, $\tau_o = 240.8$ MPa							
HSF5	104.8	192.4	349.3	0.30	351.7	0.7	203.4	−0.9
	Note: $\sigma_o = 342.7$ MPa, $\tau_o = 205.3$ MPa							
HSF6	134.9	249.8	453.2	0.30	456.8	0.8	264.1	−1.1
	Note: $\sigma_o = 443.4$ MPa, $\tau_o = 267.1$ MPa							
HSF7	134.8	239.3	435.8	0.31	441.1	1.2	253.4	−1.7
	Note: $\sigma_o = 429.1$ MPa, $\tau_o = 257.8$ MPa							

the multiaxial fatigue limit prediction equation reported in Ref.[1] and the prediction equation (34) of the material torsion fatigue limit τ_o proposed here have a high accuracy.

(*c*) From Ref.[3–6], it was found that the failure condition (for example, such as (8) and (19)), which is on the basis of the linear elastic stress field ahead of cracks or notches, can be used to well perform the fracture analysis for cracked and notched *plastic* materials. Why? Susmel and Taylor [23] said "it is much more difficult to understand why the **linear-elastic TCD** is so accurate also in estimating static strength in those situations in which notched materials undergo large-scale plastic deformations before final breakage occurs". Here, the following comments are given:

(1) Such an explanation on **"A practicability of establishing the unified prediction equation for a low/medium/high cycle fatigue of metallic materials"**, which occurred in Ref.[3] (also see the Appendix *A*), is given to perhaps help the reader to understand that it is practical that the fracture analysis for V-notch specimens made of *plastic* materials can be performed by using the local stress field failure model.

(2) Low-cycle fatigue test data of many metallic materials collected in Ref.[24] have proven that it is sure enough that the low-cycle fatigue life analysis of metallic materials is well performed by the well-known Wöhler Curve Method.

(3) The fracture analysis of notched components made of En3B performed by Susmel and Taylor [26] showed that the linear-elastic TCD can be successful in predicting static failures in notched components when the final breakage is preceded by large-scale plastic deformations.
(4) Such a fact that "A similar degree of accuracy was obtained when elasto-plastic stress analysis was used" reported by Susmel and Taylor [26] showed also that it is possible that the local stress field failure model proposed in Ref.[5,6] is used to well perform the fracture analysis for notched components made of *plastic* materials.
(5) The author considers such a fact that "A similar degree of accuracy was obtained when elasto-plastic stress analysis was used" reported by Susmel and Taylor [26] relies on the **stress intensity parameter** used in the TCD being a "stress-based" intensity parameter, not a "strain-based" and "energy-based" intensity parameter. In Appendix *A*, the author stated: at a given loading, "the *stress-based intensity parameter*" calculated using the *elastic-plastic stress analysis* almost is the same as that by the *linear-elastic stress analysis*. From a practical point of view, thus, for the LCF of metallic materials, the "*stress-based intensity parameter*" can be calculated using *the linear-elastic stress analysis* of the mechanical component.

1.6 A COMMENT ON THE UNIFIED PREDICTION EQUATION FOR A LOW/MEDIUM/HIGH CYCLE FATIGUE OF METALLIC MATERIALS (FROM PLAIN MATERIALS TO NOTCHED MATERIALS)

According to comment (c) in Section 5 and the fact that medium/high cycle fatigue life of metallic materials is well calculated by the well-known Wöhler Curve Method, obviously, the unified prediction equation for a low/medium/high cycle fatigue of metallic materials is the well-known Wöhler equation, see Ref.[24]. Note here for metallic materials including different notch geometries, the Wöhler equation is varied with notch geometries, see Ref.[25].

REFERENCES

1 Liu, B.W., Yan, X.Q. A multiaxial fatigue limit prediction equation for metallic materials. *ASME Journal of Pressure Vessel Technology* 142, 034501 (2020).
2 Liu, B.W., Yan, X.Q. A new model of multiaxial fatigue life prediction with influence of different mean stresses. *Int. J. Damage Mech.* 28(9), 1323–1343 (2019).
3 Yan, X.Q. An empirical fracture equation of mixed mode cracks. *Theoretical and Applied Fracture Mechanics* 116, 103–146 (2021).
4 Yan, X.Q. An empirical equation to assess mixed-mode fracture failure of notched components. To be published in *Journal of Testing and Evaluations*.
5 Yan, X.Q. A local approach for fracture analysis of V-notch specimens under mode I loading. To be published in *Engng. Fract. Mech.*

6 Yan, X.Q. A local stress field failure model for sharp notches. To be published in *Engng. Fract. Mech.*
7 Shetty, D.K., Rosenfield, A.R., Duckworth, W.H. Mixed-mode fracture in biaxial stress state: Application of the diametral-compression (Brazilian disk) test. *Engng. Fract. Mech.* 26, 825–840 (1987).
8 Maccagno, T.M., Knott, J.F. The mixed mode I/II fracture behavior of lightly tempered HY 130 steel at room temperature. *Engineering Fracture Mechanics* 41(6), 805–820 (1992).
9 Bhattacharjee, D., Knott, J.F. Ductile fracture in HY100 steel under mixed mode I/II loading. *Acta Metall. Mater.* 42(5), 1747–1754 (1994).
10 Seweryn, A. Brittle fracture criterion for structures with sharp notches. *Engineering Fracture Mechanics* 47, 673–681 (1994).
11 Gogotsi, G.A. Fracture toughness of ceramics and ceramic composites. *Ceram Int* 7, 777–784 (2003).
12 Torabi, A.R., Habibi, R., Hosseini, B.M. On the ability of the equivalent material concept in predicting ductile failure of U-notches under moderate-and large-scale yielding conditions. *Phys Mesomech.* 18, 337–347 (2015).
13 Cai, D.L., Qin, S.H., Gao, L., Liu, G.L., Zhang, K.S. Study on tensile fracture of pure copper plate with a center hole (in Chinese). *Journal of Guangxi University* 41(4), 1178–1186 (2016).
14 Wang, W.Z., Zhang, J.J., Sun, Y.P., Zhang, D.L. An experimental study on structural steel fracture (in Chinese). *Journal of Gansu Sciences* 19(4), 112–115 (2007).
15 Erdogan, F., Sih, G.C. On the crack extension in plates under plane loading and transverse shear. *J. Basic Engng* 85, 519–527 (1963).
16 Sih, G.C. Strain-energy-density factor applied to mixed mode crack problems. *Int. J. Fracture* 10, 305–321 (1974).
17 Gao, H., Wang, Z.Q., Yang, C.S., Zhou, A.H. An investigation on the brittle fracture of K_I-K_{II} composite mode cracks. *Aata Metallurgica Sinica* 15(3), 380–391 (1979).
18 Taylor, D. *The theory of critical distances: A new perspective in fracture mechanics.* Oxford, UK: Elsevier; 2007.
19 Susmel, L. *Multiaxial notch fatigue: From nominal to local stress/strain quantities.* Cambridge: Woodhead & CRC; 2009.
20 Berto, F., Lazzarin, P. Recent developments in brittle and quasi-brittle failure assessment of engineering materials by means of local approaches. *Materials Science and Engineering R* 75, 1–48 (2014).
21 Lazzarin, P., Zambardi, R. A finite-volume-energy based approach to predict the static and fatigue behaviour of components with sharp V-shaped notches. *International Journal of Fracture* 112, 275–298 (2001).
22 Lazzarin, P., Berto, F. Some expressions for the strain energy in a finite volume surrounding the root of blunt V-notches. *International Journal of Fracture* 135, 161–185 (2005).
23 Susmel, L., Taylor, D. The theory of critical distances to estimate the static strength of notched samples of Al6082 loaded in combined tension and torsion. Part II: Multiaxial static assessment. *Engineering Fracture Mechanics* 77, 470–478 (2010).
24 Yan, X.Q. Applicability of the Wöhler curve method for a low/medium/high cycle fatigue of metallic materials. *Submitted to Application in Science and Engineering* for publication.
25 Yan, X.Q. Notch S-N equation for a low/medium/high cycle fatigue of metallic materials. To be published in *Engng. Fract. Mech.*

26 Susmel, L., Taylor, D. On the use of the theory of critical distances to predict static failures in ductile metallic materials containing different geometrical features. *Engineering Fracture Mechanics* 75, 4410–4421, (2008).
27 Gao, Z., Zhao, T., Wang, X., Jiang, Y. Multiaxial fatigue of 16MnR steel. *ASME J Press Vess Technol* 131(2), 021403 (2009).
28 Gao, Z., Qiu, B., Wang, X., Jiang, Y. An investigation of fatigue of a notched member. *International Journal of Fatigue* 32, 1960–1969 (2010).
29 Wang, G.Y., Sun, Z.Q., Xu, J.C. Experimental method study for Mode II rock fracture toughness. *The Chinese Journal of Nonferrous Metals* 9(1), 175–179 (1999).
30 Barnard, D.J., Anderson, I.E. A Shear test method to measure shear strength of metallic materials and solder joints using small specimens. *Scripta Materialia* 40(3), 271–276 (1999).
31 Majzoobi, GH, Khodaee, F. Determination of tensile and shear strengths using a new test specimen. *Advances in Materials and Processing Technologies* 1(1–2), 254–262 (2015).

APPENDIX A: A PRACTICABILITY OF ESTABLISHING THE UNIFIED PREDICTION EQUATION FOR A LOW/MEDIUM/HIGH CYCLE FATIGUE OF METALLIC MATERIALS

In a low/medium/high cycle fatigue regime of metallic materials, does a prediction equation in which the mechanical quantity calculation is on the basis of the *linear-elastic analysis* exist? According to previous wide investigations, the prediction equation is *impossible* to exist: For the HCF of metallic materials, the mechanical quantity calculation is on the basis of the *linear-elastic analysis* of the studied component due to the fact that the studied component is subjected to elastic deformation, while for the LCF of metallic materials, the mechanical quantity calculation is on the basis of the *elastic-plastic analysis* of the studied component due to obvious plastic deformation produced in the studied component. Here, the previously mentioned *impossibility* is, no doubt, reasonable. The viewpoint that, for the LCF of metallic materials, there is large plastic deformation in the materials and therefore it's not reasonable to perform the fatigue life assessment of the materials by using the *linear-elastic analysis*, is a common view.

However, on the basis of the well-known Rambrg–Osgood relationship:

$$\varepsilon = \varepsilon_e + \varepsilon_p = \frac{\sigma}{E}\left(\frac{\sigma}{K}\right)^{\frac{1}{n}} \tag{A-1}$$

where E is Young's modulus, K is the strength coefficient and n is the strain hardening exponent, the author considers that, for the LCF of metallic materials, the way that a "stress quantity", which is calculated based on the *linear-elastic analysis* of the studied component, instead of a "strain quantity", which is calculated based on the *elastic-plastic analysis* of the studied component, is taken to be a *mechanical quantity, S*, to establish a relation of the *mechanical quantity, S,* to the fatigue life, *N*, is practicable. In order to fully indicate the author's viewpoint, here, some experimental data of low/medium/high cycle fatigue of tension–compression fatigue of plain and notched specimens made of 16MnR steel [27,28] were given in Tables 1.A-1 and 1.A-2. Comparison of experimental data and linear fitted curve of tension–compression fatigue of plain specimens in the double-logarithm frame of axes is shown in Figure 1.A-1. From the agreement between experimental data and linear fitted curve shown in Figure 1.A-1, no doubt, low/medium/high fatigue life assessment of plain specimens of the material can be carried out by using the Wöhler curve, i.e., the linear fitting relation of $\log(\sigma_a)$ to $\log(N)$, in which the fitting constants are listed in Table 1.A-3. Similarly, handling can be done for experimental data of tension–compression fatigue of notched specimens made of 16MnR steel (see Figure 1.A-2),

TABLE 1.A-1
Fully Reversed Uniaxial Fatigue Experiments of Plain Specimens Made of 16MnR Steel

No.	Strain Amplitude $\varepsilon_{x,a}$ (%)	Stress Amplitude $\sigma_{x,a}$ (MPa)	Fatigue life N (Cycles)
1	1.999	514.6	280
2	1.501	487.3	400
3	1	459.2	880
4	0.7	418.2	2220
5	0.5	380	4850
6	0.38	351	9200
7	0.3	322.9	17300
8	0.243	303.7	37600
9	0.199	285.1	79000
10	0.18	274.5	148400
11	0.16	261.9	388500
12	0.146	256	1400000

TABLE 1.A-2
Fully Reversed Uniaxial Fatigue Experiments of Notched Specimens Made of 16MnR Steel

No.	Stress Amplitude $\sigma_{x,a}$ (MPa)	Fatigue life N (Cycles)
1	438.97	800
2	345.53	2048
3	308.42	3250
4	255.81	8192
5	202.86	27,000
6	184.52	47,000
7	151.32	180,000
8	140.64	232,000
9	131.07	642,200

TABLE 1.A-3
Linear Fitted Constants in S-N Equations of 16MnR Steel

Specimen	A	C	R^2
Plain specimen	–0.08941	2.91288	0.96211
Notched specimen	–0.18268	3.1392	0.98329

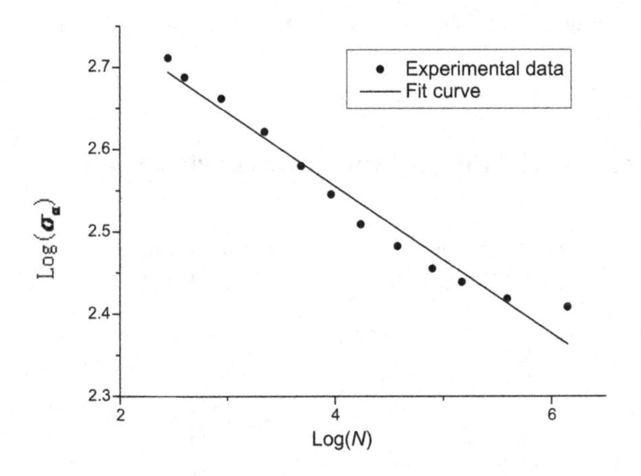

FIGURE 1.A-1 Comparison of experimental data and linear fitted curve of tension–compression fatigue of plain specimens made of 16MnR steel in the double-logarithm frame of axes.

Source: (Data from Gao, *etal* [27])

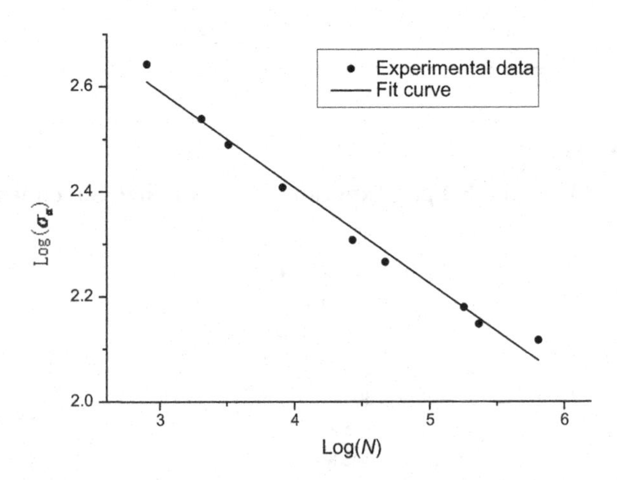

FIGURE 1.A-2 Comparison of experimental data and linear fitted curve of tension–compression fatigue of notched shaft specimens made of 16MnR steel in the double-logarithm frame of axes.

Source: (Data from Gao, *etal* [28])

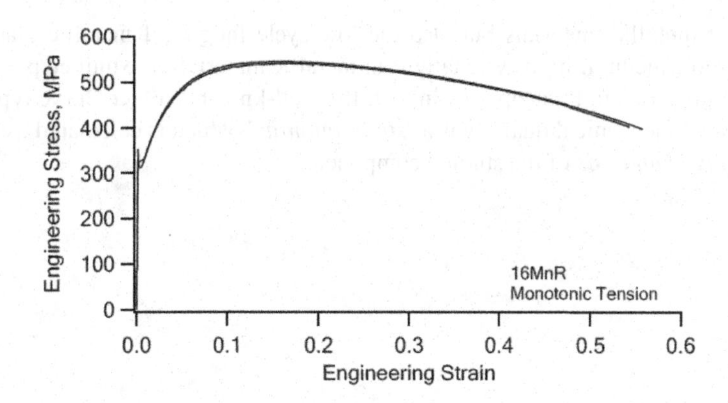

FIGURE 1.A-3 Monotonic tensile stress-strain curve of 16MnR steel.

Source: (Data from Gao, *etal* [27])

from which it can be seen that for low/medium/high cycle fatigue of notched specimens of the material, good prediction results can be obtained using the notch Wöhler Curve, i.e., the linear fitting relation of $\log(\sigma_a)$ to $\log(N)$ obtained by using experimental data of tension–compression fatigue of notched specimens shown in Table 1.A-2.

From Figure 1.A-1 and Figure 1.A-2, see also Table 1.A-3; in summary, it can be seen that, for low/medium/high cycle fatigue life assessment of 16MnR steel, a "*stress quantity*", σ_a, can be taken as a *mechanical quantity, S*, to establish a relation of the *mechanical quantity, S,* to the fatigue life, *N, i.e., a linear fitting relation between* $\log(S)$ and $\log(N)$, even though there is obvious plastic deformation in the material (see Figure 1.A-3).

Then, is a real mechanical component, which needs to be fatigue lifetime assessment, a *linear-elastic analysis* or an *elastic-plastic analysis* employed in calculating the "*stress quantity"?* Due to the fact that the magnitude of the strains involved in fatigue problems is in general not very high [19] and that the well-known Rambrg–Osgood relationship is usually employed in the *elastic-plastic analysis* of metallic materials (please here see the well-known Rambrg–Osgood relationship curve, Figure 7.7 in [19]), the author considers that, at a given loading, the "*stress quantity*" calculated using the *elastic-plastic analysis* almost is the same as that by the *linear-elastic analysis.* From a practical point of view, thus, the author considers that, for the LCF of metallic materials, the "*stress quantity" can be calculated using the linear-elastic analysis of the mechanical component.*

Thus it can be concluded that, in a low/medium/high cycle fatigue regime of metallic materials, it is practical to present a *stress invariant based equation* in which the "*stress quantity*" is calculated using the *linear-elastic analysis* of the studied component. Such a *stress invariant based equation* is, in fact, the Wöhler Curve type one usually uses to perform the fatigue life assessment of the medium/high cycle fatigue life of metallic materials. That is to say that the Wöhler Curve can be used to perform the fatigue life assessment of not only the medium/high cycle fatigue of metallic materials but also the low cycle fatigue of metallic materials. Thus, for low/medium/high cycle fatigue of metallic materials, the unified prediction equation proposed in this paper is, in fact, the well-known Wöhler Curve type one, in which the mechanical quantity is a "*stress quantity*" which is calculated using the *linear-elastic analysis* of the studied component.

2 Applicability of the Wöhler Curve Method for a Low/Medium/ High Cycle Fatigue of Metallic Materials

2.1 INTRODUCTION

It is well known that the majority of damages in real components are due to fatigue. Moreover, the problem of properly performing the fatigue assessment is further complicated due to the fact that such failures are, in general, caused by multiaxial loading: this makes it evident that engineers engaged in assessing real mechanical assemblies need sound engineering tools capable of accurately and efficiently estimating multiaxial fatigue damage.

As is well known to us, a fatigue life estimation of metallic materials is usually divided into two types, i.e., a low/medium cycle (LCF) and a high cycle fatigue (HCF). Concerning the HCF fatigue assessment, the Wöhler Curve Method is usually employed. The Modified Wöhler Curve Method (MWCM) proposed by Susmel and Lazzarin [1], for example, is used to perform the HCF fatigue assessment of metallic materials for conventional mechanical components subjected to multiaxial fatigue loading—while, for the LCF of metallic materials, no doubt, the Manson–Coffin Curve is used to perform fatigue life assessment. The Modified Manson–Coffin Curve Method by Susmel et al., [2] for example, is used to evaluate the LCF lifetime of metallic materials under multiaxial fatigue loading.

A key to fatigue life evaluation of metallic materials is, in substance, that, on the basis of experimental fatigue *data* of metallic specimens and *mechanical analyses* for those specimens, a *proper mechanical quantity*, *S*, is chosen to establish a relationship between the *mechanical quantity* and the fatigue life, *N*. For example, the Wöhler Curve and the Manson–Coffin Curve are two typical examples of the relationship. When a different *mechanical quantity*, *S*, is taken, naturally, a different *relationship equation* is obtained, which usually has different accuracy and efficiency when used to perform fatigue life estimation. Both accurate and high efficient fatigue life prediction equation is to be searched for all field investigators.

This study attempts to present a prediction equation of a low/medium/high cycle fatigue life of metallic materials. The prediction equation is required to have both accuracy and high efficiency when applied to perform the fatigue life assessment of metallic components subjected to both uniaxial and multiaxial loading. Concerning

DOI: 10.1201/9781003356721-2

fatigue assessment of a mechanical component, main factors affecting its efficiency include three aspects. One is a mechanical analysis of the studied component. The mechanical analysis, generally, is a basis of fatigue assessment of the studied component due to the fact that the mechanical analysis is a basis of calculating a chosen *mechanical quantity*, *S*. The consuming time used in the *linear-elastic analysis* of the studied component, for example, is much less than that in its *elastic-plastic analysis*. Two is to compute a chosen *mechanical quantity*, *S*. On the basis of wide research into multiaxial fatigue life predictions [1–10], it can be seen that the consuming time used in calculating the *mechanical quantity* on a stress invariant based approach is much less than that on a critical plane approach. The third aspect is to measure the material constants in the stress-stain relation used to perform the mechanical analysis of the studied component. The time and financial resources used in determining the material constants in linear stress-strain relation is, no doubt, much less than that in determining the material constants in cyclic elastic-plastic stress-strain relation.

Thus, the aim of this study is to present, on the basis of the *linear-elastic analysis* of the studied component, a *stress invariant based equation* of multiaxial fatigue life prediction. The stress invariant based equation will be both accurate and highly efficient when used in performing low/medium/high fatigue life assessment of mechanical components under both uniaxial and multiaxial loading. The systematic validation exercises given in this study, using a large number of experimental results of plain specimens in the literature, proved that the aim has been reached.

The outline of this study is follows:

2.2 Practicability of the Wöhler Curve Method for a low-cycle fatigue of metallic materials

In this section, a description on the practicability of the Wöhler Curve Method for a low-cycle fatigue of metallic materials will be given first. Then some examples from the literature, which are used here to illustrate the practicability of the Wöhler Curve Method for a low-cycle fatigue of metallic materials, are listed.

2.3 Applicability of the Wöhler Curve Method for a low/medium/high cycle fatigue of metallic materials

Medium/high cycle fatigue lives of metallic materials, as is well known to us, is well calculated by the Wöhler Curve Method. From the study of Section 2.2, naturally, it appears to be possible that, in a low/medium/high cycle fatigue of metallic materials, the Wöhler Curve Method is well suitable for fatigue life assessment. For multiaxial loading fatigue, a generalized Wöhler equation (i.e., a generalized *S-N* equation), is proposed in this study. It is a stress invariant based one, in which the computation of stress invariance is based on the *linear-elastic analysis* of the studied component. The systematic validation exercises given in this study, using a large number of experimental data of metallic materials from the literature, proved that, in a low/medium/high cycle fatigue regime of metallic materials, the generalized Wöhler equation is well suited for fatigue life assessment.

2.4 Conclusions and final comments

2.2 PRACTICABILITY OF THE WÖHLER CURVE METHOD FOR A LOW-CYCLE FATIGUE OF METALLIC MATERIALS

In this section, a description on the practicability of the Wöhler Curve Method for a low-cycle fatigue of metallic materials will be given first. Then some examples from the literature, which are used here to illustrate the practicability of the Wöhler Curve Method for a low-cycle fatigue of metallic materials, are listed.

2.2.1 A Description on the Practicability of the Wöhler Curve Method for a Low-Cycle Fatigue of Metallic Materials

On “A description on the practicability of the Wöhler Curve Method for a low-cycle fatigue of metallic materials”, “a description on the practicability of establishing the unified prediction equation for a low/medium/high cycle fatigue of metallic materials” (a quote which occurred in Appendix C in Ref.[11]), is perhaps helpful for the reader to understand it.

In Ref [11] (Yan, X. Q., An empirical fracture equation of mixed mode cracks, Theoretical and Applied Fracture Mechanics 116 (2021) 103146), recently, the author was concerned with the following question presented by a reviewer:

> *"All test data investigated in this study is obtained from brittle materials which have relatively small fracture process zone. Could the proposed approach be able to predict the fracture behavior of materials which have large fracture process zone such as rock or concrete? Please explain."*

According to this question, Appendix *B* was added in Ref.[11]. In Appendix *B* in Ref.[11], an attempt was made to test whether or not the empirical fracture equation of mixed mode cracks proposed in Ref.[11] was used to perform the fracture analysis of cracked specimens made of plastic materials. Two examples were given. From the study in Appendix *B* in Ref.[11], the author considered that the empirical fracture equation of mixed mode cracks proposed in Ref.[11] appeared to be used to perform the fracture analysis of cracked specimens made of plastic materials.

Further, such an explanation, “a description on a practicability of establishing the unified prediction equation for a low/medium/high cycle fatigue of metallic materials”, which occurred in Appendix *C* in Ref.[11], perhaps was helpful for the reader to understand it is practical that the fracture analysis of cracked specimens made of plastic materials can be performed by using the empirical fracture equation of mixed mode cracks proposed in Ref.[11].

Here, “A practicability of establishing the unified prediction equation for a low/medium/high cycle fatigue of metallic materials”, which occurred in Appendix *C* in Ref.[11], is rewritten as follows.

In a low/medium/high cycle fatigue regime of metallic materials, does a prediction equation in which the mechanical quantity calculation is on the basis of the *linear-elastic analysis* exist? According to previous wide investigations [1–10], the prediction equation is *impossible* to exist: For the HCF of metallic materials, the mechanical quantity calculation is on the basis of the *linear-elastic analysis* of the studied component due to the fact that the studied component is subjected to elastic deformation, while for the LCF of metallic materials, the mechanical quantity calculation is on the basis of the *elastic-plastic analysis* of the studied component due to obvious plastic deformation produced in the studied component. Here, the previously mentioned *impossibility* is, no doubt, reasonable. The viewpoint that, for the LCF of metallic materials, there is large plastic deformation in the materials and therefore it's not reasonable to perform the fatigue life assessment of the materials by using the *linear-elastic analysis*, is a common view.

However, on the basis of the well-known Rambrg–Osgood relationship:

$$\varepsilon = \varepsilon_e + \varepsilon_p = \frac{\sigma}{E} + \left(\frac{\sigma}{K}\right)^{\frac{1}{n}} \tag{1}$$

where E is Young's modulus, K is the strength coefficient and n is the strain hardening exponent, the author considers that, for the LCF of metallic materials, the way that a "**stress quantity**", which is calculated based on the *linear-elastic analysis* of the studied component, instead of a "strain quantity", which is calculated based on the *elastic-plastic analysis* of the studied component, is taken to be a *mechanical quantity, S*, to establish a relation of the *mechanical quantity, S*, to the fatigue life, *N*, is practicable. In order to fully indicate the author's viewpoint, here, some experimental data of low/medium/high cycle fatigue of tension–compression fatigue of plain and notched specimens made of 16MnR steel [13,14] were given in Table 2.1 and Table 2.2. Comparison of experimental data and linear fitted curve of tension–compression fatigue of plain specimens in the double-logarithm frame of axes is shown in Figure 2.1. From the agreement between experimental data and the linear fitted curve shown in Figure 2.1, no doubt, low/medium/high fatigue life assessment

TABLE 2.1
Fully Reversed Uniaxial Fatigue Experiments of Plain Specimens Made of 16MnR Steel

No.	Strain Amplitude $\varepsilon_{x,a}$ (%)	Stress amplitude $\sigma_{x,a}$ (MPa)	Fatigue life N (Cycles)
1	1.999	514.6	280
2	1.501	487.3	400
3	1	459.2	880
4	0.7	418.2	2220
5	0.5	380	4850
6	0.38	351	9200
7	0.3	322.9	17300
8	0.243	303.7	37600
9	0.199	285.1	79000
10	0.18	274.5	148400
11	0.16	261.9	388500
12	0.146	256	1400000

TABLE 2.2
Fully Reversed Uniaxial Fatigue Experiments of Notched Specimens Made of 16MnR Steel

No.	Stress amplitude $\sigma_{x,a}$ (MPa)	Fatigue life N (Cycles)
1	438.97	800
2	345.53	2048
3	308.42	3250
4	255.81	8192
5	202.86	27,000
6	184.52	47,000
7	151.32	180,000
8	140.64	232,000
9	131.07	642,200

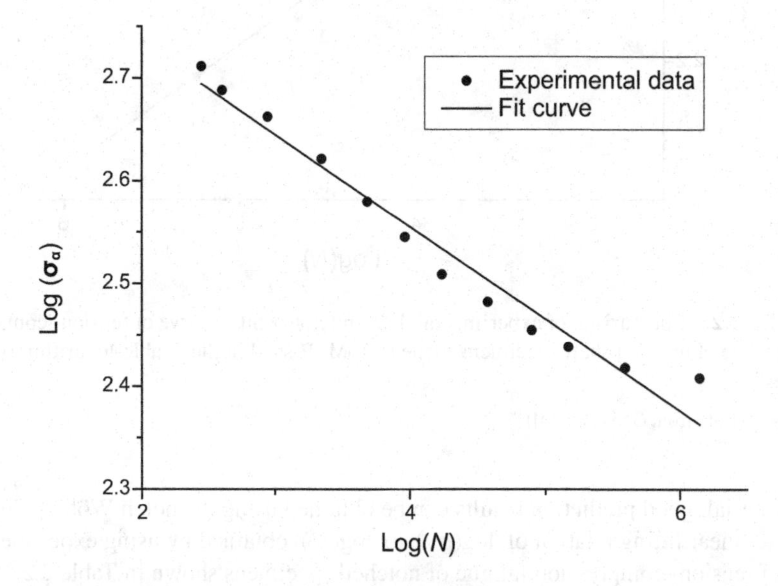

FIGURE 2.1 Comparison of experimental data and linear fitted curve of tension–compression fatigue of plain specimens made of 16MnR steel in the double-logarithm frame of axes.

Source: (Data from Gao *et al* [13])

of plain specimens of the material can be carried out by using the Wöhler curve, i.e., the linear fitting relation of $\log(\sigma_a)$ to $\log(N)$, in which the fitting constants are listed in Table 2.3. Similarly handling can be done for experimental data of tension–compression fatigue of notched specimens made of 16MnR steel (see Figure 2.2), from which it can be seen that for low/medium/high cycle fatigue of notched specimens of

TABLE 2.3
Linear Fitted Constants in the S-N Equations of 16MnR Steel

Specimen	A	C	R-square
Plain specimen	–0.08941	2.91288	0.96211
Notched specimen	–0.18268	3.1392	0.98329

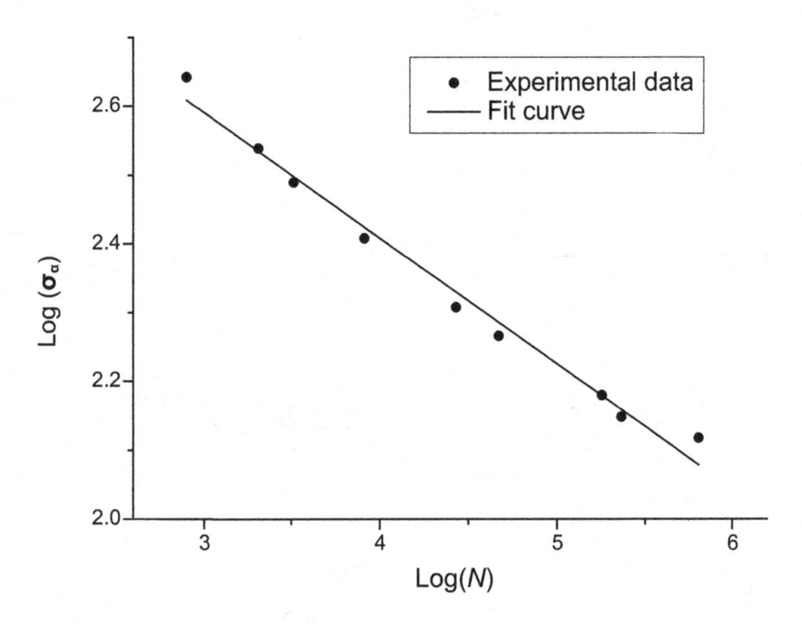

FIGURE 2.2 Comparison of experimental data and linear fitted curve of tension–compression fatigue of notched shaft specimens made of 16MnR steel in the double-logarithm frame of axes.

Source: (Data from Gao *et al* [14])

the material, good prediction results can be obtained using the notch Wöhler Curve, i.e., the linear fitting relation of $\log(\sigma_a)$ to $\log(N)$ obtained by using experimental data of tension–compression fatigue of notched specimens shown in Table 2.2.

From Figure 2.1 and Figure 2.2 (see also Table 2.3), in summary, it can be seen that, for low/medium/high cycle fatigue life assessment of 16MnR steel, a "*stress quantity*", σ_a, can be taken as a *mechanical quantity, S,* to establish a relation of the *mechanical quantity, S,* to the fatigue life, *N, i.e., a linear fitting relation between* $\log(S)$ and $\log(N)$, even though there is obvious plastic deformation in the material (see Figure 2.3).

Then, is a real mechanical component, which needs to be fatigue lifetime assessment, a *linear-elastic analysis* or an *elastic-plastic analysis* employed in calculating the "*stress quantity*"? Due to the fact that the magnitude of the strains involved in fatigue problems is in general not very high [12] and that the well-known

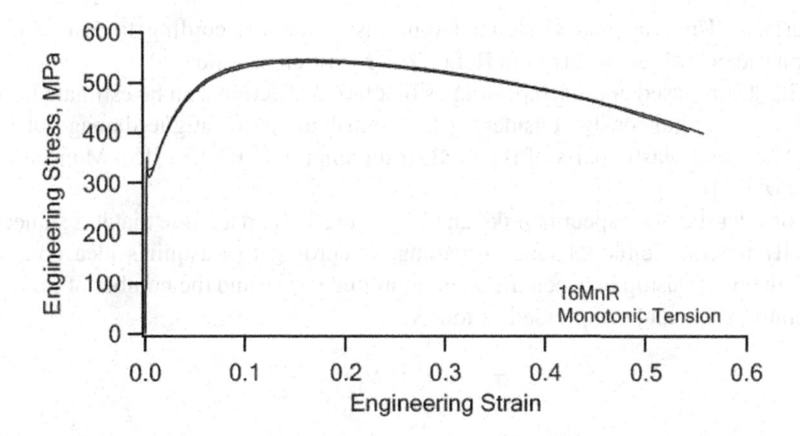

FIGURE 2.3 Monotonic tensile stress-strain curve of 16MnR steel.

Source: (Data from Gao *et al* [13])

Rambrg–Osgood relationship is usually employed in the *elastic-plastic analysis* of metallic materials (please here see the well-known Rambrg–Osgood relationship curve, Figure 7.7 in [12]), the author considers that, at a given loading, the "*stress quantity*" calculated using the *elastic-plastic analysis* almost is the same as that by the *linear-elastic analysis*. From a practical point of view, thus, the author considers that, for the LCF of metallic materials, the "*stress quantity*" *can be calculated using the linear-elastic analysis of the mechanical component.*

Thus it can be concluded that, in a low/medium/high cycle fatigue regime of metallic materials, it is practical to present a *stress invariant based equation* in which the "*stress quantity*" is calculated using the *linear-elastic analysis* of the studied component. Such a *stress invariant based equation* is, in fact, the Wöhler Curve type one usually used to perform the fatigue life assessment of the medium/high cycle fatigue life of metallic materials. That is to say that the Wöhler Curve can be used to perform the fatigue life assessment of not only the medium/high cycle fatigue of metallic materials but also the low-cycle fatigue of metallic materials. Thus, for low/medium/high cycle fatigue of metallic materials, a unified prediction equation proposed in this paper is, in fact, the well-known Wöhler Curve type one, in which the mechanical quantity is a "*stress quantity*" which is calculated using the *linear-elastic analysis* of the studied component.

2.2.2 Experimental Verifications

As seen in section 2.1, it is concluded that the Wöhler Curve Method appears to be suitable for a low cycle fatigue life analysis of metallic materials. In this subsection, some examples from the literature are given to further illustrate that the Wöhler Curve Method is suitable for a low-cycle fatigue life analysis of metallic materials.

In order to well illustrate by using some examples from the literature that the Wöhler Curve Method is suitable for a low-cycle fatigue life analysis of metallic materials, "Uniaxial and torsional fatigue assessment according to Manson and Coffin's idea", which occurred in Ref.[12], is rewritten as follows:

The strain-based approach postulates that fatigue lifetime can be estimated accurately by simultaneously considering the contributions to fatigue damage of both the elastic and plastic parts of the total strain amplitude (Coffin [15]; Manson [16]; Morrow [17]).

Consider the plain specimen sketched in Figure 2.4 and assume that it is subjected to fully reversed uniaxial fatigue loading. According to Basquin's idea (Basquin [18]), the relationship between the stress amplitude, $\sigma_{x,a}$, and the number of reversals to failure, $2N_f$, can be expressed as follows:

$$\sigma_{x,a} = \sigma_f' \, (2N_f)^b \tag{2}$$

where the fatigue strength coefficient, σ_f', and the fatigue strength exponent, b, are material constants to be determined by running appropriate experiments.

By taking full advantage of Eq. (2), it is trivial to calculate the amplitude of the elastic part of the total strain, $\varepsilon^e_{x,a}$, as the number of reversals to failure, N_f, increases, that is:

$$\varepsilon^e_{x,a} = \frac{\sigma_{x,a}}{E} = \frac{\sigma_f'}{E}(2N_f)^b \tag{3}$$

Similarly, the relationship between the plastic strain amplitude, $\varepsilon^p_{x,a}$, and $2N_f$ can be expressed as follows:

$$\varepsilon^p_{x,a} = \varepsilon_f' \, (2N_f)^c \tag{4}$$

where the fatigue ductility coefficient, ε_f', and the fatigue ductility exponent, c, are again material properties to be determined experimentally.

Finally, if the elastic contribution to the overall fatigue damage is added to the corresponding plastic contribution, the relationship between the total strain amplitude,

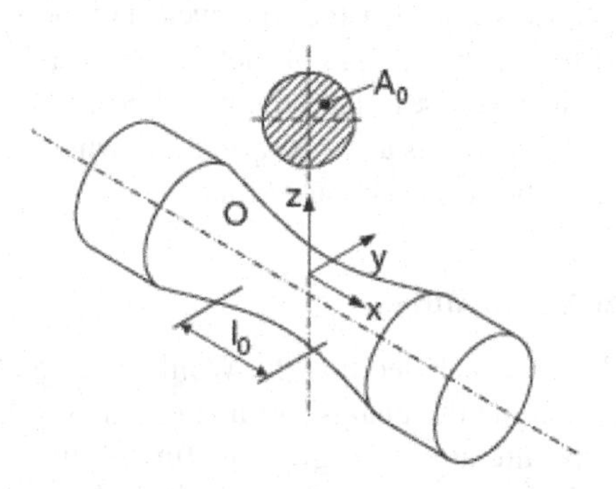

FIGURE 2.4 Plain cylindrical specimen and its gauge length.

Source: (Data from Susmel [12]).

$\varepsilon_{x,a}$, and the number of reversals to failure can be written directly in the following explicit form:

$$\varepsilon_{x,a} = \varepsilon^{e}_{x,a} + \varepsilon^{p}_{x,a} = \frac{\sigma'_f}{E}(2N_f)^b + \varepsilon'_f (2N_f)^c \tag{5}$$

This relationship is called the Manson–Coffin equation.

In this study, Eq. (2) (or (3) is called Basquin's equation.

Here, a comparison of the Wöhler equation with Basquin's equation is given below:

Eq.(2) is rewritten as:

$$\lg(\sigma_{x,a}) = b\lg(N_f) + \lg(\sigma'_f) + b\lg(2) \tag{6}$$

Here, the Wöhler equation (usually called the *S-N* equation) is written as:

$$\lg(\sigma_{x,a}) = A_1\lg(N_f) + C_1 \tag{7}$$

Obviously, Eq.(7) is the same as Eq.(2) by taking:

$$A_1 = b, \quad C_1 = \lg(\sigma'_f) + b\lg(2) \tag{8}$$

By comparing the Wöhler equation with Basquin's equation, thus, it is seen that the Wöhler equation is the same as Basquin's equation expressed in terms of (2) or (3). Based on the same, it is concluded that, in studying of low-cycle fatigue life for a metallic material by using the Manson–Coffin curve method, if the relationship between $\sigma_{x,a}$ and the number of reversals to failure, $2N_f$, is well revealed by using Basquin's equation (2) (i.e., $\lg(\sigma_{x,a})$ and $\lg(N_f)$ have good linear fitting with a high R-square value), or if the relationship between $\varepsilon^{e}_{x,a}$ and the number of reversals to failure, $2N_f$, is well revealed by using Basquin's equation (3), the low-cycle fatigue life of the metallic material can be well calculated by using the Wöhler equation. Here, a metallic material with good linear fitting relationship between $\lg(\sigma_{x,a})$ and $\lg(N_f)$ or $\lg(\varepsilon^{e}_{x,a})$ and $\lg(N_f)$ is called Basquin's material. Obviously, for Basquin's material, its low-cycle fatigue life is well calculated by the Wöhler equation, which will be illustrated by means of the following examples from the literature.

In order to illustrate the author's viewpoint that for Basquin's material, its low-cycle fatigue life is to be well predicted by the Wöhler equation, the author searched for documents on research into the low-cycle fatigue of metallic materials. Here, the documents are divided into three types:

TYPE I: Both the low-cycle fatigue test data of metallic materials and the analyzed results (strain-life curve figure) by using the Manson–Coffin equation are included in the literature. Then Basquin's curve in the strain-life curve figure reported in the literature is used to illustrate that the metallic material studied is Basquin's material, its low-cycle fatigue life is to be well predicted by the Wöhler equation and at the same time, the low-cycle fatigue test data of the material studied are analyzed by using the Wöhler equation to further illustrate that the low-cycle fatigue life of the material studied can be calculated by the Wöhler equation. Such type examples include EN AW-2007 aluminum alloy [19] and EN AW-2024-T3 aluminum alloy [20].

TYPE II: The low-cycle fatigue test data of metallic material are given in the literature. But strain-life curve figure is not reported in the literature. Then the low-cycle fatigue test data of the material studied are analyzed by using the Wöhler equation to illustrate that the low-cycle fatigue life of the material studied can be calculated by the Wöhler equation. Such type materials include A533B [21], Inconel 718 [22], SAE1045 [23], AISI304 [24], Ti-6Al-4V [25], 6061-T6 [26], 1Cr-18*N*i-9Ti [27], AISI304 [28], AISIH11 [29], 34CrNiMo6 [30], RAFM steels [31]. The details are given in Appendix *A*.

TYPE III: Strain-life curve figure of the material studied is reported in the literature. But the low-cycle fatigue test data of the material studied are not given in the literature. Then Basquin's curve in the strain-life curve figure reported in the literature is used to illustrate that the metallic material studied is Basquin's material, and its low-cycle fatigue life is to be well predicted by the Wöhler equation. Such type materials include: P92 ferritic-martensitic steel [32], Cr–Mo–V low alloy steel [33], S35C carbon steel and SCM 435 alloy steel [34,35], high strength spring steel with different heat-treatments [36], 316 L(N) stainless steel at room temperature [37], CLAM steel at room temperature [38]), Eurofer97 [39], F82H (a ferritic-martensitic steel) [40], En3B (a commercial cold-rolled low-carbon steel) [41], nodular cast iron [42] and low carbon gray cast iron [43]. The details are given in Appendix *B*.

Type I

Example 1: Low-Cycle Fatigue of EN AW-2007 Aluminum Alloy

Here, low-cycle fatigue test data (see Tables 2.4 and 2.5) of EN AW-2007 aluminum alloy reported by Szusta and Seweryn [19] is used to illustrate the author's

TABLE 2.4

Low-Cycle Fatigue Test Data of EN AW-2007 Aluminum Alloy (for Tension R = −1) and Predicted Results by the Wöhler Equation

ε_a	ε_a^e	ε_a^p	σ_a (MPa)	N_f (exp)	N_f (pre)	R.E. (%)
0.0025	0.001867	0.000632	139	7480	8090	8.2
0.0035	0.002073	0.001427	155	2066	1624	−21.4
0.005	0.002171	0.002828	162	699	847	21.1
0.008	0.002350	0.00565	175	290	271	−6.4
0.01	0.002395	0.007605	179	173	194	12.4
0.02	0.002569	0.017431	192	75	69	−7.7

Note: (a) $A_1 = -0.06784$, $C_1 = 2.40813$, R-square = 0.99; (b) M.E. = 1.0%, E.R. = [−21.4,21.1]%

TABLE 2.5
Low-Cycle Fatigue Test Data of EN AW-2007 Aluminum Alloy (for Torsion R = –1) and Predicted Results by the Wöhler Equation

γ_a	γ_a^e	γ_a^p	τ_a (MPa)	N_f (exp)	N_f (pre)	R.E. (%)
0.0290	0.0029	0.0261	97.93	63	92	46.1
0.02	0.0027	0.0173	91.20	168	196	16.4
0.0135	0.0025	0.0110	84.37	455	446	−2.1
0.0090	0.0024	0.0066	81.36	1253	655	−47.8
0.0075	0.0023	0.0052	78.12	1905	1006	−47.2
0.006	0.0021	0.0039	72.74	2964	2141	−27.8
0.0045	0.0019	0.0026	63.93	5232	8396	60.5
0.0035	0.0017	0.0018	59.43	9679	18178	87.8

Note: (a) $A_0 = -0.09449$, $C_0 = 2.41505$, R-square = 0.91; (b) M.E. = 10.7%, E.R. = [−47.7,87.8]%

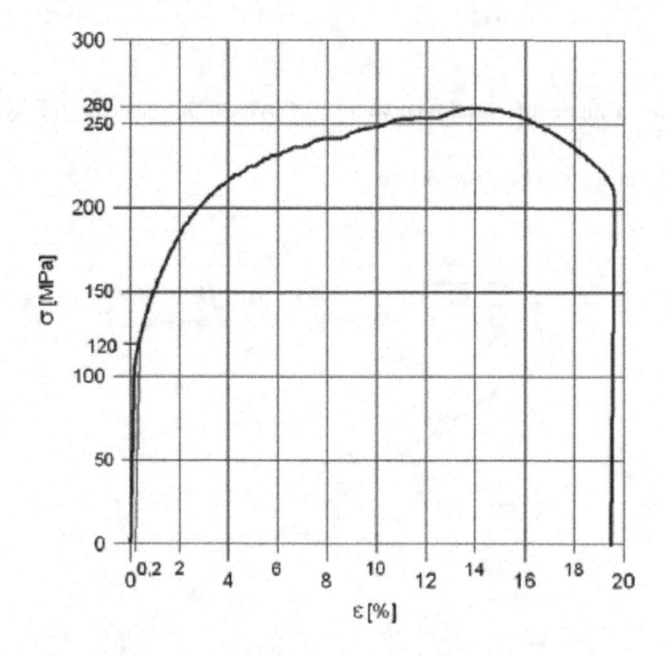

FIGURE 2.5 The monotonic tension curve of EN AW-2007 alloy.

Source: (Data from Szusta and Seweryn [19])

viewpoint that for Basquin's material, its low-cycle fatigue life is to be well predicted by the Wöhler equation. Figures 2.5 to 2.7 show the monotonic tension curve, the fatigue life strain curve for the symmetrical tension–compression, and the fatigue life curve for symmetrical torsion, respectively. From Figures 2.6 and 2.7, it is seen that the low-cycle fatigue life of the material studied is characterized by using the

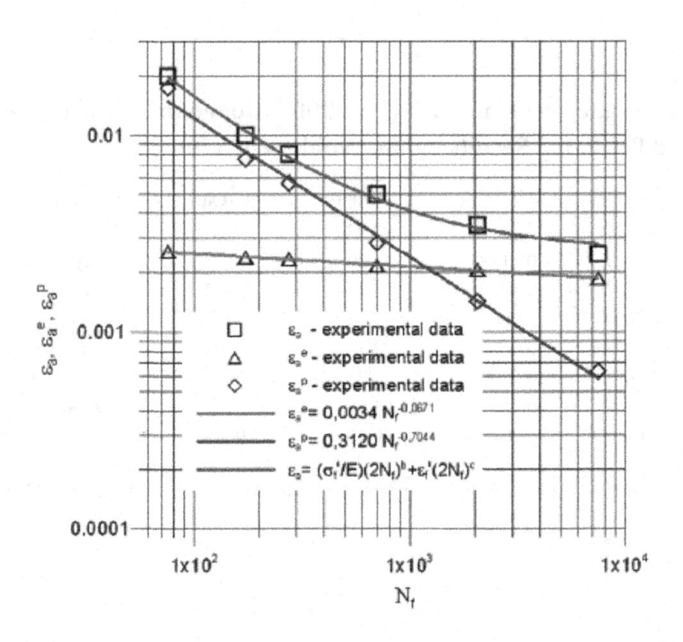

FIGURE 2.6 Fatigue life strain curve of EN AW-2007 alloy for the symmetrical tension–compression.

Source: (Data from Szusta and Seweryn [19]).

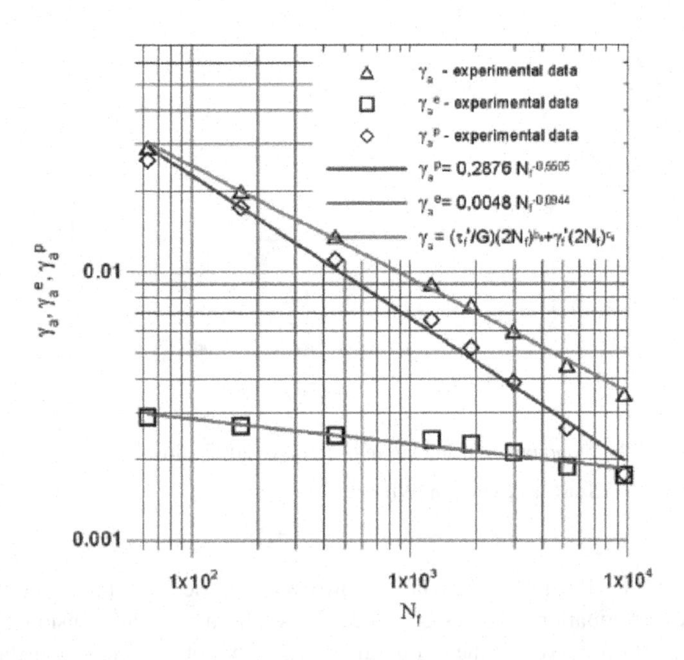

FIGURE 2.7 The fatigue life curve for EN AW-2007 alloy for symmetrical torsion.

Source: (Data from Szusta and Seweryn [19])

Manson–Coffin equation, and that the low-cycle fatigue life of the material studied is characterized also by Basquin's equation because the relationship between ε_a^e and N_f is well described by Basquin's equation. Thus the material studied, EN AW-2007 aluminum alloy, is Basquin's material.

Here, such an attempt is made that the Wöhler equation is used to analyze low-cycle fatigue test data of EN AW-2007 aluminum alloy. The obtained results are shown in Tables 2.4 to 2.6 (see also Figure 2.8). From Figure 2.8, it is found that the experimental fatigue lives are in good agreement with those predicted by the Wöhler equation, with error indexes: M.E. = 1.0%, E.R. = [−21.4,21.1]%, and M.E. = 10.7%, E.R. = [−47.7,87.8]% for tension and torsion fatigue, respectively.

Thus, it is concluded here from the obtained results shown in Tables 2.4 to 2.6 (see also Figure 2.8). that for Basquin's material, EN AW-2007 aluminum alloy, its low-cycle fatigue life is to be well predicted by the Wöhler equation.

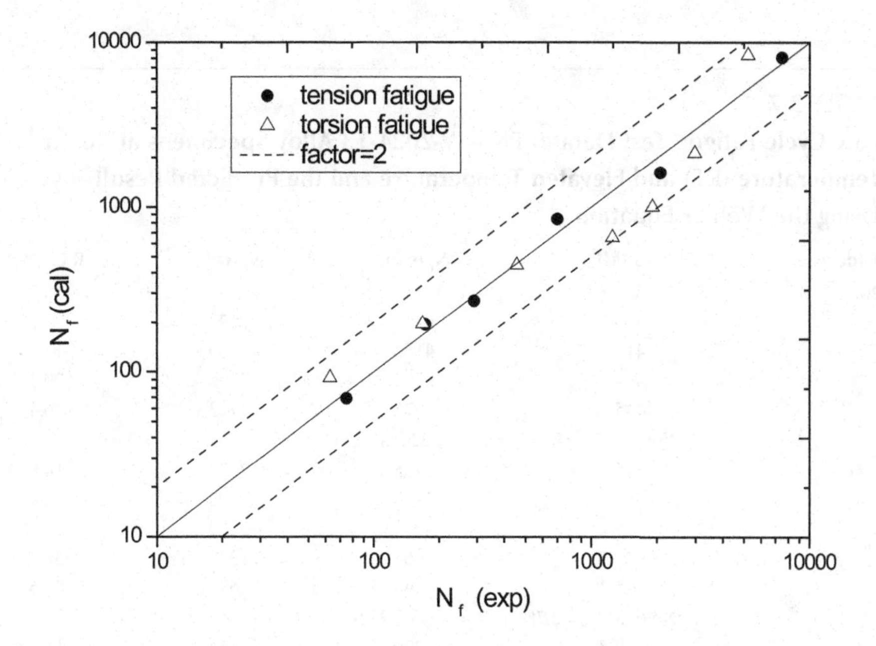

FIGURE 2.8 Comparison of experimental fatigue lives with those predicted by the Wöhler equation (EN AW-2007 alloy).

Source: (Data from Szusta and Seweryn [19])

TABLE 2.6

Fitting Constants in the Wöhler Equation of EN AW-2007 Aluminum Alloy

A_1	C_1	R-square	A_0	C_0	R-square
−0.06784	2.40813	0.99	−0.09449	2.41505	0.91

Example 2: Low-Cycle Fatigue of Aluminum Alloy EN AW-2024-T3

Here, low-cycle fatigue test data (see Table 2.7, also see Figure 2.9) of aluminum alloy EN AW-2024-T3 at room temperature (RT) and elevated temperature reported by Szusta and Seweryn [20] is used to further illustrate the author's viewpoint that for Basquin's material, its low-cycle fatigue life is to be well predicted by the Wöhler equation. Figure 2.10 shows a comparison of monotonic tension curves of EN AW-2024T3 aluminum alloy at 20°, 100°, 200°, and 300°.

From Figure 2.9, it is seen that the low-cycle fatigue life of the material studied is characterized by using the Manson–Coffin curve, and that the low-cycle fatigue life of the material studied is characterized also by Basquin's equation because the relationship between ε_a^e and N_f is well described by Basquin's equation. Thus the material studied, aluminum alloy EN AW-2024-T3, is Basquin's material.

TABLE 2.7
Low-Cycle Fatigue Test Data of EN AW-2024-T3 Alloy Specimens at Room Temperature (RT) and Elevated Temperature and the Predicted Results by Using the Wöhler Equation

T (degree)	$\sigma_{x,a}$ (MPa)	N_f (exp)	N_f (cal)	R.E. (%)
20	380.3	14597	21346	46.2
	406.7	6458	5027	−22.2
	413	4156	3609	−13.2
	423	2704	2155	−20.3
	445	572	723	26.4
	Note: M.E. = 3.4%, E.R. = [−22.2,46.2]%			
100	365	10755	12327	14.6
	379	5168	5792	12.1
	390	3209	3262	1.6
	407.7	2164	1338	−38.1
	425.3	464	573	23.5
	Note: M.E. = 2.7%, E.R. = [−38.1,23.5]%			
200	319.7	8329	7354	−11.7
	330	4435	5002	12.8
	344	2177	3018	38.7
	370.7	1679	1217	−27.5
	410.2	355	355	0.1
	Note: M.E. = 2.5%, E.R. = [−27.5,38.7]%			
300	330.7	3775	3165	−16.2
	341.7	1881	1666	−11.4
	347.7	870	1185	36.2
	356.3	464	734	58.2
	387.7	225	140	−37.7
	Note: M.E. = 5.8%, E.R. = [−37.3,58.2]%			

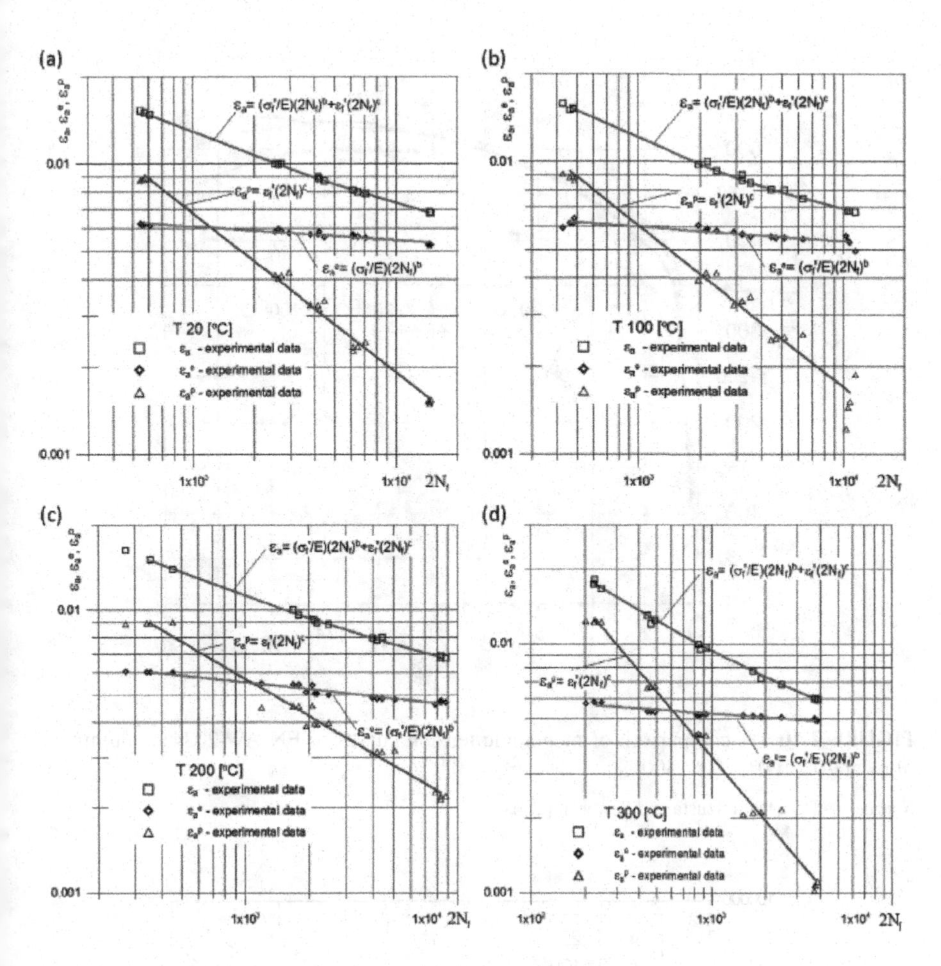

FIGURE 2.9 Fatigue life curve for EN AW-2024T3 aluminum alloy samples in the tension loading conditions and temperature of: (a) T = 20°, (b) T = 100°, (c) T = 200°, and (d) T = 300°.

Source: (Data from Szusta and Seweryn [20]).

Here, low-cycle fatigue test data (see Table 2.7) of aluminum alloy EN AW-2024-T3 at room temperature (RT) and elevated temperature reported by Szusta and Seweryn [20] are analyzed by using the Wöhler equation. The obtained results are shown in Table 2.7 (see álso Figure 2.11). From Figure 2.11, it is found that the experimental fatigue lives are in good agreement with those predicted by the Wöhler equation, with error indexes: M.E. = 3.4%, E.R. = [−22.2,46.2]%; M.E. = 2.7%, E.R. = [−38.1,23.5]%; M.E. = 2.5%, E.R. = [−27.5,38.7]% and M.E. = 5.8%, E.R. = [−37.3,58.2]% for 20°, 100°, 200° and 300°, respectively.

Table 2.8 shows the fitting constants in the Wöhler equation of aluminum alloy EN AW-2024-T3. It is seen from the various values of R square shown in Table 2.8 that the low-cycle fatigue life of the material studied is well calculated by using the Wöhler equation.

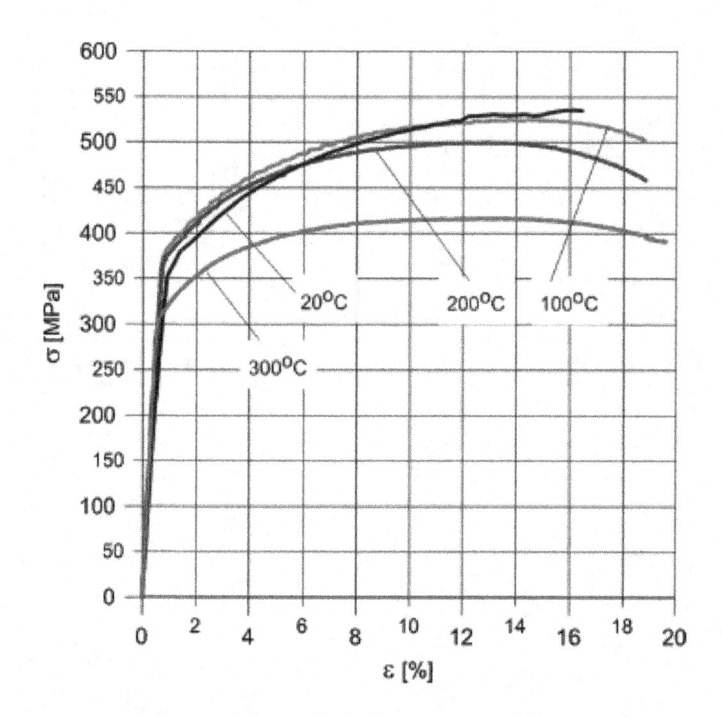

FIGURE 2.10 A comparison of monotonic tension curves of EN AW-2024T3 aluminum alloy at 20°, 100°, 200°, 300°.

Source: (Data from Szusta and Seweryn [20]).

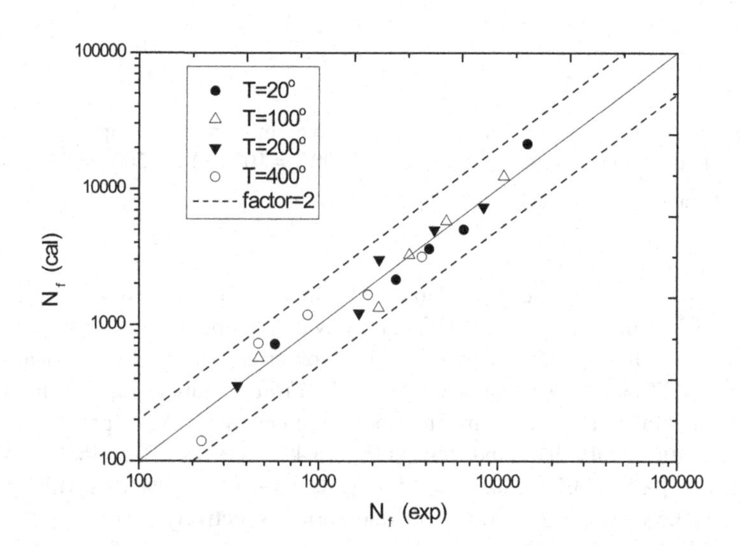

FIGURE 2.11 Comparison of the experimental fatigue lives with those predicted by the Wöhler equation.

Source: (Data from Szusta and Seweryn [20])

TABLE 2.8
Fitting Constants in the Wöhler Equation of Aluminum Alloy EN AW-2024-T3

T (degree)	A_1	C_1	R square
20	−0.04641	2.78105	0.90
100	−0.04983	2.76614	0.93
200	−0.08227	2.82284	0.94
300	−0.05102	2.69802	0.86

2.2.3 A Comment on This Section

From the study of this section, the author considers that it is sure enough that the low-cycle fatigue life analysis of metallic materials is well performed by the Wöhler Curve Method.

2.3 APPLICABILITY OF THE WÖHLER CURVE METHOD FOR A LOW/MEDIUM/HIGH CYCLE FATIGUE OF METALLIC MATERIALS

From the study of Section 2, it is sure enough that the low-cycle fatigue analysis of metallic materials is well performed by the Wöhler Curve Method. Medium/high cycle fatigue lives of metallic materials, as is well known to us, is well calculated by the Wöhler Curve Method. Thus it appears to be possible that, for a low/medium/high cycle fatigue of metallic materials, the Wöhler Curve Method is well suitable for fatigue life assessment. For multiaxial loading fatigue, a generalized Wöhler equation (i.e., a generalized *S-N* equation), is proposed in this study. It is a stress invariant based one, in which the computation of stress invariants is on the basis of the *linear-elastic analysis* of the studied component. The systematic validation exercises given here, using a large number of experimental data of metallic materials in the literature, proved that, in a low/medium/high cycle fatigue of metallic materials, the generalized Wöhler equation is well suited for fatigue life assessment.

2.3.1 A Proper Mechanical Quantity in the Fatigue Life Prediction Equation

According to the discussions in Section 2, in a low/medium/high cycle fatigue regime of metallic materials, a stress based lifetime estimation equation in which the mechanical quantity calculation is on the basis of the *linear-elastic analysis* may exist. In order to present the stress based lifetime estimation equation, is what a "stress quantity" employed under multiaxial loading? Amongst the available stress variant quantities, the most commonly adopted stress variant quantity is the von Mises equivalent stress. In this chapter, the von Mises equivalent stress will be taken as a *mechanical quantity*, *S*, to establish the relationship between the *mechanical quantity*, S, and the fatigue life, N, i.e., a stress-based lifetime estimation equation.

It is pointed out here that the von Mises equivalent stress was employed previously in fatigue of metallic materials, for example, multiaxial fatigue limits [6,7], and multiaxial fatigue life predictions [4,9,10].

In the following, the von Mises equivalent stress is called a Mises stress for short, which is expressed in term of the main stresses σ_1, σ_2 and σ_3:

$$\sigma_e = \frac{1}{\sqrt{2}}\sqrt{(\sigma_1 - \sigma_2)^2 + (\sigma_2 - \sigma_3)^2 + (\sigma_3 - \sigma_1)^2} \tag{9}$$

2.3.2 Multiaxial Fatigue Life Prediction Equation

Generally, a multiaxial fatigue life prediction equation can be expressed mathematically to be [44]:

$$F(\sigma_{ea}, \sigma_{em}, \rho, N\ , c_1, c_2, \ldots) = 0 \tag{10}$$

where σ_e is the von Mises equivalent stress, and σ_{ea} and σ_{em} are the amplitude and the mean value of the equivalent stress, respectively. A multiaxial parameter, ρ, is defined to be [44]:

$$\rho = \frac{\sigma_{11a}}{\sigma_{ea}} \tag{11}$$

where σ_{11a} is the amplitude of the first invariant of stress tensor. It is known from the formula (11) that, for the bending and torsion fatigue, the multiaxial parameter, ρ, is equal to 1 and 0, respectively. $c_1, c_2, \ldots$, in Eq. (10) are material parameters, which are usually varied with the multiaxial parameter ρ.

A concrete form of Eq.(10) will be described from a few aspects:

Multiaxial Fatigue Life Prediction Equation Under Zero Mean Stress

Under zero mean stress, a stress-based fatigue life estimation equation under multiaxial loading is assumed to be [44]:

$$\log \sigma_{ea} = A_\rho \log N + C_\rho \tag{12}$$

which, under bending and torsion fatigue, can be written to be, respectively:

$$\log \sigma_a = A_1 \log N + C_1 \tag{13}$$

and

$$\log \sqrt{3}\tau_a = A_0 \log N + C_0 \tag{14}$$

Eq. (13) is the well-known S-N curve equation expressed in terms of a stress quantity. Eq. (14) is the variant form of the S-N curve equation under torsion fatigue. A_1 and

C_1 in Eq.(13), and A_0 and C_0 in Eq.(14) are material constants determined by using bending and torsion experimental fatigue data of plain specimens, respectively.

In view of the complexity of fatigue life analysis under multiaxial loading, and also taking into account that a large number of experimental fatigue data have been accumulated in the literature, from the viewpoint of application, it is assumed here that the material parameters in Eq. (12) can be obtained by interpolating the material constants in Eqs. (13) and (14), i.e.:

$$A_\rho = A_1 \cdot \rho + A_0 \cdot (1-\rho) \tag{15}$$

$$C_\rho = C_1 \cdot \rho + C_0 \cdot (1-\rho) \tag{16}$$

Multiaxial Fatigue Life Prediction Equation with Non-Zero Mean Stress

A mean stress effect on fatigue life assessment is a classical problem. Based on the previous investigations (e.g., Tao and Xia [45]) and Marin's general equation [46]), an equivalent stress under multiaxial loading is introduced by Liu and Yan [44] to consider the effect of mean stress on multiaxial fatigue life:

$$\bar{\sigma}_{ea} = \sigma_{ea}\left(1-\alpha_\rho \frac{\sigma_{em}}{\sigma_{ea}}\right)^{n_\rho} \tag{17}$$

where α_ρ and n_ρ are material parameters dependent on the multiaxial parameter ρ.

By substituting the equivalent stress defined in formula (17) into the $\sigma_{ea} - N$ equation (12), the following $\sigma_{ea} - N$ equation with the effect of mean stress can be obtained [44]:

$$\log \sigma_{ea}\left(1-\alpha_\rho \frac{\sigma_{em}}{\sigma_{ea}}\right)^{n_\rho} = A_\rho \log N + C_\rho \tag{18}$$

which, under bending and torsion fatigue, can be written to be, respectively

$$\log \sigma_a\left(1-\alpha_1 \frac{\sigma_m}{\sigma_a}\right)^{n_1} = A_1 \log N + C_1 \tag{19}$$

and

$$\log \sqrt{3}\tau_a\left(1-\alpha_0 \frac{\tau_m}{\tau_a}\right)^{n_0} = A_0 \log N + C_0 \tag{20}$$

where α_1 and n_1 in Eq.(19), and α_0 and α_0 in Eq.(20) are material constants which are determined by using bending and torsion experimental fatigue data with mean stress effect, respectively. Meanwhile the material parameters α_ρ and n_ρ in Eq.(18), can be obtained by using the similar relations to (15) and (16).

On Dealing with Nonproportional Loading Fatigue

As to nonproportional loading fatigue, the Itoh's formula [28] is usually used to reveal the relationship of the amplitude of the equivalent nonproportional stress, σ_{ean}, and the amplitude of the equivalent proportional stress, σ_{ea}, i.e.,

$$\sigma_{ean} = \sigma_{ea}(1+\beta F) \tag{21}$$

where β is a material sensitivity constant to load-path nonproportionality defined on $\sigma - \sqrt{3}\tau$ stress plane; F is usually called a nonproportional loading factor which expresses the severity of nonproportional loading.

In this study, it is found that the prediction results of fatigue life obtained by using Itoh's formula sometimes are not satisfactory. Thus the following attempt is made:

$$\sigma_{ean} / \sigma_{ea} = \lambda + \gamma F \tag{22}$$

where λ and γ are constants to be determined by linear fitting of σ_{ean}/σ_{ea} and F. When λ in formula (22) is equal to 1, the linear fitting relation (22) becomes Itoh's formula. In this study, the formulas (21) and (22) are called Itoh's formula (I.F) and the linear formula (L.F), respectively.

Under out-of-phase loading fatigue, the nonproportional factor F (NPF) is usually decided as the ratio of the two principal axes of elliptical loading path [47]. According to this definition of NPF, two similar elliptical loading paths with different radii, in which one is, for example, much larger than the other, have a uniform value, F, which appears to be unreasonable. In order to reveal the effect of relative size of fatigue loading path on the nonproportional factor F, a new definition of NPF is proposed here as follows:

It is assumed that there are m out-of-phase loading cases, in which the nonproportional factor, F, and the equivalent proportional stress, σ_{ea}, are denoted by F_i, and $(\sigma_{ea})_i$, $(i = 1,2,..,m)$, respectively, and the maximum of $(\sigma_{ea})_i$ is denoted by $(\sigma_{ea})_{\max}$. Then a new definition of nonproportional factor (NPF') is

$$F_i' = F_i \cdot \frac{(\sigma_{ea})_i}{(\sigma_{ea})_{\max}}, \quad (i = 1,2,..,m) \tag{23}$$

Here, multiaxial fatigue life equation, Eq.(12) or Eq.(18), is called a generalized Wöhler equation, or a generalized *S-N* equation.

2.4 EXPERIMENTAL VERIFICATIONS AND DISCUSSIONS

As is well known to us, a large number of experimental fatigue data of metallic materials were assembled in the literature [12]. By using the experimental data of the low/medium/high cycle fatigue life of metallic materials from the literature, including **the stress fatigue data** (18G2A [48], Z12C*N*DV12-2 [49], 5% Cr [50], 6082-T6 [51], and **the strain control fatigue data** (S45C [52], 1Cr-18*N*i-9Ti [27], Ti-6Al-4V [25], S460N [53]), the validation and accuracy of the generalized *S-N* equation proposed in this study will be verified.

In order to quantitatively evaluate the accuracy of the multiaxial fatigue prediction equation, the following error indexes are defined:

$$ER = \frac{(N_{cal} - N_{\exp}) \times 100}{N_{\exp}} \tag{24}$$

$$MER = \frac{1}{n}\sum_{i=1}^{n}(ER)_i \tag{25}$$

$$AER = \frac{1}{n}\sum_{i=1}^{n}\left|(ER)_i\right| \tag{26}$$

where n is the number of experimental cases. The three error indexes defined in (24) to (26) are called a relative error, a mean error and an absolute mean error, respectively.

As is well known to us, there are unavoidably some *unusual experimental data* in experimental data of a physical problem, e.g., experimental fatigue life data reported in the literature, while the *unusual experimental data* cover usually the law of the studied problem. Thus the *unusual experimental data* should be eliminated in the handling of the experimental data. Concerning the experimental fatigue life data of metallic materials handled here, obviously, such experimental fatigue life data with **untimely fatigue failure** are the *unusual experimental data*, which will be eliminated in the processing of experimental fatigue life data in this study. The *unusual experimental data* are abbreviated to *unusual data* in this study.

In the study, you will see that, in the processing of experimental fatigue life data, eliminating the *unusual data* will promote the accuracy of the fatigue life prediction equation obtained. This illustrates that eliminating the *unusual data* is necessary to obtain the accurate fatigue life prediction equation. You can practice. You will find that, in the processing of experimental fatigue life data, eliminating *usual experimental data* will result in the accuracy of the fatigue life prediction equation obtained decreasing, which illustrates that your eliminating *usual experimental data* is not correct. Please note that data handled here are experimental data, from mathematical statistics, the *unusual data* should be eliminated to obtain the law of the studied problem. Of course, this is only one form of handling the *unusual data*. It is important to analyze the reason why the *unusual data* occur, which is not concerned with in this study.

Here, the author considers that the process of the fatigue life prediction equation obtained is a process of analysis and handling for experimental fatigue data, among which the main aim of the mechanical analysis of the studied specimen is to determine a *mechanical quantity*, S, which is taken to establish the relationship between the *mechanical quantity*, S and the fatigue life, N. Usually, establishing the relationship between the *mechanical quantity*, S, and the fatigue life, N, is a data curve fitting between S and N, an experimental data handling *statistically*. The *unusual data* should be not concluded mathematically in the experimental data handling *statistically*.

For the sake of clear discussions, examples given here will be described respectively. At the same time examples for **the strain control fatigue** will be reported in Appendix *C* due to limitation of length. The material constants in the S-N equation determined by using the uniaxial experimental fatigue data are listed in Table 2.9.

TABLE 2.9
Material Constants in the S-N Equation Determined by Using Uniaxial Experimental Fatigue Data

	Bending Fatigue			Torsion Fatigue		
Materials	A_1	C_1	**R-Square**	A_0	C_0	**R-Square**
Z12CNDV12-2	−0.0885	3.1748	0.9958	−0.1425	3.3788	0.8660
18G2A	−0.1318	3.2865	0.9328	−0.0749	2.9845	0.8846
6082-T6	−0.1425	3.0224	0.9906	−0.1177	2.8728	0.9273
5% Cr	−0.0714	3.1582	0.9804	−0.0790	3.5265	0.9105
S460N	−0.0960	2.9671	0.9794	−0.0747	2.8930	0.9955
S45C	−0.1153	3.0073	0.9844	−0.0983	3.0055	0.9531
1Cr-18Ni-9Ti	−0.0928	3.0344	0.9737	−0.0876	3.0212	0.9772
Ti-6Al-4V	−0.1776	3.5478	0.9673	−0.1650	3.5928	0.7339

Example 1: 18G2A Steel

According to the experimental fatigue data of 18G2A steel [48] (see Tables 2.10 to 2.11), it is a medium/high cycle fatigue problem with the influence of mean stress.

For this material, first, it is concerned with the applicability of the generalized *S-N* equation (i.e., $\sigma_{ea} - N$ equation) under zero mean stress ($R = -1$).

Under bending loading with $R = -1$, the experimental and calculated fatigue lives are shown in Table 2.10. It is seen that, by using the *S-N* equation, the prediction fatigue lives of the material are in good agreement with the experimental ones with an error range [−27.4, 104.9]%.

Under torsion loading with $R = -1$, the experimental and calculated fatigue lives are also shown in Table 2.10. It is seen that, by using the *S-N* equation, the torsion prediction fatigue lives of the material are in good agreement with experimental ones with an error range [−33.8, 80.5] %.

Under multiaxial loading with $R = -1$, the comparisons of experimental and calculated lives are also shown in Table 2.10. Because the *S-N* equation, including bending and torsion equations, can be used to reveal properly the relationships of the fatigue life, *N*, with the mechanical quantity, σ_{ea}, as discussed previously, and because the generalized *S-N* equation has an invariance of mathematical equation forms, as emphasized in Section 3, by using the generalized *S-N* equation, the prediction fatigue lives of the material under multiaxial loading are in good agreement with experimental ones (see Table 2.10, also see Figure 2.12).

Further, it is concerned with the applicability of the generalized *S-N* equation (i.e., $\sigma_{ea} - N$ equation) with mean stress ($R \neq -1$).

Under bending loading of $R \neq -1$, the experimental and calculated fatigue lives are shown in Table 2.11. It is seen that, by the *S-N* equation with the effect of mean stress, the prediction fatigue lives are in excellent agreement with the experimental ones, with an error range, [−45.4, 92.0]%. For torsion fatigue of the material with nonzero mean stress, an accurate prediction results of fatigue life are also obtained by using the *S-N* equation with the effect of mean stress (see Table 2.11), with an error range [−27.1, 53.3]%.

Under multiaxial loading with mean stress, the comparisons of experimental and calculated fatigue lives of the material are given in Table 2.12. Because the *S-N* equation with mean stress effect can be used to reveal properly the relationships of the fatigue life, *N*, with the mechanical quantity, σ_{ea}, as discussed

TABLE 2.10
Fatigue Life Prediction of 18G2A Steel with Zero Mean Stress

$\sigma_{x,a}$ (MPa)	$\sigma_{x,m}$ (MPa)	$\tau_{xy,a}$ (MPa)	$\tau_{xy,m}$ (MPa)	N_{exp} (Cycles)	ER (%)
399.6	0	0	0	128630	22.2
399.8	0	0	0	153881	1.7
399.8	0	0	0	168309	−6.9
364.1	0	0	0	327016	−2.6
365.3	0	0	0	375852	−17.3
367	0	0	0	415067	−27.7
275	0	0	0	1306495	104.9
275.4	0	0	0	2869140	−7.6
275.5	0	0	0	3364694	−21.4
0	0	223	0	224794	−10.2
0	0	222.6	0	278939	−25.9
0	0	222.7	0	310708	−33.8
0	0	195.7	0	637980	80.5
0	0	195.4	0	785106	49.7
0	0	196.2	0	874370	27.3
0	0	183.9	0	2296066	15.0
0	0	184	0	3095310	−15.2
0	0	183	0	4173870	−32.4
199.7	0	199.7	0	168497	−12.9
199.7	0	199.7	0	192588	−23.8
199.5	0	199.5	0	226543	−34.6
180.2	0	180.2	0	609639	−34.9
180.2	0	180.2	0	615487	−35.5
180	0	180	0	737954	−45.7
164.5	0	164.5	0	1303741	−26.5
164.6	0	164.6	0	1407201	−32.3
164.4	0	164.4	0	2101434	−54.1

TABLE 2.11
Fatigue Life Predictions of 18G2A Steel Under Bending and Torsion Loading with Mean Stress

$\sigma_{x,a}$ (MPa)	$\sigma_{x,m}$ (MPa)	$\tau_{xy,a}$ (MPa)	$\tau_{xy,m}$ (MPa)	N_{exp} (Cycles)	ER (%)
369.5	123.2	0	0	57172	−13.0
369.6	123.2	0	0	65725	−24.4
369.1	123	0	0	72614	−30.9
298.3	99.4	0	0	131431	92.0
298.5	99.5	0	0	168583	48.8
298.5	99.5	0	0	178962	40.2
268.8	89.6	0	0	578316	−3.8

(Continued)

TABLE 2.11 (Continued)
Fatigue Life Predictions of 18G2A Steel under Bending and Torsion Loading with Mean Stress

$\sigma_{x,a}$ (MPa)	$\sigma_{x,m}$ (MPa)	$\tau_{xy,a}$ (MPa)	$\tau_{xy,m}$ (MPa)	N_{exp} (Cycles)	ER (%)
268.9	89.6	0	0	671481	−17.4
268.9	89.6	0	0	712823	−22.2
247.3	82.4	0	0	976665	7.1
247.1	82.4	0	0	1356769	−22.4
290.2	290.2	0	0	95759	31.0
288.6	288.6	0	0	133071	−1.6
265.7	265.7	0	0	157011	56.1
274.3	274.3	0	0	173047	11.2
255	255	0	0	241679	38.5
254.5	254.5	0	0	422196	−19.5
235.3	235.3	0	0	887342	−30.5
235.4	235.4	0	0	1126890	−45.4
0	0	175	58.3	157758	−24.7
0	0	174.5	58.2	229202	−46.2
0	0	151.6	50.5	551271	46.1
0	0	151.3	50.4	640123	29.2
0	0	151.3	50.4	689753	19.9
0	0	138.4	46.1	2176479	24.7
0	0	138.4	46.1	3187921	−14.8
0	0	138.2	138.2	356608	53.3
0	0	137.8	137.8	526773	7.9
0	0	135.4	135.4	710704	1.1
0	0	135.2	135.2	860208	−14.8
0	0	131.4	131.4	1123272	−4.5
0	0	131.2	131.2	1501895	−27.1

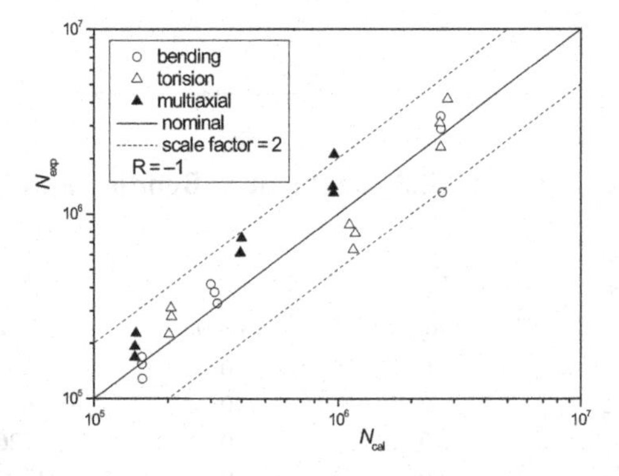

FIGURE 2.12 Comparison of the experimental fatigue lives of 18G2A steel with the ones calculated by using the $\sigma_{ea} - N$ equation under uniaxial and multiaxial loading under zero mean stress.

Source: (Data from Gasiak and Pawliczek [48])

previously, and because the generalized S-N equation has an invariance of mathematical equation forms, as emphasized in Section 3, by using the generalized S-N equation with mean stress effect, the prediction fatigue lives of the material with mean stress under multiaxial loading are in good agreement with experimental ones (see Table 2.12, also see Figure 2.13), with an error range [−48.8, 172.0]%.

At the end of this example, it can be concluded from the discussions above that the generalized S-N equation (i.e., the $\sigma_{ea} - N$ equation) has a good accuracy in performing the lifetime assessment of medium/high cycle fatigue of 18G2A steel.

TABLE 2.12
Fatigue Life Predictions of 18G2A Steel Under Multiaxial Loading with Nonzero Mean Stress

$\sigma_{x,a}$ (MPa)	$\sigma_{x,m}$ (MPa)	$\tau_{xy,a}$ (MPa)	$\tau_{xy,m}$ (MPa)	N_{exp} (Cycles)	ER (%)
146.5	48.8	146.5	48.8	87304	172.0
146.7	48.9	146.7	48.9	93327	151.0
146.5	48.8	146.5	48.8	121935	94.8
129.8	43.3	129.8	43.3	382788	99.8
129.8	43.3	129.8	43.3	417127	83.3
121.4	40.5	121.4	40.5	1408351	3.7
121.4	40.5	121.4	40.5	1505671	−2.9
118.5	39.5	118.5	39.5	2317350	−20.2
118.3	39.4	118.3	39.4	3056739	−38.5
139.4	139.4	139.4	139.4	115525	2.9
139.4	139.4	139.4	139.4	138497	−14.1
129.2	129.2	129.2	129.2	211846	17.0
129.5	129.5	129.5	129.5	233041	4.0
123.2	123.2	123.2	123.2	257221	52.6
123.2	123.2	123.2	123.2	366179	7.2
120.9	120.9	120.9	120.9	681921	−30.8
120.9	120.9	120.9	120.9	794444	−40.6
118	118	118	118	1165727	−48.8

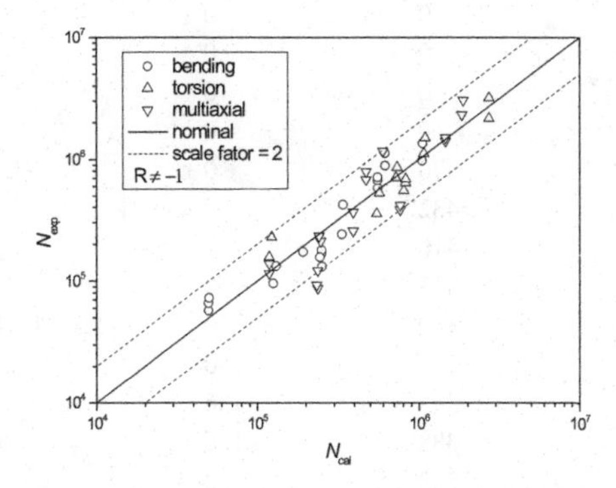

FIGURE 2.13 Comparison of the experimental fatigue lives of 18G2A steel with the ones calculated by using the $\sigma_{ea} - N$ equation under uniaxial and multiaxial loading with mean stress.

Source: (Data from Gasiak and Pawliczek [48])

Example 2: Z12CNDV12-2 Steel

According to the experimental fatigue data of Z12CNDV12-2 steel [49], it is a low/medium/high cycle fatigue problem with the influence of mean stress and out-of-phase loading.

For this material, first, the *S-N* equation is used to perform fatigue life predictions of the material under **in-phase fatigue loading**.

Under bending loading with $R = -1$, the experimental and calculated fatigue lives are shown in Table 2.13, with an error range [−10.2, 8.5]%, from which it is seen that, by using the *S-N* equation, the prediction fatigue lives of the material are in excellent agreement with the experimental ones.

Under torsion loading with $R = -1$, the comparisons of experimental and calculated fatigue lives are also shown in Table 2.13, with a larger error range [−76.8, 155.4]% because of obvious nonlinearity of torsion fatigue behavior of the material (see Figure 2.14). For the obvious nonlinearity, the satisfactory prediction fatigue lives can be obtained by using the similar *S-N* equation, which will be reported in the other study.

Under multiaxial loading with $R = -1$, the comparisons of experimental and calculated fatigue lives are also shown in Table 2.13, with an error range [−48.8, 16.6] %, from which it is seen that, by using the generalized *S-N* equation, the prediction fatigue lives are in excellent agreement with experimental ones.

For the torsion fatigue of the material with *mean stress*, very accurate prediction results of fatigue lives are obtained by using the *S-N* equation with *mean stress*, with an error range [−18.6, 22.7] %.

TABLE 2.13
Fatigue Life Predictions of Z12CNDV12-2 Steel Under In-Phase Fatigue Loading

$\sigma_{x,a}$ (MPa)	$\sigma_{x,m}$ (MPa)	$\tau_{xy,a}$ (MPa)	$\tau_{xy,m}$ (MPa)	N_{exp} (Cycles)	ER (%)
870	0	0	0	477	−4.7
775	0	0	0	1556	7.8
641	0	0	0	15951	−10.2
627	0	0	0	16929	8.5
0	0	470	0	754	155.4
0	0	432	0	3800	−8.4
0	0	410	0	6415	−21.7
0	0	373	0	42070	−76.8
0	0	160	0	1570000	135.4
0	0	323	180	21796	22.7
0	0	343	136	21591	−18.6
0	0	388	126	6692	10.4
463	0	263	0	21673	−48.8
556	0	312	0	1732	16.3

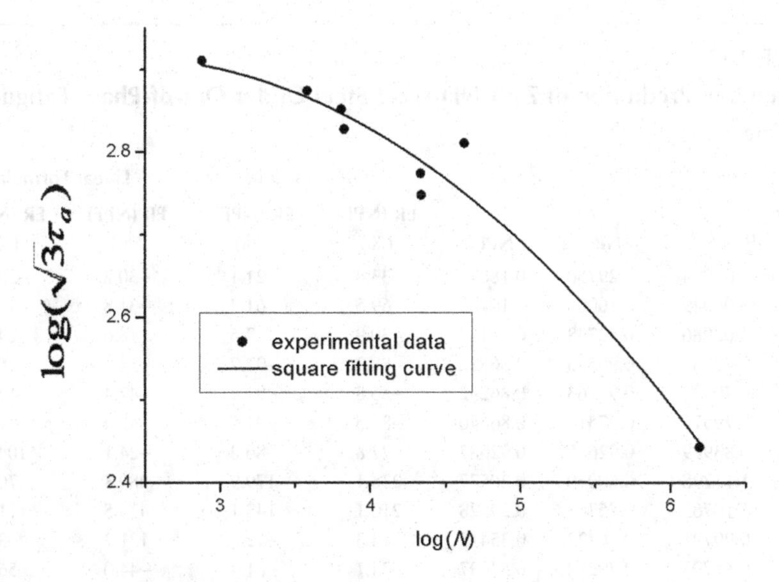

FIGURE 2.14 Torsion fatigue behavior of Z12CNDV12-2 steel.

Source: (Data from Chaudonneret [49])

TABLE 2.14
Out-of-Phase Fatigue Loading of Z12CNDV12-2 Steel

No.	$\sigma_{x,a}$ (MPa)	$\sigma_{x,m}$ (MPa)	$\tau_{xy,a}$ (MPa)	$\tau_{xy,m}$ (MPa)	δ (degree)	N_{exp}
1	219	0	425	0	90	2088
2	122	0	440	0	90	1612
3	402	0	440	0	90	780
4	791	0	441	0	90	725
5	779	0	437	0	90	724
6	778	0	438	0	90	706
7	897	0	480	0	90	130
8	316	0	159	0	63	277040
9	340	0	195	0	106	76715
10	400	0	230	0	75	43316
11	629	0	302	0	82	5240
12	716	0	300	0	64	1147
13	691	0	388	0	62	330

Second, the fatigue life prediction of the material **under out-of-phase loading** will be studied in detail. Out-of-phase loading cases are listed in Table 2.14.

A ratio, σ_{ean}/σ_{ea}, i.e., the ratio of the amplitude of the equivalent non-proportional stress, σ_{ean}, to the effective proportional stress amplitude, σ_{ea}, is briefly called an effective stress ratio. The calculated effective stress ratio,

TABLE 2.15
Fatigue Life Prediction of Z12CNDV12-2 Steel Under Out-of-Phase Fatigue Loading

				Itoh's Formula		Linear Formula	
No.	σ_{ean}/σ_{ea}	NP F	NPF′	ER (NPF) (%)	ER (NPF′) (%)	ER (NPF) (%)	ER (NPF′) (%)
1	1.03094	0.29750	0.18682	33.4	21.1	−30.2	−23.3
2	1.06979	0.16008	0.10102	69.5	61.4	−31.8	−18.4
3	1.02086	0.52748	0.37162	30.9	7.5	18.6	24.5
4	0.78387	0.96565	0.86822	−88.3	−92.7	−54.5	−20.7
5	0.79388	0.97163	0.86292	−86.8	−91.7	−47.4	−12.5
6	0.79571	0.97511	0.86640	−86.5	−91.5	−45.4	−9.18
7	0.83979	0.92685	0.92687	−77.8	−86.2	−24.1	104.0
8	1.11798	0.60040	0.20577	238.1	179.9	280.8	70.2
9	1.10763	0.75353	0.29528	210.1	143.1	473.5	111.5
10	0.99791	0.76727	0.35417	14.3	−12.2	121.3	−3.5
11	0.85793	0.79288	0.53037	−74.1	−81.4	−44.0	−56.6
12	0.92160	0.56236	0.40679	−50.7	−61.7	−50.8	−47.0
13	0.97064	0.60034	0.47316	−15.3	−35.2	−4.8	16.1

σ_{ean}/σ_{ea}, and the nonproportional loading factors, F, and F', are given in Table 2.15 with the out-of-phase loading cases. At the same time, the fatigue life prediction results are also shown in Table 2.15. The fatigue life prediction results by Itoh's formula have an error range [−88.8, 238.1]% and an absolute mean error 82.8%, which are much more than those of in-phase loading. ***This results in the author's deep analysis for it***. First it is found that, within the range [0.78, 1.1] of σ_{ean}/σ_{ea}, it is hard to establish the suitable relationship between σ_{ean}/σ_{ea} and F by Itoh's formula. Then such an attempt is made that a linear relation between σ_{ean}/σ_{ea} and F is proposed. The fatigue life prediction results obtained by the linear relation between σ_{ean}/σ_{ea} and F are also given in Table 2.15, with an error range [−54.5,473.6]% and an absolute mean error 94.4%, which are also very poor. Further, a new definition, F', of the nonproportional loading factor, as expressed in formula (23), is proposed, and in addition, the new definition is employed in the linear relation between σ_{ean}/σ_{ea} and F' to calculate the fatigue lives of the material under out-of-phase loading. The obtained results are given in Table 2.15, with an error range, [−55.6,111.5]% and an absolute mean error 39.8%, from which it is seen that the prediction results obtained on the basis of the new definition F' are much better than those on the basis of the original definition F.

The error indexes calculated under out-of-phase loading are listed in Table 2.16.

By the way, the new definition, F', of the nonproportional loading factor, as expressed in formula (23), and the linear relation, as expressed in formula (22), proposed in this study, is originated from the fatigue life analysis of the material under the out-of-phase loading.

TABLE 2.16
Error Indexes of Fatigue Life Analysis of Z12CNDV12-2 Steel Under Out-of-Phase Fatigue Loading

	ER (NPF)				ER (NPF′)			
	Min ER (%)	Max ER. (%)	AER (%)	MER (%)	Min ER (%)	Max ER. (%)	AER (%)	MER (%)
Itoh's F	−88.8	238.1	82.8	8.9	−92.7	179.9	74.3	−10.7
Linear F	−54.5	473.6	94.4	43.1	−56.6	111.5	39.8	10.4

Example 3: 5% Cr Steel

According to the experimental fatigue data of 5% Cr steel [50], it is a low/medium/high cycle fatigue problem with the influence of mean stress and out-of-phase loading.

The comparisons of experimental and calculated fatigue lives of the material under **in-phase loading** are shown in Figure 2.15, from which it is seen that the predicted results are in good agreement with experimental ones.

The fatigue life prediction of the material under **out-of-phase loading** will be studied in detail in the following.

Out-of-phase loading cases are listed in Table 2.17. On the basis of the Itoh's formula, and the original and new definitions of the nonproportional loading factor, the fatigue life prediction results are given in Table 2.18, from which it is found that there is an experimental case (No. 8) whose relative error (ER(NPF)) reaches to 775.3%. The case failure obviously is **untimely fatigue failure**. The experimental fatigue data of this case is the *unusual data*, which will be eliminated in the fatigue life prediction of the material.

After eliminating the *unusual data*, the calculated results, including the variation of $\sigma_{ean} / \sigma_{ea}$, NPF, NPF′, and relative errors of fatigue life prediction with out-of-phase loading cases, are given in Table 2.19. The calculated error indexes are listed in Table 2.20, in which error indexes including the *unusual data* are also listed.

On the basis of Itoh's formula, and the original definitions of the nonproportional loading factor, the comparisons of two sets of error indexes including and excluding the unusual data illustrate that it is important to exclude the unusual data in the fatigue life prediction of the material: error ranges, [−88.8, 775.3]% and [−78.1, 206.2]%, mean errors, 96.3% and 50.9%.

While on the basis of Itoh's formula, and the original and new definitions of the nonproportional loading factor, the comparisons of two sets of error indexes show that the calculated results of fatigue lives of the material by using the new definition of the nonproportional loading factor are better than those by using the original definition of the nonproportional loading factor: the absolute mean errors, 78.3% and 98.6%, the means errors, 29.6% and 50.9%.

Also on the basis of the original definition of the nonproportional loading factor, the fact that the calculated results of fatigue lives of the material by the linear formula are better than those by Itoh's formula—the absolute mean errors, 79.9% and 98.6%, the means errors, 41.06% and 50.9%—illustrates that it is necessary to employ the linear formula (22) to the fatigue life analysis of the material.

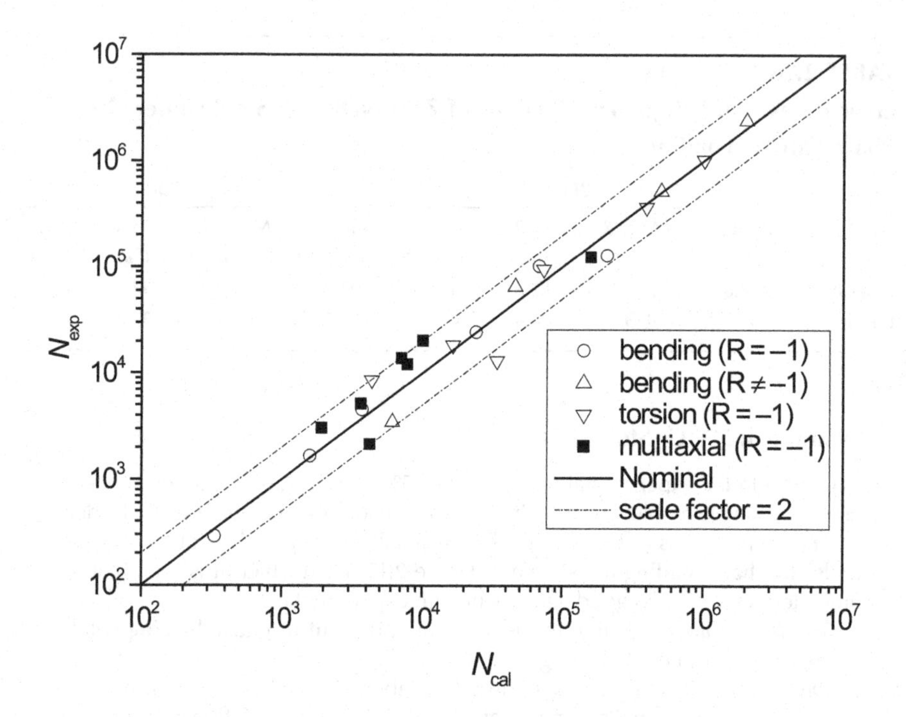

FIGURE 2.15 Comparison of experimental and calculated fatigue lives of 5% Cr steel under in-phase loading.

Source: (Data from Kim, *et al* [50])

TABLE 2.17
Out-of-Phase Fatigue Loading of 5% Cr Steel

No.	$\sigma_{x,a}$ (MPa)	$\sigma_{x,m}$ (MPa)	$\tau_{xy,a}$ (MPa)	$\tau_{xy,m}$ (MPa)	δ (degree)	N_{exp} (Cycles)
1	748	0	433	0	58	46794
2	735	0	432	0	94	45358
3	589	0	605	0	58	6040
4	594	0	597	0	101	20609
5	434	0	751	0	50	13727
6	430	0	745	0	101	6890
7	690	0	394	0	72	45731
8	680	0	390	0	108	13842
9	550	0	549	0	101	279968
10	549	0	543	0	94	110207
11	642	0	791	0	79	8889

TABLE 2.18
Fatigue Life Prediction Results of 5% Cr Steel Under Out-of-Phase Loading by Itoh's Formula (Including the Unusual Experimental Data)

No.	$\left(\frac{\sigma_{ean}}{\sigma_{ea}}-1\right)/F$	$\left(\frac{\sigma_{ean}}{\sigma_{ea}}-1\right)/F'$	ER(NPF) (%)	ER(NPF′) (%)
1	−0.37979	−0.5425	−66.6	−73.6
2	−0.21232	−0.30627	114.9	33.3
3	−0.09835	−0.12378	170.4	178.2
4	−0.22602	−0.28676	33.1	35.8
5	−0.39856	−0.43976	−38.1	−28.6
6	−0.12301	−0.13683	85.6	123.0
7	−0.19278	−0.30055	120.1	26.6
8	−0.0661	−0.10435	775.3	388.3
9	−0.3966	−0.54625	−71.1	−75.3
10	−0.2747	−0.38164	−12.8	−27.5
11	−0.48768	−0.48768	−80.8	−68.2

TABLE 2.19
Fatigue Life Prediction Results of 5% Cr Steel Under Out-of-Phase Loading (Excluding the Unusual Experimental Data)

				Itoh's Formula		Linear Formula	
No.	σ_{ean}/σ_{ea}	NP F	NPF′	ER(NPF) (%)	ER(NPF′) (%)	ER(NPF) (%)	ER(NPF′) (%)
1	0.7894	0.5542	0.3880	−60.4	−69.7	−63.8	−69.2
2	0.8024	0.9304	0.6450	197.7	72.3	44.5	79.7
3	0.9578	0.4288	0.3407	206.2	212.7	235.4	215.9
4	0.8747	0.5542	0.4368	57.3	58.7	44.1	61.7
5	0.9028	0.2436	0.2208	−33.9	−23.4	−8.2	−23.4
6	0.9599	0.3256	0.2927	103.2	145.6	154.5	147.0
7	0.8599	0.7263	0.4659	179.1	50.3	94.2	53.8
9	0.7787	0.5578	0.4050	−65.8	−71.5	−68.9	−71.1
10	0.8404	0.5808	0.4180	4.0	−15.9	−8.4	−14.4
11	0.7779	0.4553	0.4553	−78.1	−62.6	−76.9	−61.8

From Table 2.20, it is seen that, on the basis of the new definition of the non-proportional loading factor, the fatigue life prediction results obtained by means of Itoh's formula are nearly the same as those by the linear formula (22), which illustrates that the linear formula (22) becomes Itoh's formula to the material because the constant λ in the linear formula (22) is nearly equal to 1 ($\lambda = 1.0012$).

TABLE 2.20
Error Indexes of Fatigue Life Analysis of 5% Cr Steel Under Out-of-Phase Loading

	ER(NPF)				ER(NPF′)			
	Min ER (%)	Max ER. (%)	AER (%)	MER (%)	Min ER (%)	Max ER. (%)	AER (%)	MER (%)
Itoh's F	−78.1	206.2	98.6	50.9	−71.5	212.7	78.3	29.6
Linear F	−68.9	235.4	79.9	41.0	−71.1	215.9	79.8	31.8
Itoh's F (Including unusual data)	−80.8	775.3	135.3	93.6	−75.3	388.3	96.2	46.5

Example 4: 6082-T6 Steel

According to the experimental fatigue data of 6082-T6 steel [51], it is a medium/high cycle fatigue problem.

The comparisons of experimental and calculated fatigue lives of the material under **in-phase loading** are shown in Figure 2.16, from which it is seen that the predicted results are in good agreement with experimental ones.

The fatigue life prediction of 6082-T6 steel under **out-of-phase loading** will be discussed in detail in the following.

Out-of-phase loading cases are listed in Table 2.21. On the basis of Itoh's formula, and the original and new definitions of the nonproportional loading factor, the fatigue life prediction results of the material are given in Table 2.22, from which it is found that there is an experimental case (No.7) whose relative error (ER(NPF′) reaches to 916.9%. The case failure obviously is **untimely fatigue failure**. The experimental fatigue data of this case is the *unusual data*, which will be eliminated in the fatigue life prediction of the material.

After eliminating the *unusual data*, the calculated results are given in Table 2.23. The calculated error indexes are listed in Table 2.24, in which error indexes including the *unusual data* are also listed.

On the basis of Itoh's formula, and the original definition of the nonproportional loading factor, the comparisons of two sets of error indexes including and excluding the *unusual data* (see Table 2.24) again illustrate that ***it is important to exclude the unusual data in the processing of experimental fatigue data***: error ranges, [−63.7, 632.9]% and [−62.9, 237.8]%, and mean errors, 62.2% and 27.99%.

On the basis of the original definition of the nonproportional loading factor, the fact that the calculated results of fatigue life of 6082-T6 steel by the linear formula are better than those by Itoh's formula—error ranges, [−58.3, 162.9]% and [−62.9, 237.8]%, absolute mean errors, 43.3% and 54.2%, and means errors, 11.3% and 27.9%—illustrates that it is necessary to employ the linear formula (22) in the fatigue life analysis of the material.

It is seen from Table 2.24 that the fatigue life prediction results of the material obtained on the basis of the new definition of the nonproportional loading factor and the linear formula are best in all calculated results.

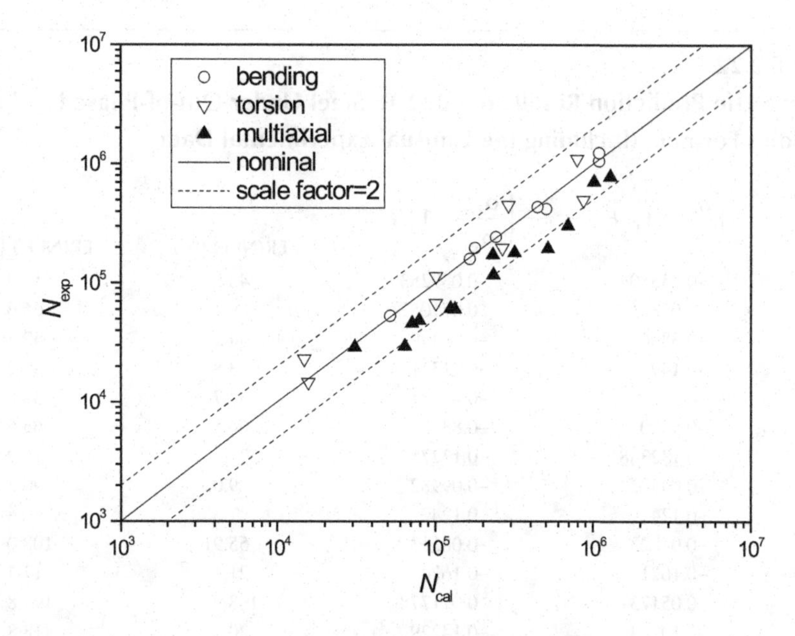

FIGURE 2.16 Comparison of experimental and calculated fatigue lives of 6082-T6 steel under in-phase loading.

Source: (Data from Susmel and Petrone [51])

TABLE 2.21
Out-of-Phase Loading of 6082-T6 Steel

No.	$\sigma_{x,a}$ (MPa)	$\sigma_{x,m}$ (MPa)	$\tau_{xy,a}$ (MPa)	$\tau_{xy,m}$ (MPa)	δ (degree)	N_{exp} (Cycles)
1	79	0	129	0	129	20730
2	79	0	116	0	125	41490
3	69	0	110	0	126	188882
4	68	0	99	0	128	234725
5	68	0	99	0	125	368080
6	60	0	94	0	126	1016280
7	188	0	106	0	89	5590
8	189	0	106	0	94	27420
9	189	0	106	0	88	34015
10	171	0	99	0	91	44750
11	190	0	105	0	91	47020
12	149	0	68	0	93	114845
13	151	0	67	0	94	273325
14	155	0	72	0	92	445560
15	152	0	47	0	91	456725

TABLE 2.22
Fatigue Life Prediction Results of 6082-T6 Steel Under Out-of-Phase Loading by Itoh's Formula (Including the Unusual Experimental Data)

No.	$\left(\frac{\sigma_{ean}}{\sigma_{ea}}-1\right)/F$	$\left(\frac{\sigma_{ean}}{\sigma_{ea}}-1\right)/F'$	ER(NPF) (%)	ER(NPF′) (%)
1	0.035304	0.039253	42.3	51.9
2	0.058219	0.071057	59.2	65.6
3	−0.38065	−0.49499	−44.5	−43.7
4	−0.14933	−0.21331	−3.8	−5.2
5	−0.31527	−0.45035	−37.7	−38.6
6	−0.55003	−0.83527	−64.5	−65.6
7	0.132388	0.132752	577.5	916.9
8	−0.09982	−0.09982	29.0	90.3
9	−0.1242	−0.1242	7.3	60.6
10	−0.07823	−0.08513	55.91	108.0
11	−0.1621	−0.1624	−21.1	17.4
12	0.051737	0.071778	198.9	194.8
13	−0.10284	−0.14229	20.1	18.8
14	−0.21841	−0.28929	−44.1	−42.6
15	−0.09256	−0.14144	17.9	11.2

TABLE 2.23
Fatigue Life Prediction Results of 6082-T6 Steel Under Out-of-Phase Loading (Excluding the Unusual Experimental Data)

				Itoh's Formula		Linear Formula	
No.	σ_{ean}/σ_{ea}	NP F	NPF′	ER(NPF) (%)	ER(NPF′) (%)	ER(NPF) (%)	ER(NPF′) (%)
1	1.0092	0.2608	0.2346	48.2	58.7	69.8	74.2
2	1.0177	0.3049	0.2498	67.0	73.5	85.0	87.1
3	0.8937	0.2792	0.2147	−42.1	−41.4	−34.5	−34.2
4	0.9562	0.2932	0.2053	0.6	−1.5	12.4	11.6
5	0.9031	0.3072	0.2150	−34.6	−36.1	−27.7	−28.4
6	0.8439	0.2836	0.1867	−62.9	−64.4	−58.3	−58.8
8	0.9074	0.9271	0.9271	49.8	129.4	4.0	10.8
9	0.8813	0.9555	0.9555	25.3	95.0	−14.8	−9.2
10	0.9231	0.9825	0.9029	83.0	149.4	21.5	24.1
11	0.8453	0.9539	0.9522	−8.0	42.4	−37.4	−33.3
12	1.0406	0.7858	0.5664	237.8	227.2	162.9	147.6
13	0.9217	0.7613	0.5502	35.2	31.4	7.2	1.3
14	0.8247	0.8022	0.6057	−36.6	−35.7	−51.3	−53.5
15	0.9504	0.5354	0.3503	27.6	18.1	19.6	13.9

TABLE 2.24
Error Indexes of Fatigue Life Analysis of 6082-T6 Steel Under Out-of-Phase Loading

	ER(NPF)				ER(NPF′)			
	Min ER (%)	Max ER. (%)	AER (%)	MER (%)	Min ER (%)	Max ER. (%)	AER (%)	MER (%)
Itoh's F	−62.9	237.8	54.2	27.9	−64.4	227.2	71.7	46.1
Linear F	−58.3	162.9	43.3	11.3	−58.8	147.6	42.0	10.9
Itoh's F (Including unusual data)	−63.7	632.9	88.9	62.2	−65.6	916.9	115.4	89.3

2.4 CONCLUSIONS AND FINAL COMMENTS

From this study, the following conclusions can be made:

(1) The Wöhler Curve Method is well suitable for the low-cycle fatigue life analysis of metallic materials.
(2) In a low/medium/high cycle fatigue regime of metallic materials, the Wöhler Curve Method and the generalized Wöhler Curve Method (the generalized *S-N* equation) proposed in this study are well suited to perform the fatigue life assessment of metallic materials under uniaxial and multiaxial loading, respectively.
(3) On dealing with nonproportional loading fatigue, a new definition of nonproportional factor (NPF') (see formula (23)), and the linear formula (see formula (22)), proposed in this study, has been proven to have high accuracy with respect to Itoh's formula (21) in the fatigue life assessment of metallic materials.

The three comments given here are as follows:

(1) Although it appears to be sure enough by using a number of the experimental fatigue data of metallic materials that, in a low/medium/high cycle fatigue regime of metallic materials, the Wöhler Curve Method (including the generalized Wöhler Curve Method) are well suited to perform the fatigue life assessment of metallic materials, further experimental verifications are needed.
(2) In this study, it is important to illustrate that, for the LCF of metallic materials, the way that a "stress quantity" calculated based on the *linear-elastic analysis* of the studied component is taken to be a *mechanical quantity, S*, to establish a relationship between the *mechanical quantity, S*, and the fatigue life, *N*, is practicable. Based on the practicability, the generalized Wöhler equation (or the generalized *S-N* equation), for a low/medium/ high cycle fatigue life assessment of metallic materials, is proposed. The

prediction equation is a stress invariant based one, in which the computation of stress invariants is on the basis of the *linear-elastic analysis* of the studied component.

(3) It is worth here highlighting that, by using *linear-elastic analysis,* the LCF life assessment of metallic materials can be performed by the Wöhler Curve Method without the need for carrying out complex and time-consuming elastic-plastic analyses.

REFERENCES

1 Susmel, L., Lazzarin, P. A bi-parametric modified Wöhler curve for high cycle multiaxial fatigue assessment. *Fatigue and Fracture of Engineering Materials and Structures* 25, 63–78 (2002).

2 Susmel, L., Meneghetti, G., Atzori, B. A simple and efficient reformulation of the classical Manson-Coffin curve to predict lifetime under multiaxial fatigue loading. Part I: Plain materials. *Transactions of the ASME, Journal of Engineering Materials and Technology* 131, 021009-1 to 9 (April 2009).

3 Socie, D. F. Multiaxial fatigue damage models. Transactions of the ASME, Journal of Engineering Materials and Technology 109, 293–298 (1987).

4 Cristofori, A., Susmel, L., Tovo, R. A stress invariant based criterion to estimate fatigue damage under multiaxial loading. *International Journal of Fatigue* 30, 1646–1658 (2008).

5 Papadopoulos, I.V., Davoli, P., Gorla, C., Filippini, M., Bernasconi, A. A comparative study of multiaxial highcycle fatigue criteria for metals. *Int J Fatigue* 19, 219–235 (1997).

6 Sines, G. Behaviour of metals under complex static and alternating stresses. In: *Metal fatigue*. New York: McGraw Hill; 1959. p. 145–169.

7 Crossland, B. Effect of large hydroscopic pressures on the torsional fatigue strength of an alloy steel. In: Proceedings of international conference on fatigue of metals, London, New York; 1956. p. 138–149.

8 Li, B., De Freitas, M. A procedure for fast evaluation of high cycle fatigue under multiaxial random loading. *J Mech Design* 124, 558–563 (2002).

9 Wang, C.H., Brown, M.W. Multiaxial random load fatigue: Life prediction techniques and experiments. Fourth international conference on biaxial/multiaxial fatigue, Paris; 1994, May 31–June 3.

10 Bishop, N.W.M., Sherratt, F. *Finite element based fatigue calculations.* Farnham UK: NAFEMS Ltd; 2000.

11 Yan, X.Q. An empirical fracture equation of mixed mode cracks. *Theoretical and Applied Fracture Mechanics* 116, 103146 (2021).

12 Susmel, L. *Multiaxial notch fatigue, from nominal to stress/strain quantities.* Woodhead Publishing Limited, CRC Press; Cambridge, UK 2009.

13 Gao, Z., Zhao, T., Wang, X., Jiang, Y. Multiaxial fatigue of 16MnR steel. *ASME J Press Vess Technol* 131(2), 021403 (2009).

14 Gao, Z., Qiu, B., Wang, X., Jiang, Y. An investigation of fatigue of a notched member. *International Journal of Fatigue* 32, 1960–1969 (2010).

15 Coffin, L.F. A study of the effects of cyclic thermal stresses on a ductile metal. *Transactions of the ASME* 76, 931–950 (1954).

16 Manson, S.S. *Behaviour of materials under conditions of thermal stress.* NACA TN-2933, National Advisory Committee for Aeronautics; 1954.

17 Morrow, J.D. Cyclic plastic strain energy and fatigue of metals. In: Internal friction, damping and cyclic plasticity, ASTM STP 378, American Society for Testing and Materials, Philadelphia, PA; 1965. p. 45–84.
18 Basquin, O.H. The exponential law of endurance tests. *Proceedings of American Society for Testing and Materials* 10, 625–630 (1910).
19 Szusta, J., Seweryn, A. Fatigue damage accumulation modelling in the range of complex low-cycle loadings: The strain approach and its experimental verification on the basis of EN AW-2007 aluminum alloy. *Int J Fatigue* 33, 255–264 (2011).
20 Szusta, J., Seweryn, A. Experimental study of the low-cycle fatigue life under multiaxial loading of aluminum alloy EN AW-2024-T3 at elevated temperatures. *Int J Fatigue* 33, 255–264 (2011).
21 Nelson, D.V., Rostami, A. Biaxial fatigue of A533B pressure vessel steel. *Transactions of the ASME, Journal of Pressure Vessel Technology* 119, 325–331 (1997). doi: 10.1115/1.2842312
22 Socie, D.F., Kurath, P., Koch, J. A multiaxial fatigue damage parameter. In: Brown, M.W., Miller, K.J., editors. *Biaxial and Multiaxial Fatigue*, EGF 3. London: Mechanical Engineering Publications; 1989. p. 535–550.
23 Kurath, P., Downing, S.D., Galliart, D.R. Summary of non-hardened notched shaft-round robin program. In: Leese, G.E., Socie, D.F., editors. *Multiaxial fatigue-analysis and experiments*, SAE AE-14. Warrendale, PA: Society of Automotive Engineers; (1989). p. 13–32.
24 Socie, D.F. Multiaxial fatigue damage models. *Transactions of the ASME, Journal of Engineering Materials and Technology* 109, 293–298 (1987).
25 Kallmeyer, A.R., Krgo, A., Kurath, P. Evaluation of multiaxial fatigue life prediction methodologies for Ti-6Al-4V. *Transactions of the ASME, Journal of Engineering Materials and Technology* 124, 229–237 (2002). doi: 10.1115/1.1446075
26 Lin, H., Nayeb-Hashemi, H., Pelloux, R.M. Constitutive relations and fatigue life prediction for anisotropic Al-6061-T6 rods under biaxial proportional loadings. *International Journal of Fatigue* 14, 249–259 (1992). doi: 10.1016/0142-1123(92)90009-2
27 Chen, X., An, K., Kim, K.S. Low-cycle fatigue of 1Cr–18Ni–9Ti stainless steel and related weld metal under axial, torsional and 90 out-of-phase-loading. *Fatigue and Fracture of Engineering Materials and Structures* 27, 439–448 (2004). doi: 10.1111/j.1460-2695.2004.00740.x
28 Itoh, T., Sakane, M., Ohnami, M., Socie, D.F. Nonproportional low cycle fatigue criterion for type 304 stainless steel. *Transactions of the ASME, Journal of Engineering Materials and Technology* 117, 285–292 (1995).
29 Du, W., Luo, Y., Wang, Y., Chen, S., Yu, D. A new energy-based method to evaluate low-cycle fatigue damage of AISI H11 at elevated temperature. *Fatigue Fract Engng Mater Struct* 40, 994–1004 (2017).
30 Branco, R., Costa, J.D., Antunes, F.V. Low-cycle fatigue behaviour of 34CrNiMo6 high strength steel. *Theoretical and Applied Fracture Mechanics* 58, 28–34 (2012).
31 Shankar, V., Mariappan, K., Nagesha, A., Reddy, G.V. P., Sandhya, R., Mathew, M.D., Jayakumar, T. Effect of tungsten and tantalum on the low cycle fatigue behavior of reduced activation ferritic/martensitic steels. *Fusion Engineering and Design* 87, 318–324 (2012).
32 Zhang, Z., Hu, Z.F., Schmauder, S., Zhang, B.S., Wang, Z.Z. Low cycle fatigue properties and microstructure of P92 ferritic-martensitic steel at room temperature and 873 K. *Materials Characterization* 157, 109923 (2019).
33 Li, Z.Q., Han, J.M., Li., W.J., Pan, L.K. Low cycle fatigue behavior of Cr–Mo–V low alloy steel used for railway brake discs. *Materials and Design* 56, 146–157 (2014).

34 Hatanaka, K. Cyclic stress-strain responds and low-cycle fatigue life in metallic materials. *JSME, International Journal*, Series 1 33(1), 13–25 (1990).
35 Hatanaka, K., Fujimisu, T. The cyclic stress-strain response and strain life behavior of metallic materials. *Proc. of Fatigue '84* 1, 93 (1984).
36 Li, D.M., Kim, K.W., Lee, C.S. Low cycle fatigue data evaluation for a high strength spring steel. *Int. J. Fatigue* 19(8–9), 607-612 (1997).
37 Roy, S.C., Goyal, S., Sandhy, R., Ray, S.K. Low cycle fatigue life prediction of 316 L(N) stainless steel based on cyclic elasto-plastic response. *Nuclear Engineering and Design* 253, 219–225 (2012).
38 Hu, X., Huang, L.X., Wang, W.G., Yang, Z.G., Sha, W., Wang. W.Y., Shan, Y.Y. Low cycle fatigue properties of CLAM steel at room temperature. *Fusion Engineering and Design* 88, 3050–3059 (2013).
39 Marmy, P., Kruml, T. Low cycle fatigue of Eurofer 97. *Journal of Nuclear Materials* 377, 52–58 (2008).
40 Stubbins, J.F., Gelles, D.S. Fatigue performance and cyclic softening of F82H, a ferritic-martensitic steel. *Journal of Nuclear Materials* 233–237, 331–335 (1996).
41 Atzori, B., Meneghetti, G., Susmel, L., Taylor, D. The modified Manson-Coffin method to estimate low-cycle fatigue damage in notched cylindrical bars. In: U. S. Fernando, editor. Proceedings of the 8th international conference on multiaxial fatigue and fracture, Sheffield, UK; 2007 July.
42 Blaž Šamec, Iztok Potrč, Matjaž Šraml. Low cycle fatigue of nodular cast iron used for railway brake discs. *Engineering Failure Analysis* 18, 1424–1434 (2011).
43 Pevec, M., Oder, G., Potrč, I., Šraml, M. Elevated temperature low cycle fatigue of grey cast iron used for automotive brake discs. *Engineering Failure Analysis* 42, 221–230 (2014).
44 Liu, B, Yan, X. A new model of multiaxial fatigue life prediction with influence of different mean stresses. *International Journal of Damage Mechanics* 28(9), 1323–1343 (2019).
45 Tao, G., Xia, Z. Mean stress/strain effect on fatigue behavior of an epoxy resin. *Int J Fatigue* 29(12), 2180–2190 (2007).
46 Marin, J. Interpretation of fatigue strength for combined stresses. In: Proceedings of international conference on fatigue of metals, London; 2005. p. 184–192.
47 Freitas, M., Li, B., Santos, J.L.T. Multiaxial fatigue and deformation: Testing and prediction. *ASTM STP* 1387 (2000).
48 Gasiak, G., Pawliczek, R. The mean loading effect under cyclic bending and torsion of 18G2A steel. In: de Freitas, M., editor. Proceedings of 6th international conference on biaxial/multiaxial fatigue and fracture, Lisbon, Portugal; 2001, p. 213–222.
49 Chaudonneret, M. A simple and efficient multiaxial fatigue damage model for engineering applications of macro-crack initiation. *Transactions of the ASME, Journal of Engineering Materials and Technology* 115, 373–379 (1993).
50 Kim, K.S., Nam, K.M., Kwak, G.J., Hwang, S.M. A fatigue life mode for 5% chrome work roll steel under multiaxial loading. *International Journal of Fatigue* 26, 683–689 (2004).
51 Susmel, L., Petrone, N. Multiaxial fatigue life estimations for 6082-T6 cylindrical specimens under in-phase and out-of-phase biaxial loadings. In: Biaxial and Multiaxial Fatigue and Fracture, edited by A. Carpinteri, M. de Freitas and A. Spagnoli, Elsevier and ESIS, 83–104. DOI: 10.1016/S1566-1369(03)80006-7.
52 Kim, K.S., Park, J.C., Lee, J.W. Multiaxial fatigue under variable amplitude loads. *Transaction of the ASME, Journal of Engineering Materials and Technology* 121, 286–293 (1999).
53 Jiang, Y., Hertel, O., Vormwald, M. An experimental evaluation of three critical plane multiaxial fatigue criteria. *International Journal of Fatigue* 29, 1490–1502 (2007).

APPENDIX A: EXPERIMENTAL INVESTIGATIONS: THE WÖHLER CURVE METHOD IS WELL SUITED FOR THE LOW-CYCLE FATIGUE LIFE ANALYSIS OF METALLIC MATERIALS BY THE LOW-CYCLE FATIGUE TEST DATA OF METALLIC MATERIALS FROM THE LITERATURE

Here, the low-cycle fatigue test data of metallic materials reported in the literature are analyzed by using the Wöhler equation to illustrate that the low-cycle fatigue life of the metallic materials studied can be calculated by the Wöhler equation. The obtained results are listed in Tables 2.A-2 to 2.A-14. A summary of these results is given in Table 2.A-1. A comparison of the experimental low-cycle fatigue lives with those calculated by the Wöhler equation is shown in Figures 2.A-1 and 2.A-2. The type materials include A533B by Nelson and Rostami [21] (Table 2.A-2), Inconel 718 by Socie et al. [22] (Table 2.A-3), SAE 1045 by Kurath et al.[23] (Table 2.A-4), AISI 304 by Socie [24] (Table 2.A-45), Ti-6Al-4V by Kallmeyer et al.[25] (Table 2.A-6), 6061-T6 by Lin et al.[26] (Table 2.A-7, Table A.2-8), 1Cr-18*N*i-9Ti by Chen et al.[27] (Table 2.A-9, Table 2.A-10), AISI 304 by Itoh et al. [28] (Table 2.A-11), AISI H11 by Du et al. [29] (Table 2.A-12), 34CrNiMo6 by Branco et al. [30] (Table 2.A-13), and RAFM steels by Shankar et al.[31] (Table 2.A-14).

From Tables 2.A-1 to 2.A-14 and Figures 2.A-1 to 2.A-2, it is seen that the experimental low-cycle fatigue lives of the metallic materials studied are in good agreement with those calculated by the Wöhler equation, which illustrates that the low-cycle fatigue life of the metallic materials studied can be calculated by the Wöhler equation.

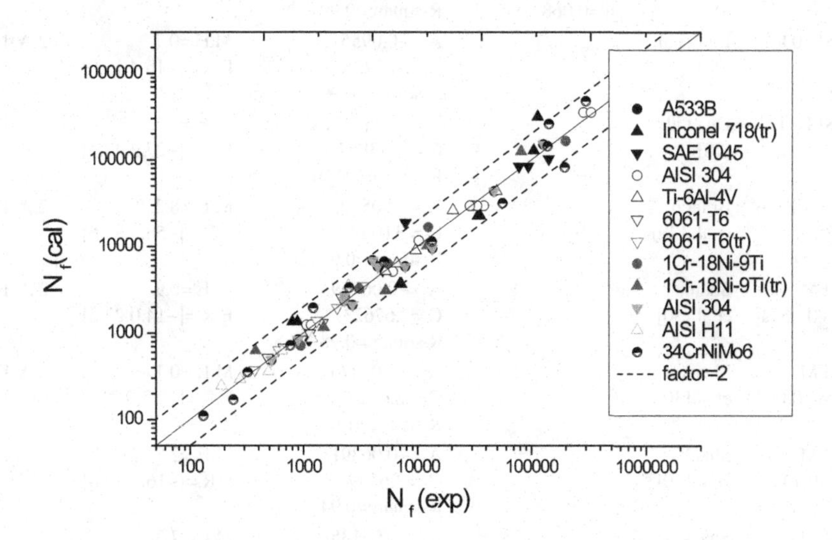

FIGURE 2.A-1 Comparison of the experimental low-cycle fatigue lives of metallic materials studied with those calculated by the Wöhler equation.

TABLE 2.A-1
Summary of the Obtained Results That the Low-Cycle Fatigue Test Data of Metallic Materials Reported in the Literature Are Analyzed by Using the Wöhler Equation or Basquin's Equation

Material	Refs	Constants in Rambrg–Osgood Relationship	Constants in Basquin's Equation or the Wöhler Equation	Error Indexes (%)	Tables
A533B	Nelson and Rostami [21]	$E = 198000$, $K' = 827$, $n' = 0.13$	$A = -0.08206$, $C = -2.39763$, R-square = 0.94	M.E. = 3.3 E.R. = [−30.1,68.5]	2.A-2
Inconel 718	Socie et al. [22]	G = 77800, K′ = 860, n′ = 0.079	B = −0.12807, D = −1.71402, R-square = 0.89	M.E. = 19.3, E.R. = [−48.8,176.5]	2.A-3
SAE 1045	Kurath et al. [23]	E = 204000, K = 1185, N = 0.23	A = −0.10394, C = −2.4121, R-square = 0.95	M.E. = 9.4, E.R. = [−27.8,139.6]	2.A-4
AISI 304	Socie [24]	E=183000, K= 1210, n=0.193	A=−0.09292, C=−2.30296, R-square=0.99	M.E.=1.0 E.R.=[−23.2,23.1]	2.A-5
Ti-6Al-4V	Kallmeyer et al.[25]	E=116000, K′= 854, n′=0.0149	A_1=−0.17767, C_1=3.5478, R-square=0.99	M.E.=0.9, E.R.=[−12.0,28.2]	2.A-6
6061-T6	Lin et al. [26]	E=71500, K′= 436, N′=0.069	A_1=−0.03142, C_1=2.55881, R-square=0.94	M.E.=0.7, E.R.=[−12.8,18.0]	2.A-7
		G=28200, K′=292, n′=0.068	A_0=−0.04613, C_0=2.43043, R-square=0.99	M.E.=0.2, E.R.=[−12.6,6.4]	2.A-8
AISI 304	Itoh et al. [27]		A_1=−0.09551, C_1=2.87225, R-square=0.99	M.E.=0.7, E.R.=[−16.4,15.9]	2.A-11
AISI H11	Du et al. [28]		A_1=−0.11517, C_1= 3.43867, R-square=0.90	M.E.=3.3, E.R.=[−27.2,42.3]	2.A-12
34CrNiMo6	Branco et al. [29]		A_1=−0.0559, C_1= 3.07246, R-square=0.97	M.E.=8.2, E.R.=[−58.4,81.6]	2.A-13
RAFM 2W–0.06Ta	Shankar et al.[30]		A_1=−0.06959, C_1=2.67623, R-square=0.98	M.E.=0.9 E.R.=[−14.0,23.6]	2.A-14
RAFM 1.4W–0.06Ta	Shankar et al.[30]		A_1=−0.06476, C_1= 2.66297, R-square=1.0	M.E.=0.1, E.R.=[−2.2,5.8]	2.A-14
RAFM 1W–0.14Ta	Shankar et al.[30]		A_1=−0.06391, C_1= 2.62494, R-square=0.94	M.E.=2.9, E.R.=[−16.1,49.0]	2.A-14
RAFM 1W–0.06Ta	Shankar et al.[30]		A_1=−0.06135, C_1= 2.62391, R-square=0.84	M.E.=7.3, E.R.=[−37.1,64.4]	2.A-14

Note: Basquin's equation: (a) $\lg(\varepsilon_a^e) = A\lg(N_f) + C$ or $\lg(\gamma_a^e) = B\lg(N_f) + D$; (b) Wöhler equation: $\lg(\sigma_a) = A_1\lg(N_f) + C_1$ or $\lg(\tau_a) = A_0\lg(N_f) + C_0$

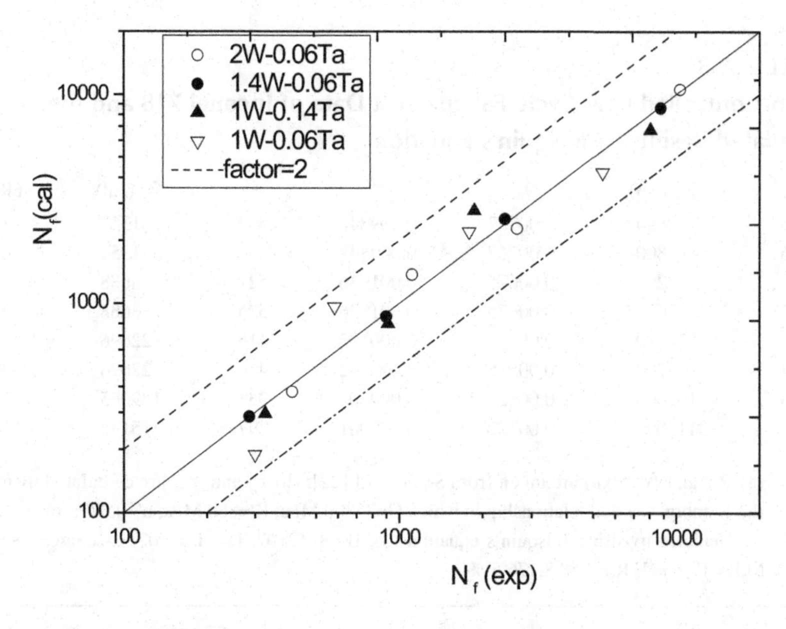

FIGURE 2.A-2 Comparison of the experimental low-cycle fatigue lives of AFM steels with those calculated by the Wöhler equation.

TABLE 2.A-2
Strain-Controlled Low-Cycle Fatigue Test Data of A533B and the Calculated Results by Basquin's Equation

ε_a	N_f (exp)	ε_a^e	ε_a^p	σ_a	N_f (cal)	ER(%)
0.0168	520	0.00241	0.01451	477	487	−6.4
0.0126	900	0.0023	0.01026	456	842	−6.4
0.0079	2700	0.00214	0.00576	423	2104	−22.1
0.0075	2300	0.00212	0.00535	419	2363	2.7
0.0071	2300	0.0021	0.00497	415	2656	15.5
0.005	4500	0.00196	0.00302	389	5842	29.8
0.005	7800	0.00196	0.00302	389	5842	−25.1
0.0047	4060	0.00194	0.00274	384	6840	68.5
0.0041	13500	0.00189	0.00223	374	9434	−30.1
0.0025	47300	0.00167	0.00085	330	43362	−8.3
0.0019	127500	0.00151	0.00039	298	150290	17.9

Note: (a) ε_a and N_f (exp) are taken from Nelson and Rostami [21]; (b) σ_a, ε_a^e and ε_a^p are calculated by using the Rambrg–Osgood relationship in which E=198000 MPa, K′=827 MPa, n′=0.13; (c) N_f (cal) is calculated by using Basquin's equation (3): A=−0.08206, C=−2.39763, R-square=0.94; (e) M.E.=3.3%, E.R.=[−30.1,68.5]%.

TABLE 2.A-3
Strain-Controlled Low-Cycle Fatigue Test Data of Inconel 718 and the Calculated Results by Basquin's Equation

γ_a	N_f (exp)	γ_a^e	γ_a^p	τ_a	N_f (cal)	ER(%)
0.0176	890	0.00767	0.009848	597	1352	51.9
0.0176	800	0.00767	0.009848	597	1352	69
0.0087	7200	0.00675	0.001936	525	3688	−48.8
0.0087	7000	0.00675	0.001936	525	3688	−47.3
0.0054	34000	0.00535	0.000102	416	22696	−33.2
0.0054	35700	0.00535	0.000102	416	22696	−36.4
0.0043	105000	0.00428	0.000006	333	129005	22.9
0.0038	114000	0.00382	0.000001	297	315196	176.5

Note: (a) γ_a and N_f(exp) are taken from Socie et al.[22]; (b) τ_a and γ_a^p are calculated by using the Rambrg–Osgood relationship in which G=77800 MPa, K′=860 MPa, n′=0.079; (c) N_f (cal) is calculated by using Basquin's equation (3): B=−0.12807, D=−1.71402, R-square=0.89; (e) M.E.=19.3%, E.R.=[−48.8,176.5]%.

TABLE 2.A-4
Strain-Controlled Low-Cycle Fatigue Test Data of SAE 1045 and the Calculated Results by Basquin's Equation

ε_a	N_f (exp)	ε_a^e	ε_a^p	σ_a	N_f (cal)	ER(%)
0.01	1107	0.00192	0.00815	392	845	−23.6
0.01	1137	0.00192	0.00815	392	845	−25.6
0.0051	4959	0.00159	0.00356	324	5286	6.6
0.0034	7839	0.00139	0.00201	284	18782	139.6
0.0022	78270	0.00119	0.00102	243	84176	7.5
0.0022	94525	0.00119	0.00102	243	84176	−10.9
0.0021	142500	0.00117	0.00093	238	102815	−27.8

Note: (a) ε_a and N_f(exp) are taken from Kurath et al.[23]; (b) σ_a, ε_a^e and ε_a^p are calculated by using the Rambrg–Osgood relationship in which E= 204000 MPa, K= 1185 MPa, n=0.23; (c) N_f (cal) is calculated by using Basquin's equation (3): A= −0.10394, C= −2.4121, R-square=0.95; (e) M.E.=9.4%, E.R.=[−27.8,139.6]%.

TABLE 2.A-5
Strain-Controlled Low-Cycle Fatigue Test Data of AISI 304 and the Calculated Results by Basquin's Equation

ε_a	N_f (exp)	ε_a^e	ε_a^p	σ_a	N_f (cal)	ER(%)
0.0035	38500	0.00191	0.00162	350	29564	−23.2
0.0046	10300	0.00209	0.00254	382	11531	11.9
0.01	1070	0.00257	0.00745	470	1239	15.8
0.01	1167	0.00257	0.00745	470	1239	6.1
0.006	6080	0.00225	0.00376	412	5111	−15.9
0.0035	30700	0.00191	0.00162	350	29564	−3.7
0.0035	33530	0.00191	0.00162	350	29564	−11.8
0.0035	29000	0.00191	0.00162	350	29564	1.9
0.002	286400	0.00152	0.00049	278	352537	23.1
0.002	333100	0.00152	0.00049	278	352537	5.8

Note: (a) ε_a and N_f (exp) are taken from Socie [24]; (b) $\sigma_a, \varepsilon_a^e$ and ε_a^p are calculated by using the Rambrg–Osgood relationship in which E= 183000 MPa, K= 1210 MPa, n=0.193; (c) N_f (cal) is calculated by using Basquin's equation (3): A=−0.09292, C=−2.30296, R-square=0.99; (e) M.E.=1.0%, E.R.=[−23.2,23.1]%.

TABLE 2.A-6
Strain-Controlled Low-Cycle Fatigue Test Data of Ti-6Al-4V and the Calculated Results by the Wöhler Equation

ε_a	N_f (exp)	ε_a^e	ε_a^p	σ_a	N_f (cal)	ER(%)
0.0078	5246	0.00666	0.00118	772.4	5182	−1.2
0.0065	6608	0.00642	0.0001	744.7	6365	−3.7
0.006	9640	0.00604	0	700.5	8982	−6.8
0.005	20515	0.00499	0	578.8	26292	28.2
0.0046	49518	0.00456	0	529.1	43580	−12

Note: (a) ε_a and N_f (exp) are taken from Kallmeyer et al.[25]; (b) $\sigma_a, \varepsilon_a^e$ and ε_a^p are calculated by using the Rambrg–Osgood relationship in which E=116000 MPa, K'=854 MPa, n'=0.0149; (c) N_f(cal) is calculated by using the Wöhler equation: A_1= −0.17767, C_1= 3.5478, R-square=0.99; (e) M.E.=0.9%, E.R.=[−12.0,28.2]%.

TABLE 2.A-7
Strain-Controlled Low-Cycle Fatigue Test Data of 6061-T6 and the Calculated Results by the Wöhler Equation (Tension)

ε_a	N_f (exp)	ε_a^e	ε_a^p	σ_a	N_f (cal)	ER(%)
0.0059	2160	0.00396	0.0019	283	2548	18
0.0066	1470	0.00404	0.00258	289	1307	−11.1
0.007	1140	0.00408	0.00292	291.5	994	−12.8
0.0076	670	0.00413	0.00348	295	680	1.4
0.0081	480	0.00416	0.00393	297.5	519	8.2

Note: (a) ε_a and N_f (exp) are taken from Lin et al.[26]; (b) σ_a, ε_a^e and ε_a^p are calculated by using the Rambrg–Osgood relationship in which E=71500 MPa, K= 436 MPa, n=0.069; (c) N_f (cal) is calculated by using the Wöhler equation: A_1=−0.03142, C_1=2.55881, R-square=0.94; (e) M.E.=0.7%, E.R.=[−12.8,18.0]%.

TABLE 2.A-8
Strain-Controlled Low-Cycle Fatigue Test Data of 6061-T6 and the Calculated Results by the Wöhler Equation (Torsion)

γ_a	N_f (exp)	γ_a^e	γ_a^p	τ_a	N_f (cal)	ER(%)
0.0078	3360	0.00656	0.00122	185	3460	3
0.0085	1980	0.00674	0.0018	190	1941	−2
0.0092	1310	0.00684	0.00227	193	1382	5.5
0.0102	960	0.007	0.00318	197.5	839	−12.6
0.0109	600	0.00709	0.00383	200	638	6.4
0.0129	370	0.00727	0.0055	205	374	1

Note: (a) γ_a and N_f (exp) are taken from Lin et al. [26]; (b) τ_a, γ_a^e and γ_a^p are calculated by using the Rambrg–Osgood relationship in which G=28200 MPa, K'=292 MPa, n'=0.068; (c) N_f (cal) is calculated by using the Wöhler equation: A_0=−0.04613, C_0=2.4304, 3 R-square=0.99; (e) M.E.=0.2%, E.R.=[−12.6,6.4]%.

TABLE 2.A-9
Strain-Controlled Low-Cycle Fatigue Test Data of1Cr-18Ni-9Ti and the Calculated Results by the Wöhler Equation (Tension)

ε_a	N_f (exp)	ε_a^e	ε_a^p	σ_a	N_f (cal)	ER(%)
0.002	200000	0.00183	0.00015	354	164195	−17.9
0.003	12410	0.00226	0.00076	437	16681	34.4
0.004	5500	0.00248	0.00151	478	6301	14.6
0.005	3100	0.00263	0.00241	508	3254	5
0.01	950	0.00303	0.00702	584	716	−24.6

Note: (a) ε_a and N_f (exp) are taken from Chen et al.[27]; (b) σ_a, ε_a^e and ε_a^p are calculated by using the Rambrg–Osgood relationship in which E=193000 MPa, K'=1115 MPa, n'=0.1304; (c) N_f (cal) is calculated by using the Wöhler equation: A_1=−0.09211, C_1=3.02939, R-square=0.98; (e) M.E.=2.3%, E.R.=[−24.6,34.4]%.

TABLE 2.A-10
Strain-Controlled Low-Cycle Fatigue Test Data of1Cr-18Ni-9Ti and the Calculated Results by the Wöhler Equation (Torsion)

γ_a	N_f (exp)	γ_a^e	γ_a^p	τ_a	N_f (cal)	ER(%)
0.0043	81376	0.00424	0.00009	315	125803	54.6
0.0069	12188	0.00585	0.00111	435	10243	−16
0.0104	5283	0.00682	0.00362	507	3115	−41
0.0173	1500	0.00774	0.00952	575	1172	−21.9
0.026	376	0.00838	0.01764	623	628	67.1

Note: (a) γ_a and N_f (exp) are taken from Chen et al.[27]; (b) τ_a, γ_a^e and γ_a^p are calculated by using the Rambrg–Osgood relationship in which G=74300 MPa, K'=1053 MPa, n'=0.13; (c) N_f (cal) is calculated by using the Wöhler equation: A_0=−0.12869, C_0=3.15459, R-square=0.94; (e) M.E.=8.6%, E.R.=[−41.0,67.1]%.

TABLE 2.A-11
Strain-Controlled Low-Cycle Fatigue Test Data of AISI 304 and the Calculated Results by the Wöhler Equation

σ_a	N_f (exp)	N_f (cal)	R.E.
265	49000	50248	2.5
290	23400	19552	–16.4
315	7100	8226	15.9
365	1500	1759	17.3
365	1700	1759	3.5
402.5	690	632	–8.4
412.5	540	489	–9.5

Note: (a) A_1=–0.09551, C_1=2.87225, R-square=0.99; (b) M.E.=0.7%, E.R.=[–16.4,15.9]%

TABLE 2.A-12
Strain-Controlled Low-Cycle Fatigue Test Results of AISI H11 Hot-Work Tool Steel and the Predicted Results by Using the Wöhler Equation

ε_a (%)	σ_a (MPa)	N_f (exp)	N_f (cal)	R.E.(%)
0.6	1071.5	2484	3536	42.3
0.7	1260	1189	866	–27.2
0.8	1308.5	649	624	–3.9
0.9	1394.5	491	359	–26.9
1.0	1428	275	292	6.2
1.1	1457.5	189	245	29.4

Note: (a) A_1 = –0.11517, C_1 = 3.43867, R-square=0.90; (b) M.E.=3.3%, E.R.=[–27.2,42.3]%

TABLE 2.A-13
Low-Cycle Fatigue Life Test Data of 34CrNiMo6 High-Strength Steel and the Predicted Results by Using the Wöhler Equation

ε_a (%)	ε_a^e	ε_a^p	σ_a	N_f (exp)	N_f (cal)	R.E.(%)
2.003	0.425	1.578	891.8	131	110	–16.3
1.503	0.414	1.089	869.0	240	169	–29.6
1.254	0.396	0.858	831.6	321	352	9.7
1.004	0.380	0.624	796.8	767	719	–6.3
0.806	0.358	0.448	750.6	1219	1948	59.8
0.607	0.346	0.261	726.6	2523	3352	32.9
0.512	0.332	0.18	697.5	5140	6632	29
0.413	0.322	0.091	675.3	13,378	11380	–14.9
0.303	0.303		635.0	56,181	31789	–43.4
0.286	0.286		600.0	196724	81910	–58.4
0.277	0.277		580.0	138769	144255	4
0.267	0.267		560.0	142690	259153	81.6
0.257	0.257		540.0	299787	475590	58.6

Note: (a) A_1 = –0.0559, C_1= 3.07246, R-square=0.97; (b) M.E.=8.2%, E.R.=[–58.4,81.6]%

TABLE 2.A-14
Low-Cycle Fatigue Life Test Data of RAFM Steels and the Predicted Results by Using the Wöhler Equation

Materials	$\Delta\varepsilon_t$ (%)	$\Delta\varepsilon_p$ (%)	σ_a	N_f (exp)	N_f (cal)	R.E.(%)
2W–0.06Ta	0.5	0.19	249	10347	10568	2.1
	0.8	0.4	277	2658	2285	−14
	1.2	0.75	287	1111	1373	23.6
	2	1.52	314	410	377	−8
	Note: (a) A_1 =−0.06959, C_1=2.67623, R-square=0.98; (b) M.E.=0.9%, E.R.=[−14.0,23.6]%					
1.4W–0.06Ta	0.5	0.18	256	8770	8579	−2.2
	0.8	0.41	277	2401	2539	5.8
	1.2	0.81	297	898	865	−3.6
	2	1.6	319	286	287	0.4
	Note: (a) A_1 = −0.06476, C_1 = 2.66297, R-square=1.0; (b) M.E.=0.1%, E.R.=[−2.2,5.8]%					
1W–0.14Ta	0.5	0.22	240	8048	6750	−16.1
	0.8	0.47	254	1866	2780	49
	1.2	0.84	275	912	802	−12.1
	2	1.62	293	327	297	−9
	Note:(a) A_1 = −0.06391, C_1 = 2.62494, R-square=0.94; (b) M.E.=2.9%, E.R.=[−16.1,49.0]%					
1W–0.06Ta	0.5	0.2	252	5407	4235	−21.7
	0.8	0.42	262.3	1788	2204	23.3
	1.2	0.84	276	584	961	64.6
	2	1.6	305	300	189	−37.1
	Note: (a) A_1 = −0.06135, C_1 = 2.62391, R-square=0.84; (b) M.E.=7.3%, E.R.=[−37.1,64.4]%					

APPENDIX B: EXPERIMENTAL INVESTIGATIONS: THE WÖHLER CURVE METHOD IS WELL SUITED FOR THE LOW-CYCLE FATIGUE LIFE ANALYSIS OF METALLIC MATERIALS BY BASQUIN'S CURVE IN THE STRAIN-LIFE CURVE FIGURE OF METALLIC MATERIAL FROM THE LITERATURE

Here, Basquin's curve in the strain-life curve figure of metallic material reported in the literature is used to illustrate that the metallic material studied is Basquin's material, its low-cycle fatigue life is to be well predicted by the Wöhler equation. Such type materials include: P92 ferritic-martensitic steel [32] (Figure 2.B-1), Cr–Mo–V low alloy steel [33] (Figure 2.B-2), S35C carbon steel and SCM 435 alloy steel [34,35] (Figure 2.B-3), high-strength spring steel with different heat treatments [36] (Figure 2.B-4), 316 L(N) stainless steel at room temperature [37] (Figure 2.B-5), CLAM steel at room temperature [38] (Figure 2.B-6), Eurofer97 [39] (Figure 2.B-7), F82H (a ferritic-martensitic steel) [40] (Figure 2.B-8), En3B (a commercial cold-rolled low-carbon steel) [41] (Figure 2.B-9), nodular cast iron [42] (Figure 2.B-10) and low carbon grey cast iron [43] (Figure 2.B-11).

From Figures 2.B-1 to 2.B-11, it is seen that all Basquin's curves have good linear fitting relationships between $\lg(\varepsilon_{x,a}^{e})$ and $\lg(N_f)$. Thus all metallic materials studied

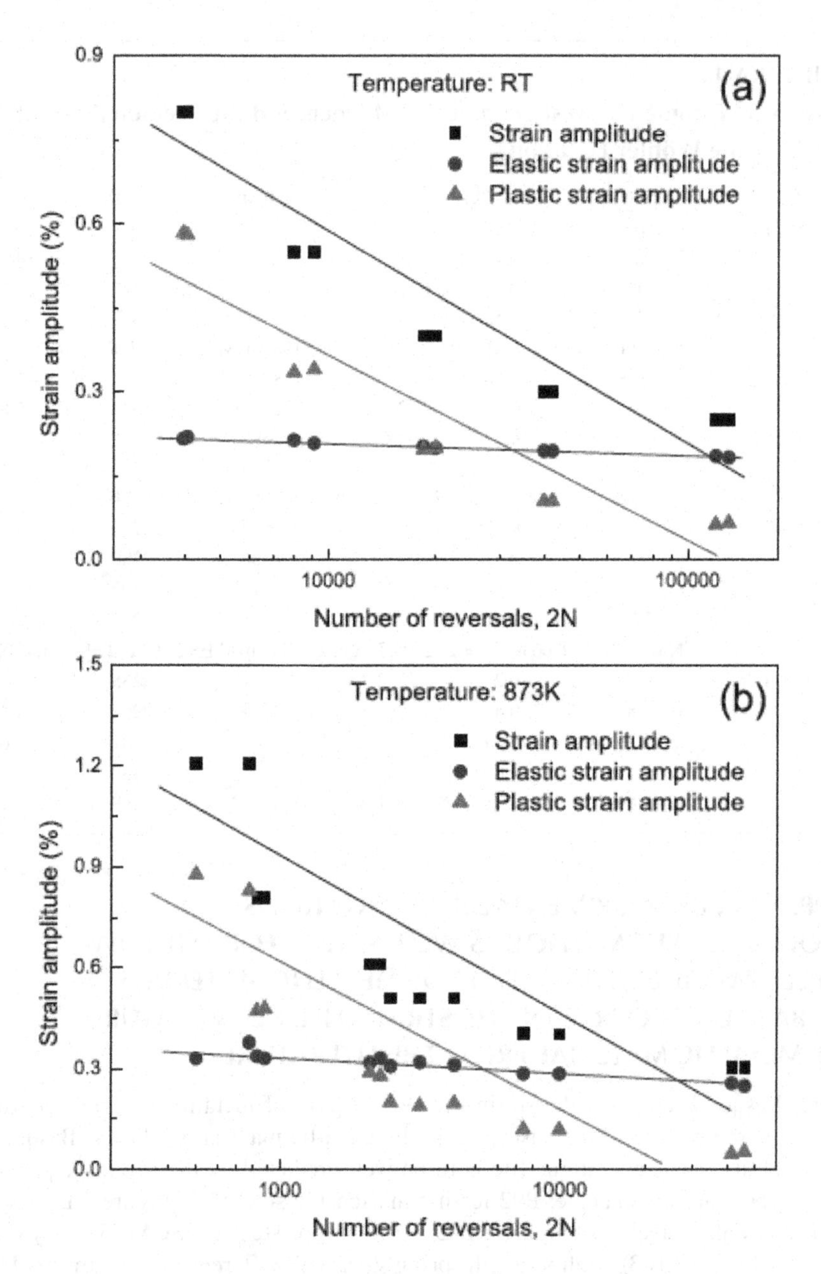

FIGURE 2.B-1 Strain-life curve of P92 ferritic-martensitic steel (a) RT, (b) 873 K.

Source: (Data from Zhang et al.[32])

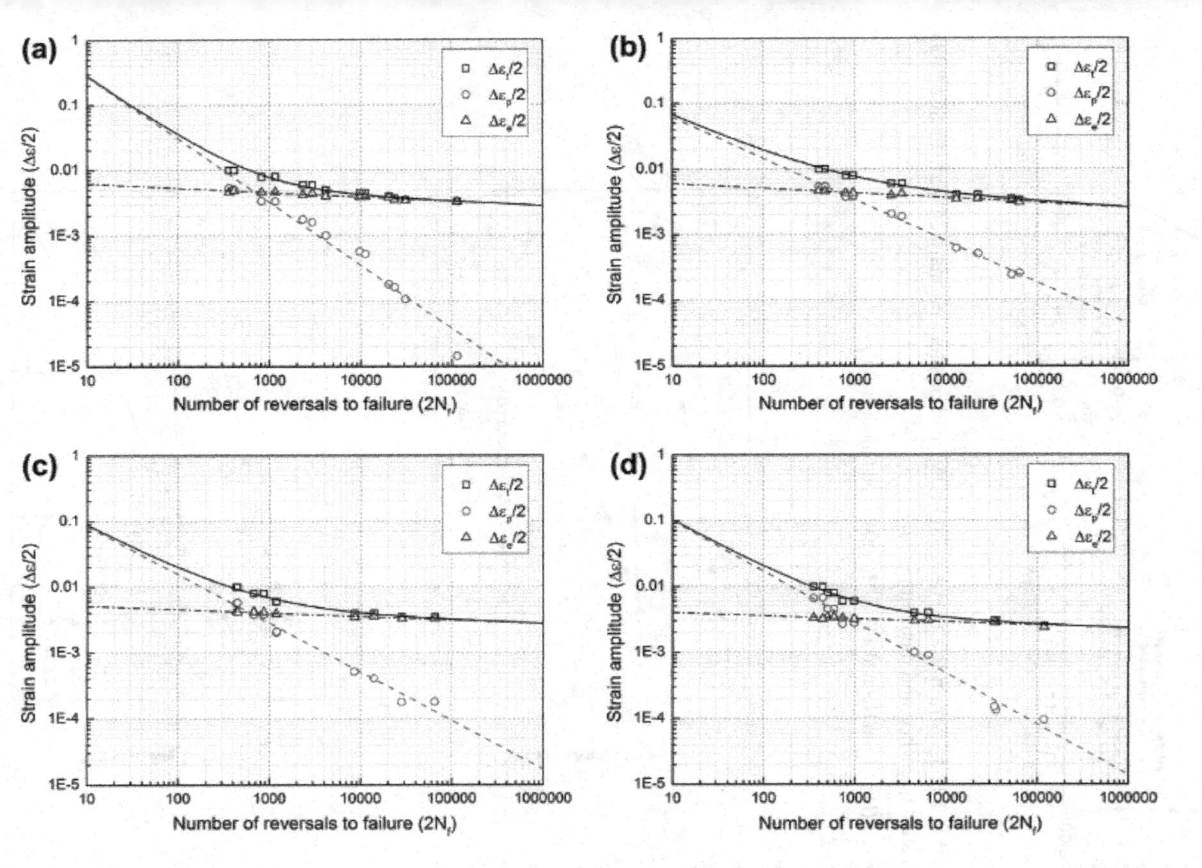

FIGURE 2.B-2 Strain-life curves of Cr–Mo–V low alloy steel at (a) RT, (b) 200°, (c) 400°, and (d) 600°.

Source: (Data from Li et al.[33])

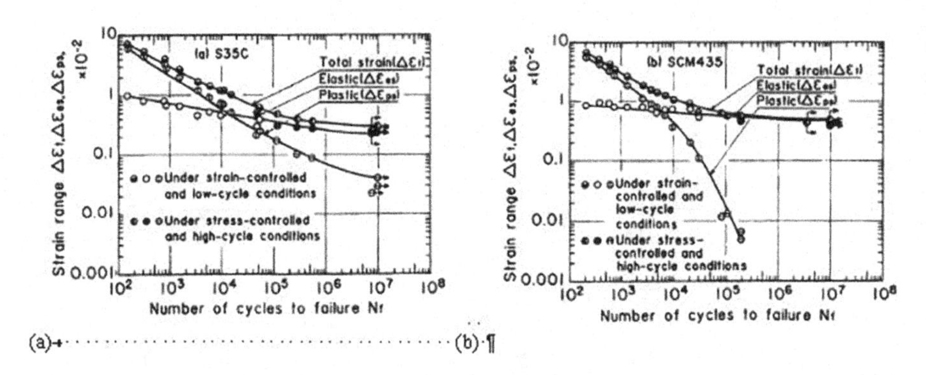

FIGURE 2.B-3 Strain-life curves of (a) S35C carbon steel normalized and (b) SCM 435 alloy steel quenched and then tempered.

Source: (Data from Hatanaka [34], Hatanaka and Fujimisu [35])

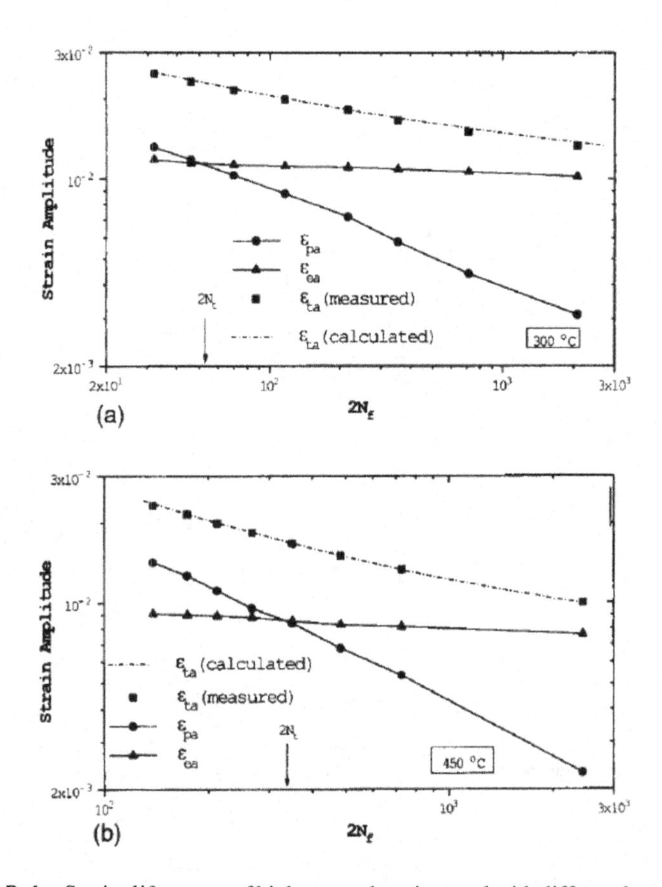

FIGURE 2.B-4 Strain–life curves of high-strength spring steel with different heat treatments.

Source: (Data from Li et al.[36])

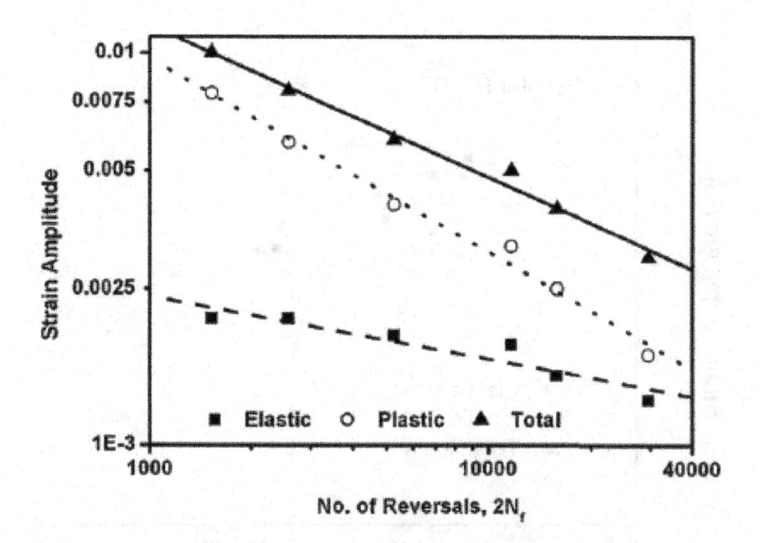

FIGURE 2.B-5 Strain-life plots of 316 L(N) stainless steel at room temperature.

Source: (Data from Roy et al.[37])

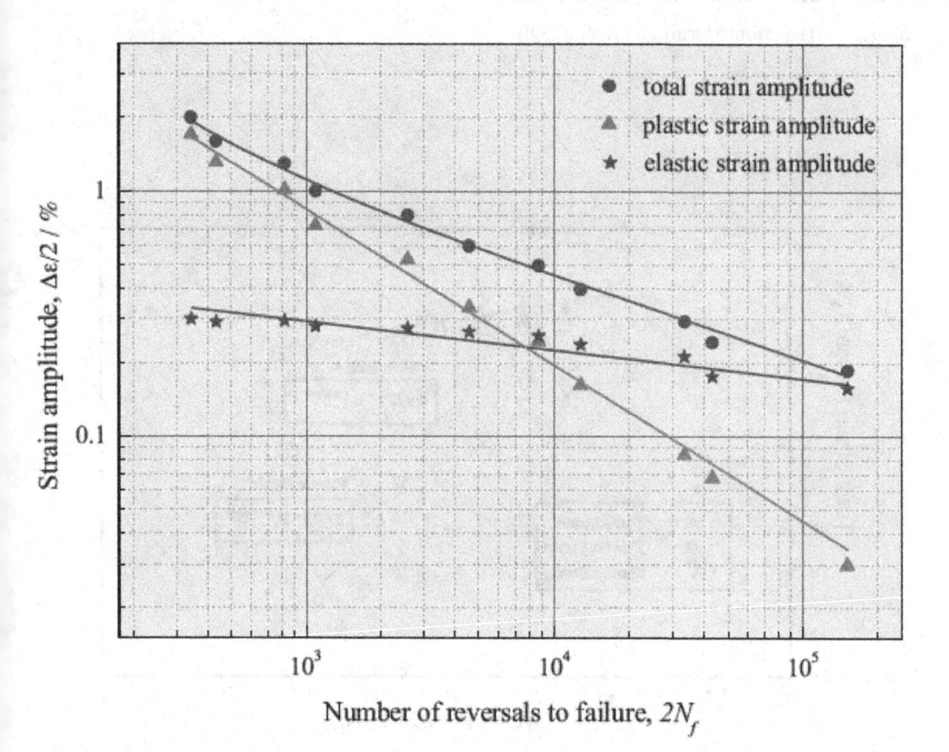

FIGURE 2.B-6 Strain-life plots of CLAM steel at room temperature.

Source: (Data from Hu, et al.[38])

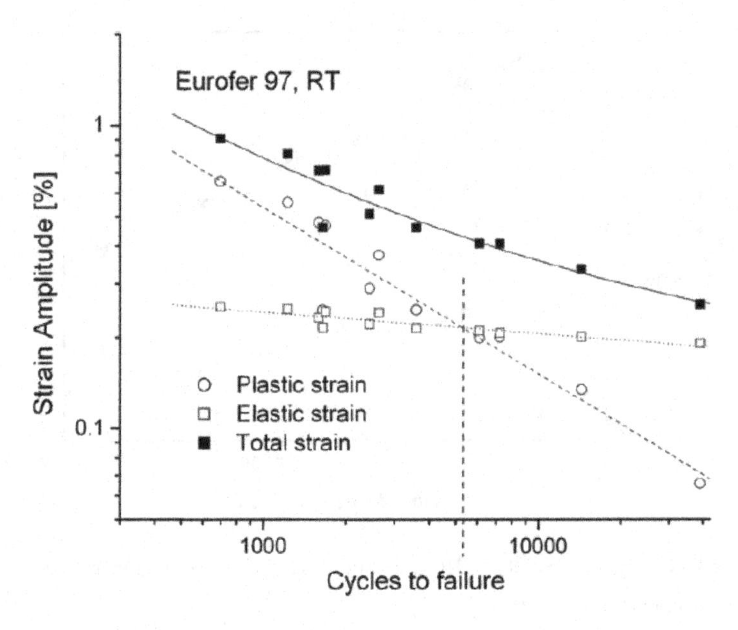

FIGURE 2.B-7 Strain-life plots of Eurofer 97.

Source: (Data from Marmy and Kruml [39])

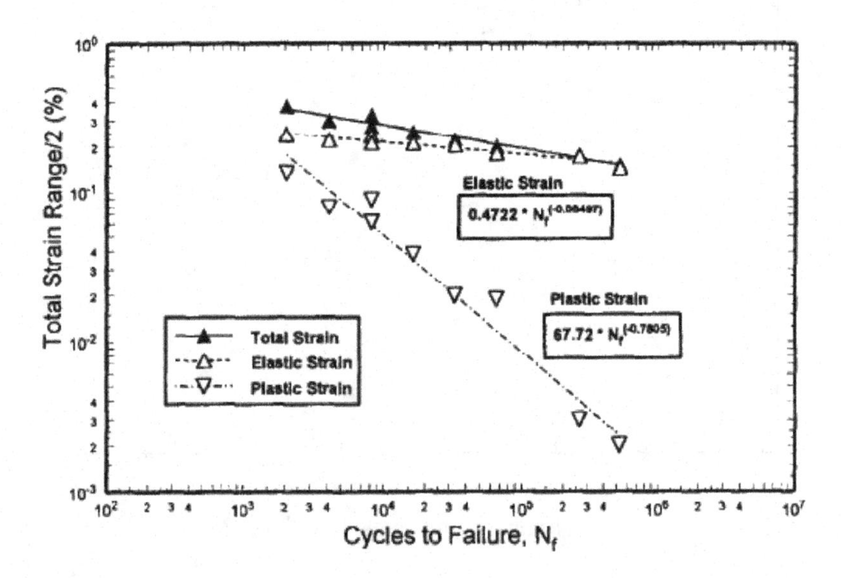

FIGURE 2.B-8 Strain-life plots of F82H (a ferritic-martensitic steel).

Source: (Data from Stubbins and Gelles [40])

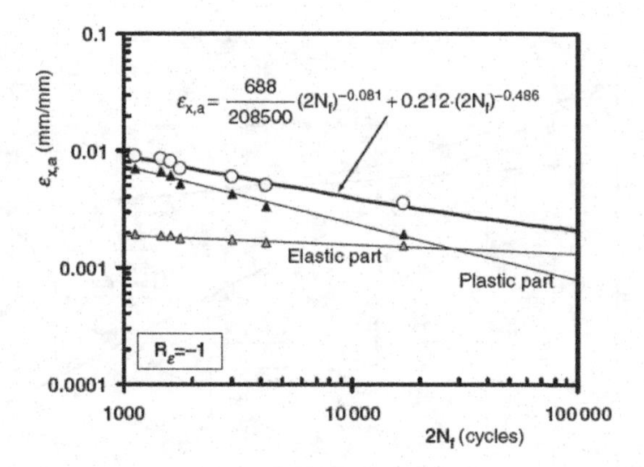

FIGURE 2.B-9 Strain-life plots of En3B (a commercial cold-rolled low-carbon steel).

Source: (Data from Atzori et al.[41])

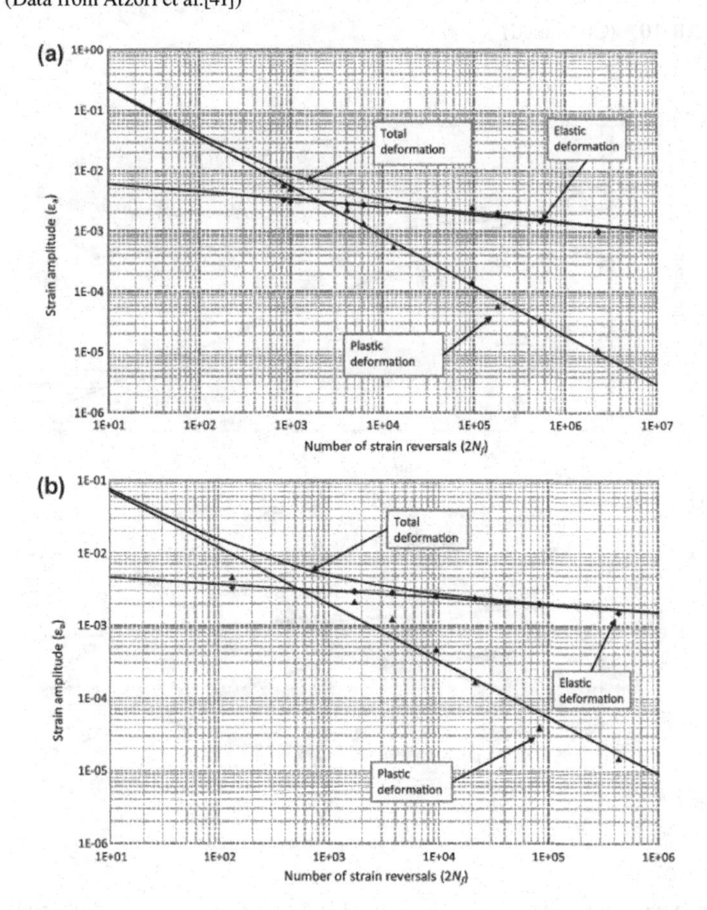

FIGURE 2.B-10 Strain–life curves of nodular cast iron at (a) RT, (b) 300°, (c) and 400°.

Source: (Data from Blazˇ et al.[42])

(Continued)

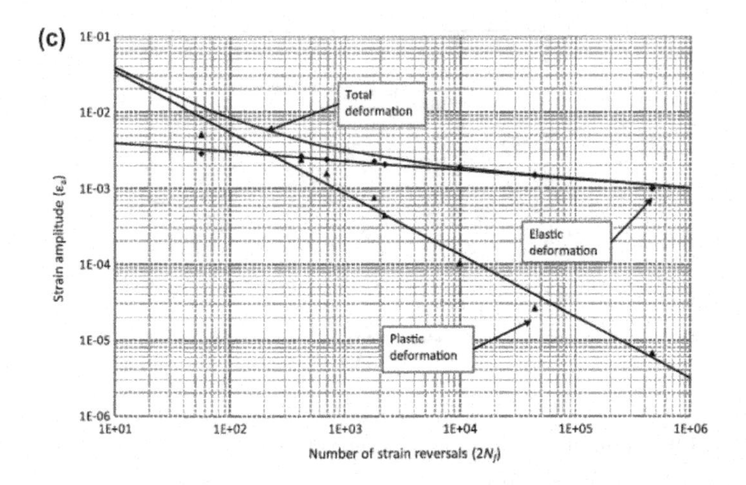

FIGURE 2.B-10 (Continued)

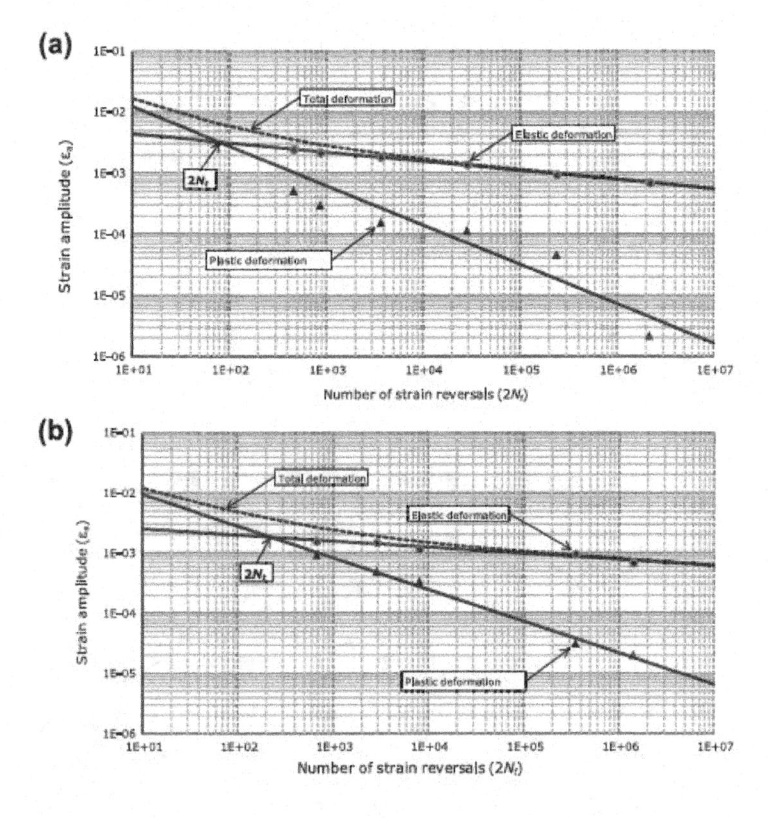

FIGURE 2.B-11 Strain–life curves of low-carbon grey cast iron at (a) RT, (b) 500°, (c) 600° and (d) 700°.

Source: (Data from Miha et al.[43]).

(Continued)

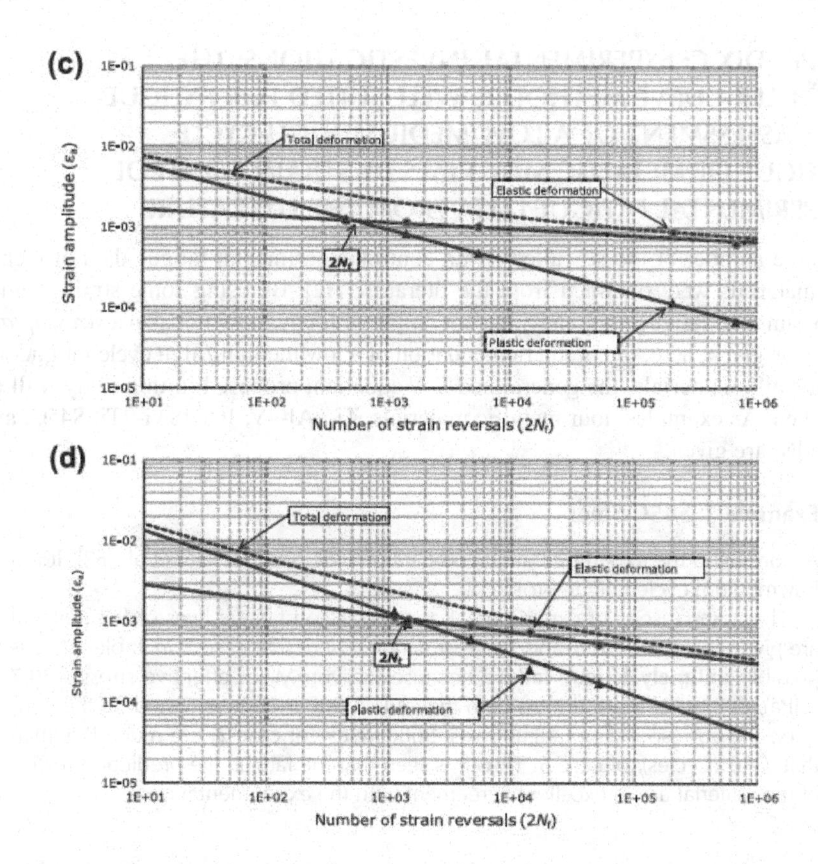

FIGURE 2.B-11 (Continued)

are Basquin's materials, and their low-cycle fatigue lives can be calculated by the Wöhler equation.

Form Figure 2.B-1 (Strain-life curve of P92 ferritic-martensitic steel (a) RT, (b) 873 K (from Zhang et al.[32])), it is seen that there is no linear fitting relationship between $\lg(\varepsilon^p_{x,a})$ and $\lg(N_f)$. The author considers that the low-cycle fatigue life of P92 ferritic-martensitic steel appears to be not well calculated by the Manson–Coffin equation.

From Figure 2.B-3 (Strain-life curves (a) S35C carbon steel normalized and (b) SCM 435 alloy steel quenched and then tempered), it is found that, from low-cycle to high-cycle fatigue of the two materials, the relationship between ε^e_a and N_f is well expressed by Basquin's equation. Thus the author considers that it is very possible that the fatigue life of the two materials, from low-cycle to high-cycle fatigue, is well calculated by the Wöhler equation.

In addition, it is found from Figure 2.B-10 (Strain–life curves of nodular cast iron) that, from low-cycle to high-cycle fatigue of the material at room temperature, the relationship between ε^e_a and N_f is well expressed by Basquin's equation. Thus it is very possible that the fatigue life of the material at room temperature, from low-cycle to high-cycle fatigue, is well calculated by the Wöhler equation.

APPENDIX C: EXPERIMENTAL INVESTIGATIONS: THE WÖHLER CURVE METHOD IS WELL SUITED FOR FATIGUE LIFE ASSESSMENT OF A LOW/MEDIUM/HIGH CYCLE FATIGUE OF METALLIC MATERIALS BY STRAIN CONTROL EXPERIMENTAL FATIGUE DATA FROM THE LITERATURE

A large number of the so-called strain control experimental fatigue data of metallic materials were obtained from the literature [12]. By using some strain control experimental fatigue data, in which not only a strain quantity but also a *stress quantity* was given, here, the prediction equation of a low/medium/high cycle fatigue life of metallic materials, the generalized *S-N* equation, proposed in this study, will be verified. As examples, four metallic materials, Ti-6Al-4V, 1Cr-18*N*i-9Ti, S45C, and S460N, are given.

Example 1: S45C Steel

According to the experimental fatigue data of S45C steel by Kim et al. [52], it is a low/medium cycle fatigue problem.

The comparisons of experimental and calculated fatigue lives of the material are given in Tables 2.C-1 and 2.C-2. It is noted here that case 2 in Table 2.C-2 is possibly **untimely fatigue failure** because the von Mises effective stress, 439.7 (MPa), of case 2, is a little less than that (441.8 (MPa), of case 5, and on the contrary, the corresponding fatigue lives, 568(Cycles), of case 2, are much less than that, (1181(Cycles), of case 5. Thus it is seen that the fatigue life prediction results of the material are in excellent agreement with the experimental ones.

TABLE 2.C-1
Fatigue Life Prediction of S45C Steel Under Uniaxial Loading

$\sigma_{x,a}$ (MPa)	$\sigma_{x,m}$ (MPa)	$\tau_{xy,a}$ (MPa)	$\tau_{xy,m}$ (MPa)	N_{exp} (Cycles)	ER (%)
595.2	0	0	0	110	−5.3
480.2	0	0	0	852	−21.3
389.1	0	0	0	3383	22.7
361.9	0	0	0	5514	41.1
509	0	0	0	421	−3.9
365.5	0	0	0	8933	−20.04
325.9	0	0	0	22071	−12.5
500.2	0	0	0	407	15.5
0	0	287.1	0	1151	20.6
0	0	283.8	0	1761	−11.3
0	0	286.3	0	1771	−19.3
0	0	244.3	0	5644	27.0
0	0	229.4	0	14930	−8.8

TABLE 2.C-2
Fatigue Life Prediction of S45C Steel Under Multiaxial Loading

No.	$\sigma_{x,a}$ (MPa)	$\sigma_{x,m}$ (MPa)	$\tau_{xy,a}$ (MPa)	$\tau_{xy,m}$ (MPa)	N_{exp} (Cycles)	ER (%)
1	370.3	0	103.6	0	2278	25.7
2	395.2	0	111.4	0	568	181.2
3	391.1	0	108.9	0	1366	30.5
4	285.1	0	164.4	0	4647	–6.2
5	430.6	0	57.3	0	1181	19.7

Example 2: 1Cr-18Ni-9Ti

According to the experimental fatigue data of 1Cr-18Ni-9Ti steel by Chen et al. [27], it is a low/medium cycle fatigue problem.

The comparisons of experimental and calculated fatigue lives of the material, **under in-phase loading**, are given in Table 2.C-3. According to similar analysis to **untimely fatigue failure** of S45C steel, the multiaxial loading case with a relative error, 347.7%, in Table 2.C-3, is possibly **untimely fatigue failure**. Thus it is seen that the fatigue life prediction results of the material are in excellent agreement with the experimental ones.

The comparisons of experimental and calculated fatigue lives of the material, **under out-of-phase loading**, are given in Table 2.C-4, from which it is seen that the prediction results obtained by the new definition of the nonproportional loading factor and the linear formula are much better than those by the original definition of the nonproportional loading factor and Itoh's formula, with the error ranges [–21.9, 24.2]% and [–59.1, 228.4]%, respectively.

TABLE 2.C-3
Fatigue Life Prediction of 1Cr-18Ni-9Ti Steel Under In-Phase Loading

$\sigma_{x,a}$ (MPa)	$\sigma_{x,m}$ (MPa)	$\tau_{xy,a}$ (MPa)	$\tau_{xy,m}$ (MPa)	N_{exp} (Cycles)	ER (%)
350	0	0	0	200000	–4.3
448	0	0	0	12410	7.9
472	0	0	0	5500	38.8
535	0	0	0	3100	–36.1
568	0	0	0	950	9.4
0	0	221.7	0	81376	19.2
0	0	271.4	0	12188	–20.8
0	0	282.9	0	5283	13.6
0	0	329.1	0	1500	–28.8
0	0	352.2	0	376	30.9
438.4	0	253.1	0	646	–37.1
497.1	0	287	0	184	–44.2
258.1	0	149	0	30028	347.7
338	0	195.1	0	3648	92.3

TABLE 2.C-4
Fatigue Life Prediction of 1Cr-18Ni-9Ti Steel Under Out-of-Phase Loading

$\sigma_{x,a}$ (MPa)	$\sigma_{x,m}$ (Mpa)	$\tau_{xy,a}$ (MPa)	$\tau_{xy,m}$ (MPa)	δ (degree)	N_{exp} (Cycles)	Itoh's F ER(NPF) (%)	Linear F ER(NPF′) (%)
438.4	0	253.1	0	90	646	−53.8	−21.9
497.1	0	287	0	90	184	−59.1	24.2
258.1	0	149	0	90	30028	228.4	9.5
338	0	195.1	0	90	3648	41.1	−6.2

Example 3: Ti-6Al-4V Steel

According to the experimental fatigue data of Ti-6Al-4V steel by Kallmeyer et al. [25], it is a low/medium cycle fatigue problem. The comparisons of experimental and calculated fatigue lives of the material are given in Figure 2.C-1, from which it is seen that the experimental fatigue lives are in good agreement with those calculated by the *S-N* equation.

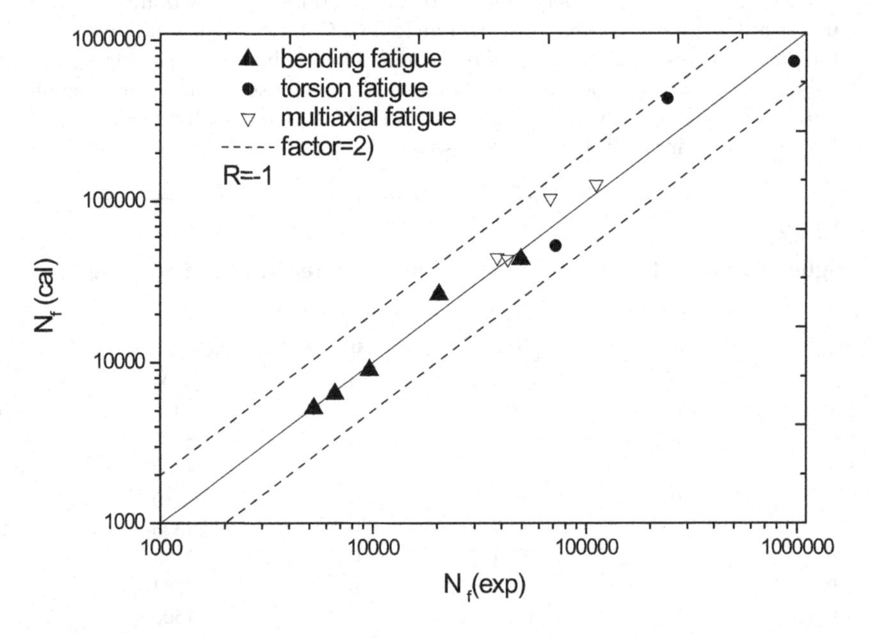

FIGURE 2.C-1 Comparisons of experimental and calculated fatigue lives of Ti-6Al-4V steel.

Source: (Data from Chen et al. [25])

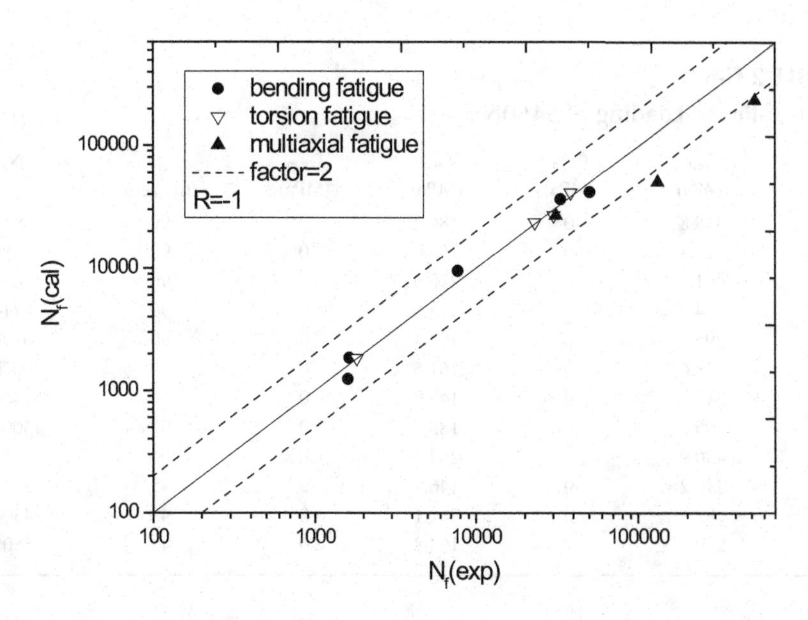

FIGURE 2.C-2 Comparisons of experimental and calculated fatigue lives of S460N steel.

Source: (Data from Jiang et al. [53])

Example 4: S460N Steel

According to the experimental fatigue data of S460N steel by Jiang et al. [53], it is a low/medium cycle fatigue problem.

The fatigue life predictions of the material under *in-phase loading* are shown in Figure 2.C-2, from which it is seen that, under bending and torisional loading, the predicted fatigue lives are in excellent agreement with the experimental ones, but under multiaxial loading, the predicted fatigue lives are conservative, with an error range [−61.0,−13.2]%.

The fatigue life predictions of S460N steel under **out-of-phase loading** will be studied in detail in the following.

Out-of-phase loading cases are listed in Table 2.C-5. On the basis of Itoh's formula, and the original and new definitions of the nonproportional loading factor, the fatigue life prediction results of the material under out-of-phase loading are given in Table 2.C-6, from which it is found that there is a case (No. 11) whose $\left(\frac{\sigma_{ean}}{\sigma_{ea}}-1\right)/F$ (or $\left(\frac{\sigma_{ean}}{\sigma_{ea}}-1\right)/F'$) are largely different from its mean value. Here, the experimental fatigue data of this case is the *unusual data*, which will be eliminated in the fatigue life prediction of the material.

After eliminating the *unusual data*, the calculated results are given in Table 2.C-7. The calculated error indexes are listed in Table 2.C-8, in which error indexes including the *unusual data* are also listed.

On the basis of Itoh's formula, and the original definition of the nonproportional loading factor, the comparisons of two sets of error indexes *including and*

TABLE 2.C-5
Out-of-Phase Loading of S460N Steel

No.	$\sigma_{x,a}$ (MPa)	$\sigma_{x,m}$ (MPa)	$\tau_{xy,a}$ (MPa)	$\tau_{xy,m}$ (MPa)	δ (degree)	N_{exp} (Cycles)
1	318.8	0	184.2	0	90	39670
2	325.1	0	199.6	0	90	22800
3	391.5	0	230.9	0	90	6570
4	284.3	0	195.5	0	90	47140
5	295	0	193.4	0	90	51900
6	228.6	0	161.6	0	90	90700
7	215	0	149.2	0	90	574600
8	295.3	0	183	0	90	30000
9	480.9	0	269.3	0	90	540
10	216.2	0	146.3	0	45	218400
11	282.1	0	282.1	0	45	43000
12	280.4	0	164.8	0	45	55000

TABLE 2.C-6
Fatigue Life Prediction of S460N Steel Under Out-of-Phase Loading by Itoh's Formula (Including the Unusual Data)

No.	$\left(\frac{\sigma_{ean}}{\sigma_{ea}}-1\right)/F$	$\left(\frac{\sigma_{ean}}{\sigma_{ea}}-1\right)/F'$	Itoh's F ER(NPF) (%)	Itoh's F ER(NPF') (%)
1	−0.24435	−0.3629	341.8	286.7
2	−0.25983	−0.3668	218.4	268.0
3	−0.29021	−0.3474	115.3	643.1
4	−0.28344	−0.4295	103.7	68.4
5	−0.28659	−0.4301	104.3	74.2
6	−0.14625	−0.2711	830.8	275.9
7	−0.23585	−0.4700	258.0	13.4
8	−0.20619	−0.3188	542.2	368.1
9	−0.25918	−0.2592	235.1	5235.8
10	−0.28805	−0.5793	30.6	−20.1
11	−1.14933	−1.3648	−98.7	−98.2
12	−0.41617	−0.6968	−34.5	−48.5

excluding the unusual data (see Table 2.C-8), again illustrate that *it is important to exclude the unusual data in the processing of experimental fatigue data*: error ranges [−98.8, 830.9]% and [−55.6, 276.0]%, mean errors 220.6% and 31.5%.

While on the basis of the original definition of the nonproportional loading factor, the fact that the calculated results of fatigue life of S460N steel by the linear

TABLE 2.C-7
Fatigue Life Prediction of S460N Steel Under Out-of-Phase Loading (Excluding the Unusual Data)

				Itoh's Formula		Linear Formula	
No.	σ_{ean}/σ_{ea}	NP F	NPF′	ER(NPF) (%)	ER(NPF′) (%)	ER(NPF) (%)	ER(NPF′) (%)
1	0.7558	0.9992	0.6726	36.3	64.4	−0.7	−16.5
2	0.7556	0.9403	0.6660	7.6	57.6	−15.6	−18.3
3	0.7158	0.9789	0.8177	−31.3	135.2	−48.7	−30.8
4	0.7620	0.8396	0.5540	−20.2	−13.3	−29.5	−34.7
5	0.7476	0.8806	0.5867	−24.5	−14.7	−36.5	−42.2
6	0.8805	0.8167	0.4405	276.0	129.1	240.9	145.8
7	0.8037	0.8319	0.4174	41.6	−28.5	26.1	−17.9
8	0.8078	0.9316	0.6024	120.0	124.0	74.1	44.2
9	0.7486	0.9699	0.9698	8.9	1060.9	−17.5	79.1
10	0.8827	0.4069	0.2023	−11.2	−34.6	23.2	34.6
12	0.8276	0.4141	0.2473	−55.6	−59.7	−39.0	−26.7

TABLE 2.C-8
Error Indexes of Fatigue Life Analysis of S460N Steel Under Out-of-Phase Loading

	ER(NPF)				ER(NPF′)			
	Min ER (%)	Max ER. (%)	AER (%)	MER (%)	Min ER (%)	Max ER. (%)	AER (%)	MER (%)
Itoh's F	−55.6	276.0	57.6	31.5	−34.6	1060.9	129.1	156.5
Linear F	−48.7	240.9	50.1	16.0	−42.2	145.8	44.6	10.5
Itoh's F (Including unusual data)	−98.8	830.9	242.8	220.6	−98.3	5235.9	616.7	588.9

formula are better than those by Itoh's formula—error ranges [−48.7, 240.9]%, and [−55.6, 276.0]%; mean errors 16.03% and 31.59%—illustrates that it is necessary to employ the linear formula to the fatigue life analysis of the material.

For the material, the fatigue life prediction results obtained on the basis of the new definition of the nonproportional loading factor and the linear formula are best in all calculated results (see Table 2.C-8).

3 Notch *S-N* Equation for a Low/Medium/ High Cycle Fatigue of Metallic Materials

3.1 INTRODUCTION

It is well known that the majority of damages in real components are due to fatigue. Moreover, the problem of properly performing the fatigue assessment is further complicated due to the fact that such failures are, in general, caused by multiaxial loading: this makes it evident that engineers engaged in assessing real mechanical assemblies need sound engineering tools capable of accurately and efficiently estimating multiaxial fatigue damage.

As is well known to us, a fatigue life estimation of metallic materials is usually divided into two types, i.e., a low/medium cycle (LCF) and a high-cycle fatigue (HCF). Concerning the HCF fatigue assessment, the Wöhler Curve Method is usually employed. The Modified Wöhler Curve Method (MWCM) proposed by Susmel and Lazzarin [1], for example, is used to perform the HCF fatigue assessment of metallic materials for conventional mechanical components subjected to multiaxial fatigue loading. While for the LCF of metallic materials, no doubt, the Manson–Coffin Curve is used to perform fatigue life assessment. The Modified Manson–Coffin Curve Method by Susmel et al. [2], for example, is used to evaluate LCF lifetime of metallic materials under multiaxial fatigue loading.

A key to fatigue life evaluation of metallic materials is, in substance, that, on the basis of experimental fatigue *data* of metallic specimens and *mechanical analyses* for those specimens, a *proper mechanical quantity*, *S*, is chosen to establish a relationship between the *mechanical quantity* and the fatigue life, *N*. For example, the Wöhler Curve and the Manson–Coffin Curve are two typical examples of the relationship. When a different *mechanical quantity*, *S*, is taken, naturally, a different *relationship equation* is obtained, which has usually different accuracy and efficiency when used to perform fatigue life estimation. Both an accurate and highly efficient fatigue life prediction equation is to be searched for by all field investigators.

Recently, the author attempted to present a prediction equation of a low/medium/high cycle fatigue life of metallic materials. The prediction equation is required to have both accuracy and high efficiency when applied to perform the fatigue life assessment of metallic components subjected to both uniaxial and multiaxial loading. Concerning fatigue assessment of a mechanical component, main factors affecting its efficiency include three aspects. One is a mechanical analysis of the studied

DOI: 10.1201/9781003356721-3

component. The mechanical analysis, generally, is a basis of fatigue assessment of the studied component due to the fact that the mechanical analysis is a basis of calculating a chosen *mechanical quantity*, *S*. The time consumption involved in the *linear-elastic analysis* of the studied component, for example, is much less than that in its *elastic-plastic analysis*. Two is to compute a chosen *mechanical quantity*, *S*. On the basis of wide research into multiaxial fatigue life predictions [1–10], it can be seen that the time consumption involved in calculating the *mechanical quantity* based on a stress invariant based approach is much less than that on a critical plane approach. The third aspect is to measure the material constants in the stress-stain relation used to perform the mechanical analysis of the studied component. The consumption of time and financial resources involved in determining the material constants in linear stress-strain relations is, no doubt, much less than that in determining the material constants in cyclic elastic-plastic stress-strain relations.

Thus, the aim of the author's study was to present, on the basis of the *linear-elastic analysis* of the studied component, a *stress invariant based equation* of multiaxial fatigue life prediction. The stress invariant based equation will be both accurate and highly efficient when used in performing low/medium/high fatigue life assessment of mechanical components under both uniaxial and multiaxial loading. The systematic validation exercises made by the author, by using a large number of experimental results of plain specimens in the literature, prove that the author's aim has been reached. This success was published partly in Ref.[11].

In this study, by using concepts of the critical stress, the critical distance and the local approach in dealing with the fracture failure of notches, the Notch *S-N* equation for a low/medium/high cycle fatigue of metallic materials is proposed and is verified to have high accuracy in performing the fatigue failure assessment of a low/medium/high cycle fatigue of metallic materials. No doubt, the Notch *S-N* curve method, which is on the basis of the linear elastic stress analysis of the notched components, is very efficient because there is no need for elastic-plastic analysis even for low-cycle fatigue of metallic materials.

Moreover, an attempt is made to show that the Notch *S-N* equations proposed in this study are applied in multiaxial fatigue limit analysis of notched components.

3.2 A BRIEF DESCRIPTION OF THE UNIFIED LIFETIME ESTIMATION EQUATION OF A LOW/MEDIUM/HIGH CYCLE FATIGUE OF METALLIC MATERIALS

In this section, a description on the practicability of the Wöhler Curve Method for a low-cycle fatigue of metallic materials is given first. Then the unified lifetime estimation equation of a low/medium/high cycle fatigue of metallic materials is described.

3.2.1 A Description of the Practicability of the Wöhler Curve Method for a Low-Cycle Fatigue of Metallic Materials

On "A description of the practicability of the Wöhler Curve Method for a low-cycle fatigue of metallic materials", "a description on a practicability of establishing the unified prediction equation for a low/medium/high cycle fatigue of metallic materials", which occurred in Appendix C in Ref.[11], quoted here perhaps is suitable and is further helpful for the reader to understand it.

In Ref [11], recently, the author was concerned with the following question presented by a reviewer:

"All test data investigated in this study is obtained from brittle materials which have relatively small fracture process zone. Could the proposed approach be able to predict the fracture behavior of materials which have large fracture process zone such as rock or concrete? Please explain."

According to this question, Appendix *B* was added in Ref.[11]. In Appendix *B* in Ref.[11], an attempt was made to test whether or not the empirical fracture equation of mixed mode cracks proposed in Ref.[11] was used to perform the fracture analysis of cracked specimens made of plastic materials. Two examples were given. From the study in Appendix *B* in Ref.[11], the author considered that the empirical fracture equation of mixed mode cracks proposed in Ref.[11] appeared to be used to perform the fracture analysis of cracked specimens made of plastic materials.

Further, such an explanation, "a description of the practicability of establishing the unified prediction equation for a low/medium/high cycle fatigue of metallic materials", which occurred in Appendix *C* in Ref.[11], perhaps was helpful for the reader to understand it is practical that the fracture analysis of cracked specimens made of plastic materials can be performed by using the empirical fracture equation of mixed mode cracks proposed in Ref.[11].

Here, "The practicability of establishing the unified prediction equation for a low/medium/high cycle fatigue of metallic materials", which occurred in Appendix *C* in Ref.[11], is rewritten as follows.

In a low/medium/high cycle fatigue regime of metallic materials, does a prediction equation, in which the mechanical quantity calculation is on the basis of the *linear-elastic analysis*, exist? According to previous wide investigations [1–10], the prediction equation is *impossible* to exist: For the HCF of metallic materials, the mechanical quantity calculation is on the basis of the *linear-elastic analysis* of the studied component due to the fact that the studied component is subjected to elastic deformation, while for the LCF of metallic materials, the mechanical quantity calculation is on the basis of the *elastic-plastic analysis* of the studied component due to obvious plastic deformation produced in the studied component. Here, the previously mentioned *impossibility* is, no doubt, reasonable. The viewpoint that, for the LCF of metallic materials, there is large plastic deformation in the materials and therefore it's not reasonable to perform the fatigue life assessment of the materials by using the *linear-elastic analysis*, is a common view.

However, on the basis of the well-known Rambrg–Osgood relationship:

$$\varepsilon = \varepsilon_e + \varepsilon_p = \frac{\sigma}{E} + (\frac{\sigma}{K})^{\frac{1}{n}} \tag{1}$$

where E is Young's modulus, K is the strength coefficient and n is the strain hardening exponent, the author considers that, for the LCF of metallic materials, the way that a "**stress quantity**", which is calculated based on the *linear-elastic analysis* of the studied component, instead of a "strain quantity", which is calculated based on the *elastic-plastic analysis* of the studied component, is taken to be a *mechanical quantity, S*, to establish a relation of the *mechanical quantity, S,* to the fatigue life, *N*, is practicable. In order to fully indicate the author's viewpoint, here, some experimental data of low/medium/high cycle fatigue of tension–compression fatigue of plain and notched specimens made of 16MnR steel [13,3] were given in Table 3.1 and Table 3.2. Comparison of experimental data and the linear fitted curve of tension–compression fatigue of plain specimens in the double-logarithm frame of

TABLE 3.1
Fully Reversed Uniaxial Fatigue Experiments of Plain Specimens Made of 16MnR Steel

No.	Strain Amplitude $\varepsilon_{x,a}$ (%)	Stress Amplitude $\sigma_{x,a}$ (MPa)	Fatigue Life N (Cycles)
1	1.999	514.6	280
2	1.501	487.3	400
3	1	459.2	880
4	0.7	418.2	2220
5	0.5	380	4850
6	0.38	351	9200
7	0.3	322.9	17300
8	0.243	303.7	37600
9	0.199	285.1	79000
10	0.18	274.5	148400
11	0.16	261.9	388500
12	0.146	256	1400000

TABLE 3.2
Fully Reversed Uniaxial Fatigue Experiments of Notched Specimens Made of 16MnR Steel (from [3])

No.	Stress Amplitude $\sigma_{x,a}$ (MPa)	Fatigue Life N (Cycles)
1	438.97	800
2	345.53	2048
3	308.42	3250
4	255.81	8192
5	202.86	27,000
6	184.52	47,000
7	151.32	180,000
8	140.64	232,000
9	131.07	642,200

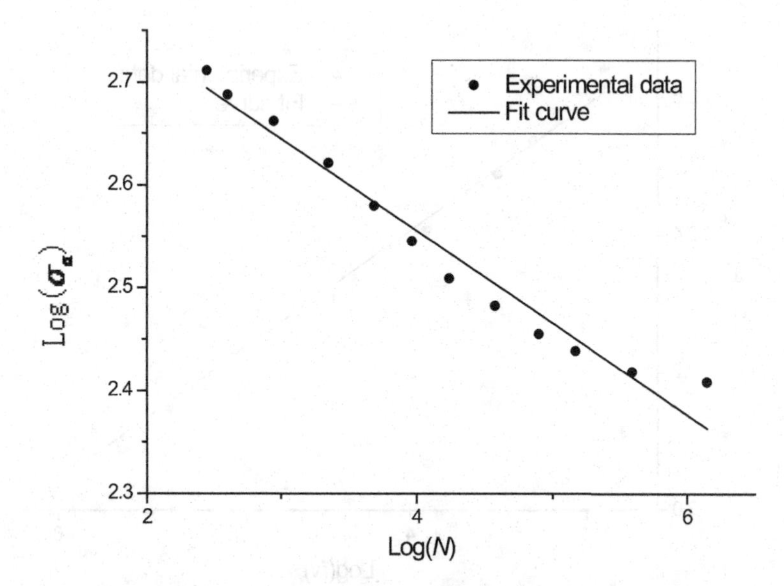

FIGURE 3.1 Comparison of experimental data and linear fitted curve of tension–compression fatigue of plain specimens made of 16MnR steel in the double-logarithm frame of axes.
Source: (Data from Gao *et al.*[13])

axes is shown in Figure 3.1. From the good agreement between experimental data and the linear fitted curve shown in Figure 3.1, no doubt, low/medium/high fatigue life assessment of plain specimens of the material can be carried out by using the Wöhler curve, i.e., the linear fitting relation of $\log(\sigma_a)$ to $\log(N)$, in which the fitting constants are listed in Table 3.3. Similarly, handling can be done for experimental data of tension–compression fatigue of notched specimens made of 16MnR steel (see Figure 3.2), from which it can be seen that for low/medium/high cycle fatigue of notched specimens of the material, good prediction results can be obtained using the notch Wöhler Curve, i.e., the linear fitting relation of $\log(\sigma_a)$ to $\log(N)$ obtained by using experimental data of tension–compression fatigue of notched specimens shown in Table 3.2.

From Figure 3.1 and Figure 3.2 (see also Table 3.3), in summary, it can be seen that, for low/medium/high cycle fatigue life assessment of 16MnR steel, a "*stress quantity*", σ_a, can be taken as a *mechanical quantity, S*, to establish a relation of the *mechanical quantity, S*, to the fatigue life, N, i.e., a linear fitting relation between $\log(S)$ and $\log(N)$, even though there is obvious plastic deformation in the material (see Figure 3.3).Then, for a real mechanical component which needs to be fatigue lifetime assessment, is either a *linear-elastic analysis* or an *elastic-plastic analysis* employed in calculating the "*stress quantity*"? Due to the fact that the magnitude of the strains involved in fatigue problems is in general not very high [12] and that the well-known Rambrg–Osgood relationship is usually employed in the *elastic-plastic analysis* of metallic materials (please here see the well-known Rambrg–Osgood relationship curve, Figure 7.7 in [12]), the author considers that, at a given loading,

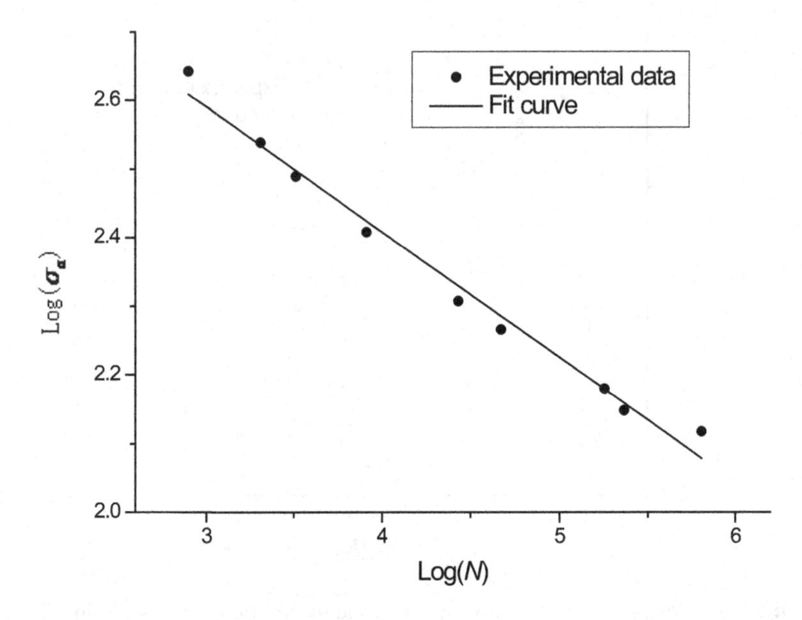

FIGURE 3.2 Comparison of experimental data and linear fitted curve of tension–compression fatigue of notched shaft specimens made of 16MnR steel in the double-logarithm frame of axes.

Source: (Data from Gao *et al.*[3])

TABLE 3.3
Linear Fitted Constants in the *S-N* Equations of 16MnR Steel

Specimen	A	C	R-square
Plain specimen	−0.08941	2.91288	0.96211
Notched specimen	−0.18268	3.1392	0.98329

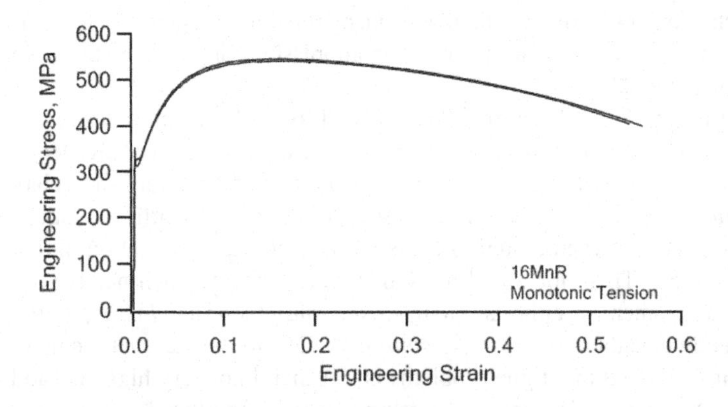

FIGURE 3.3 Monotonic tensile stress-strain curve of 16MnR steel.

Source: (Data from Gao *et al.* [13])

the "*stress quantity*" calculated using the *elastic-plastic analysis* almost is the same as that by the *linear-elastic analysis*. From a practical point of view, thus, the author considers that, for the LCF of metallic materials, the "*stress quantity*" *can be calculated using the linear-elastic analysis of the mechanical component.*

Thus it can be concluded that, in a low/medium/high cycle fatigue regime of metallic materials, it is practical to present a *stress invariant based equation* in which the "*stress quantity*" is calculated using the *linear-elastic analysis* of the studied component. Such a *stress invariant based equation* is, in fact, the Wöhler Curve type one usually used to perform the fatigue life assessment of the medium/high cycle fatigue life of metallic materials. That is to say that the Wöhler Curve can be used to perform the fatigue life assessment of not only the medium/high cycle fatigue of metallic materials but also the low-cycle fatigue of metallic materials. Thus, for low/medium/high cycle fatigue of metallic materials, a unified prediction equation proposed in this study is, in fact, the well-known Wöhler Curve type one, in which the mechanical quantity is a "*stress quantity*" which is calculated using the *linear-elastic analysis* of the studied component.

In order to let the reader be sure that the Wöhler Curve can be used to perform the fatigue life assessment of the low-cycle fatigue of metallic materials, two examples are given below.

Example 1: Low-Cycle Fatigue of EN AW-2007 Aluminum Alloy

Here, such an attempt is made that the Wöhler equation is used to analyze low-cycle fatigue test data (see Tables 3.4 and 3.5) of EN AW-2007 aluminum alloy reported by Szusta and Seweryn [14]. The obtained results are shown in Tables 3.4 to 3.6 (see also Figure 3.4). From Figure 3.4, it is found that the experimental fatigue lives are in good agreement with those predicted by the Wöhler equation, with error indexes: M.E. = 1.0%, E.R. = [-21.4,21.1]%, and M.E. = 10.7%, E.R. = [-47.7,87.8]% for tension and torsion fatigue, respectively.

Figure 3.5 shows the monotonic tension curve of EN AW-2007 aluminum alloy.

TABLE 3.4
Low-Cycle Fatigue Test Data of EN AW-2007 Aluminum Alloy [14] (for Tension R = −1) and Predicted Results by the Wöhler Equation

ε_a	ε_a^e	ε_a^p	σ_a (MPa)	N (exp)	N (cal)	R.E.(%)
0.0025	0.001867	0.000632	139	7480	8090	8.2
0.0035	0.002073	0.001427	155	2066	1624	−21.4
0.005	0.002171	0.002828	162	699	847	21.1
0.008	0.002350	0.00565	175	290	271	−6.4
0.01	0.002395	0.007605	179	173	194	12.4
0.02	0.002569	0.017431	192	75	69	−7.7

Note: (a) $A_1 = -0.06784$, $C_1 = 2.40813$, R-square = 0.99; (b) M.E. = 1.0%, E.R. = [−21.4,21.1]%

TABLE 3.5
Low-Cycle Fatigue Test Data of EN AW-2007 Aluminum Alloy [14] (for Torsion R = −1) and Predicted Results by the Wöhler Equation

γ_a	γ_a^e	γ_a^p	τ_a (MPa)	N (exp)	N (cal)	R.E. (%)
0.0290	0.0029	0.0261	97.93	63	92	46.1
0.02	0.0027	0.0173	91.20	168	196	16.4
0.0135	0.0025	0.0110	84.37	455	446	−2.1
0.0090	0.0024	0.0066	81.36	1253	655	−47.8
0.0075	0.0023	0.0052	78.12	1905	1006	−47.2
0.006	0.0021	0.0039	72.74	2964	2141	−27.8
0.0045	0.0019	0.0026	63.93	5232	8396	60.5
0.0035	0.0017	0.0018	59.43	9679	18178	87.8

Note: (a) $A_0 = -0.09449$, $C_0 = 2.41505$, R-square = 0.91; (b) M.E. = 10.7%, E.R. = [−47.7,87.8]%

TABLE 3.6
Fitting Constants in the Wöhler Equation of EN AW-2007 Aluminum Alloy

A_1	C_1	R-square	A_0	C_0	R-square
−0.06784	2.40813	0.99	−0.09449	2.41505	0.91

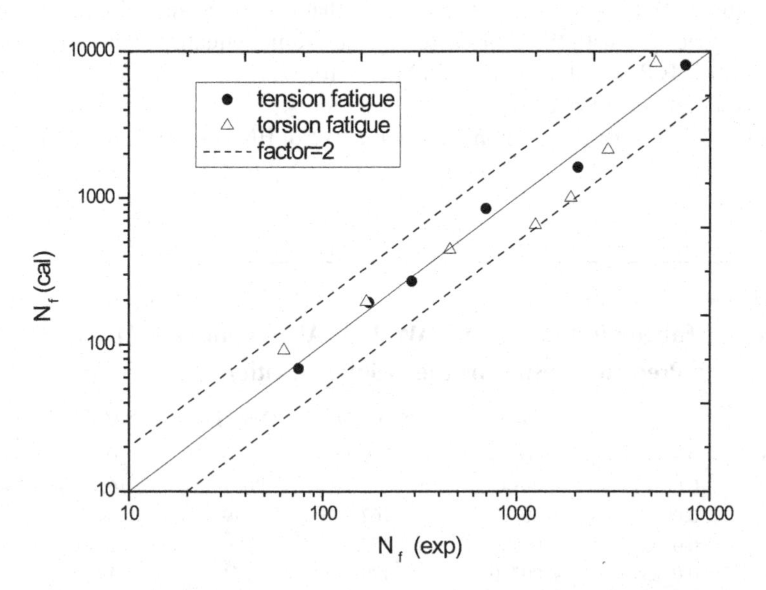

FIGURE 3.4 Comparison of experimental fatigue lives with those predicted by the Wöhler equation (EN AW-2007 alloy.

Source: (Data from Szusta and Seweryn [14])

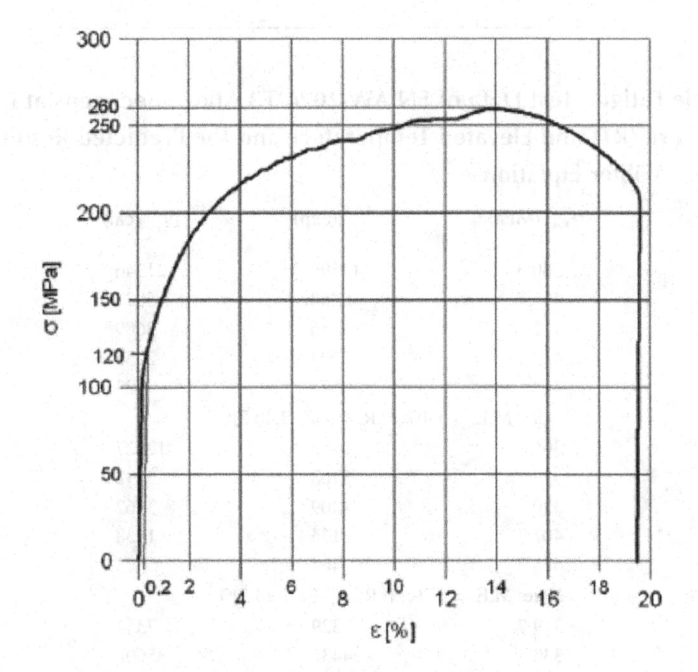

FIGURE 3.5 The monotonic tension curve of EN AW-2007 alloy.

Source: (Data from Szusta and Seweryn [14])

Thus, it is concluded here from the obtained results shown in Tables 3.4 to 3.6 (see also Figure 3.4), that for EN AW-2007 aluminum alloy, its low-cycle fatigue life is to be well predicted by the Wöhler equation.

Example 2: Low-Cycle Fatigue of Aluminum Alloy EN AW-2024-T3

Here, low-cycle fatigue test data (see Table 3.7) of aluminum alloy EN AW-2024-T3 at room temperature (RT) and elevated temperature reported by Szusta and Seweryn [15] are analyzed by using the Wöhler equation. The obtained results are shown in Table 3.7 (see also Figure 3.6). From Figure 3.6, it is found that the experimental fatigue lives are in good agreement with those predicted by the Wöhler equation, with error indexes: M.E. = 3.4%, E.R. = [−22.2,46.2]%; M.E. = 2.7%, E.R. = [−38.1,23.5]%; M.E. = 2.5%, E.R. = [−27.5,38.7]% and M.E. = 5.8%, E.R. = [−37.3,58.2]% for 20°, 100°, 200° and 300°, respectively.

Table 3.8 shows the fitting constants in the Wöhler equation of aluminum alloy EN AW-2024-T3. It is seen from the various values of R square shown in Table 3.8 that the low-cycle fatigue life of the material studied is well calculated by using the Wöhler equation.

Figure 3.7 shows a comparison of monotonic tension curves of EN AW-2024T3 aluminum alloy at 20°, 100°, 200°, and 300°.

TABLE 3.7
Low-Cycle Fatigue Test Data of EN AW-2024-T3 Alloy Specimens at Room Temperature (RT) and Elevated Temperature and the Predicted Results by Using the Wöhler Equation

T (degree)	$\sigma_{x,a}$ (MPa)	N_f (exp)	N_f (cal)	R.E. (%)
20	380.3	14597	21346	46.2
	406.7	6458	5027	−22.2
	413	4156	3609	−13.2
	423	2704	2155	−20.3
	445	572	723	26.4
	Note: M.E. = 3.4%, E.R. = [−22.2,46.2]%			
100	365	10755	12327	14.6
	379	5168	5792	12.1
	390	3209	3262	1.6
	407.7	2164	1338	−38.1
	425.3	464	573	23.5
	Note: M.E. = 2.7%, E.R. = [−38.1,23.5]%			
200	319.7	8329	7354	−11.7
	330	4435	5002	12.8
	344	2177	3018	38.7
	370.7	1679	1217	−27.5
	410.2	355	355	0.1
	Note: M.E. = 2.5%, E.R. = [−27.5,38.7]%			
300	330.7	3775	3165	−16.2
	341.7	1881	1666	−11.4
	347.7	870	1185	36.2
	356.3	464	734	58.2
	387.7	225	140	−37.7
	Note: M.E. = 5.8%, E.R. = [−37.3,58.2]%			

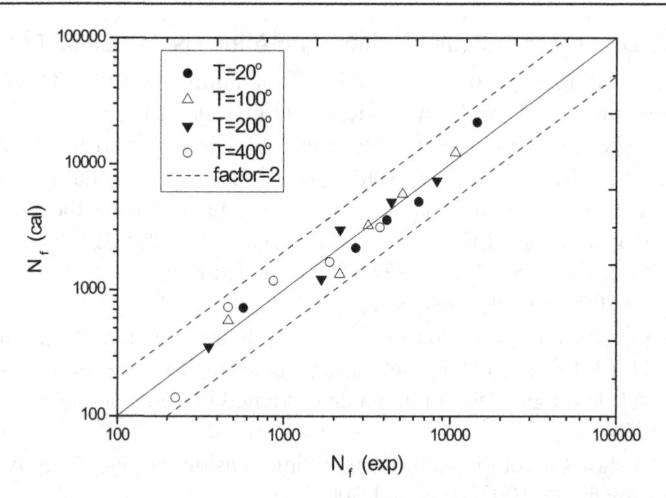

FIGURE 3.6 Comparison of the experimental fatigue lives with those predicted by the Wöhler equation.

Source: (Data from Szusta and Seweryn [15])

TABLE 3.8
Fitting Constants in the Wöhler Equation of Aluminum Alloy EN AW-2024-T3

T(degree)	A_1	C_1	R square
20	−0.04641	2.78105	0.90
100	−0.04983	2.76614	0.93
200	−0.08227	2.82284	0.94
300	−0.05102	2.69802	0.86

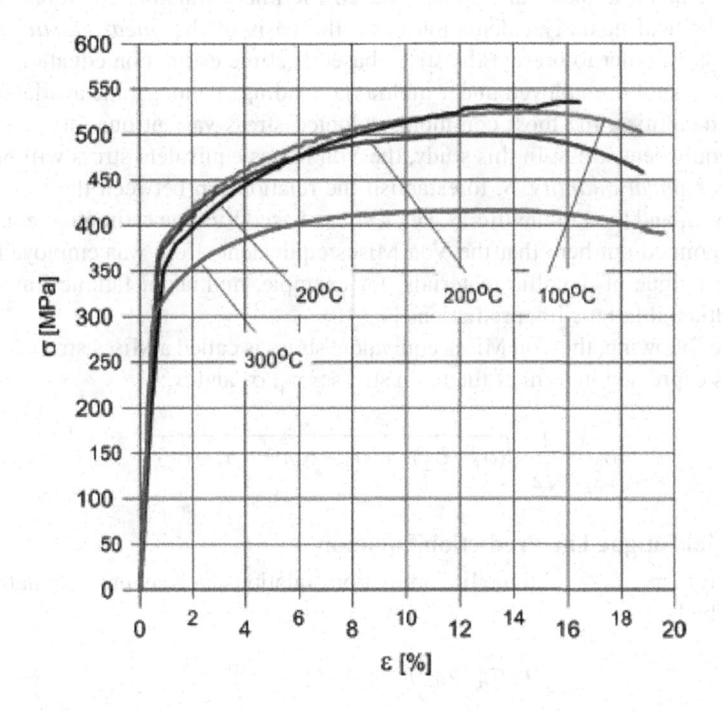

FIGURE 3.7 A comparison of monotonic tension curves of EN AW-2024T3 aluminum alloy at 20°, 100°, 200°, 300°.

Source: (Data from Szusta and Seweryn [15])

3.2.2 The Unified Lifetime Estimation Equation of a Low/Medium/High Cycle Fatigue of Metallic Materials

From the study of Section 2.1, it is sure enough that the low-cycle fatigue life analysis of metallic materials is well performed by the Wöhler Curve Method. Medium/high cycle fatigue lives of metallic materials, as is well known to us, is well calculated by the Wöhler Curve Method. Thus it appears to be possible that, in a low/medium/high cycle fatigue regime of metallic materials, the Wöhler Curve Method is well suited for fatigue life assessment. For multiaxial loading fatigue, a generalized Wöhler equation (i.e., a generalized *S-N* equation) was proposed recently in Ref.

[54]. It is a stress invariant based one, in which the computation of stress invariants is on the basis of the *linear-elastic analysis* of the studied component. The systematic validation exercises given in Ref.[54] by using a large number of experimental data of metallic materials in the literature proved that, in a low/medium/high cycle fatigue regime of metallic materials, the generalized Wöhler equation is well suited for fatigue life assessment.

A Proper Mechanical Quantity in the Fatigue Life Prediction Equation

According to the discussions in Section 2.1, in a low/medium/high cycle fatigue regime of metallic materials, a stress based lifetime estimation equation in which the mechanical quantity calculation is on the basis of the *linear-elastic analysis* may exist. In order to present the stress based lifetime estimation equation, what is a "stress quantity" employed under multiaxial loading? Amongst the available stress variant quantities, the most commonly adopted stress variant quantity is the Von Mises equivalent stress. In this study, the Von Mises equivalent stress will be taken as a *mechanical quantity, S*, to establish the relationship between the *mechanical quantity, S*, and the fatigue life, *N*, i.e., a stress-based lifetime estimation equation.

It is pointed out here that the Von Mises equivalent stress was employed previously in fatigue of metallic materials, for example, multiaxial fatigue limits [6,7], and multiaxial fatigue life predictions [4,9,10].

In the following, the Von Mises equivalent stress is called a Mises stress for short, which is expressed in term of the main stresses σ_1, σ_2 and σ_3:

$$\sigma_e = \frac{1}{\sqrt{2}}\sqrt{(\sigma_1-\sigma_2)^2+(\sigma_2-\sigma_3)^2+(\sigma_3-\sigma_1)^2} \tag{2}$$

Multiaxial Fatigue Life Prediction Equation

Generally, a multiaxial fatigue life prediction equation can be expressed mathematically to be [16]:

$$F(\sigma_{ea},\sigma_{em},\rho,N\ ,c_1,c_2,...)=0 \tag{3}$$

where σ_e is the Mises equivalent stress, σ_{ea} and σ_{em} are the amplitude and the mean value of the equivalent stress, respectively. A multiaxial parameter, ρ, is defined to be [16]:

$$\rho = \frac{\sigma_{kk,a}}{\sigma_{ea}} \tag{4}$$

where $\sigma_{kk,a}$ is the amplitude of the first invariant of stress tensor. It is known from the formula (4) that, for the bending and torsion fatigue, the multiaxial parameter, ρ, is equal to 1 and 0, respectively. $c_1, c_2, ...$, in Eq. (3) are material parameters, which are usually varied with the multiaxial parameter ρ.

Under zero mean stress, a stress-based fatigue life estimation equation under multiaxial loading is assumed to be [16]:

$$\log\sigma_{ea} = A_\rho \log N + C_\rho \tag{5}$$

which, under bending and torsion fatigue, can be written to be, respectively:

$$\log \sigma_a = A_1 \log N + C_1 \tag{6}$$

and

$$\log \sqrt{3}\tau_a = A_0 \log N + C_0 \tag{7}$$

Eq. (6) is the well-known *S-N* curve equation expressed in terms of a stress quantity. Eq. (7) is the variant form of the *S-N* curve equation under torsion fatigue. A_1 and C_1 in Eq.(6), and A_0 and C_0 in Eq.(7) are material constants determined by using bending and torsion experimental fatigue data of plain specimens, respectively.

In view of the complexity of fatigue life analysis under multiaxial loading, and also taking into account that a large number of experimental fatigue data have been accumulated in the literature, from the viewpoint of application, it is assumed here that the material parameters in Eq. (5) can be obtained by interpolating the material constants in Eqs. (6) and (7), i.e.:

$$A_\rho = A_1 \cdot \rho + A_0 \cdot (1-\rho) \tag{8}$$

$$C_\rho = C_1 \cdot \rho + C_0 \cdot (1-\rho) \tag{9}$$

Here, multiaxial fatigue life equation, Eq.(5) is called a generalized Wöhler equation, or a generalized *S-N* equation.

3.3 *S-N* EQUATION OF NOTCH SPECIMENS

V-notches, which are widely used in mechanical components such as bolts, nuts and screws, decrease dramatically the load-bearing capacity of components due to the concentration of stress at the vicinity of their tips. A reliable prediction of the mechanical failure like crack formation and growth in the vicinity of V-notches has been a topic of great interest to researchers.

Since the beginning of the last century by Neuber [17,18] and by Peterson [19], a lot of advance of local approaches to deal with brittle failure of notched components has been made, in particular, the CD approach [20,21] (often called TCD, Theory of Critical Distance) and the volume-based SED (strain energy density) approach [22–24]. In the two well-known local approaches, the size of Neuber's elementary volume is taken to be a material constant. For brittle materials (such as ceramics [25]), for example, the critical distance according to Taylor's TCD is determined by

$$L = \frac{1}{\pi}\left(\frac{K_{Ic}}{\sigma_0}\right)^2 \tag{10}$$

where K_{Ic} is the plane strain fracture toughness and σ_0 is taken to be the ultimate tensile stress, σ_{UTS}. Whereas when static failures are preceded by a certain amount of plasticity, σ_0 takes a value which is higher than the plain material strength and both parameters in TCD (i.e., the critical stress σ_0 and the critical distance L) can be determined only carrying out ad hoc experimental investigations [26,27]. For example, the critical stress and the critical distance of Perspex are the values of the point of intersection of stress-distance curves for the sharp notch and blunt notch (3 mm hole) at the failure load, as shown in Figure 3.8. According to Taylor's PM, thus, failure is predicted to occur when the critical stress σ_0 is present at a critical distance $r = r_c = \frac{L}{2}$ from the notch root (see Figure 3.9). This failure condition is expressed as:

$$\sigma_\theta(r = r_c, \theta = 0) = \sigma_0 \tag{11}$$

where σ_θ is the circumference stress, the coordinate r is the distance from the notch tip.

Here, it is worthy to mention the fatigue life prediction of notched specimens reported in Refs [28-31]. In 2010, using the concept of the TCD and on the basis of notch elastic-plastic stress analysis, the Manson-Coffin equation together with S WTparameter was employed by Susmel and Taylor [28] to perform LCF assessment of notched components. In 2011, using the approach by Susmel and Taylor [28], Yang et al. [31] performed high-temperature LCF life prediction of notched DS Ni-based superalloy. Yang et al. [31] found that, using two calibration fatigue failure curves of a smooth specimen and a notched specimen, the critical distance obtained is different rather than taking a determinate value at a given fatigue life N_f. Thus it

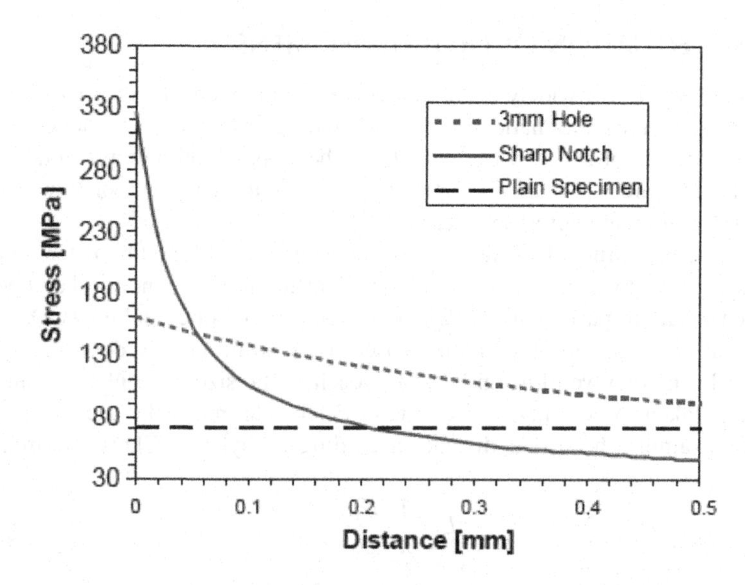

FIGURE 3.8 Stress-distance curves at the failure load for the sharp notch, 3 mm hole and plain specimen, in Perspex.

Source: (Data from Taylor et al.[26])

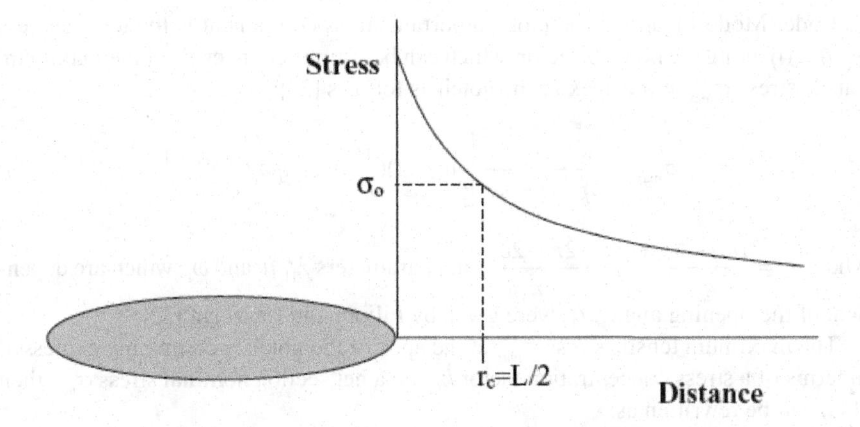

FIGURE 3.9 The theory of critical distances—the point method: failure is predicted to occur when the critical stress σ_0 is present at the critical distance *L*/2 from the notch root.

Source: (Data from Taylor *et al.*[26])).

could be speculated that the equation between the critical distance and fatigue life N_f obtained according to the approach by Susmel and Taylor [29,30],

$$L(N_f) = A \cdot N_f^B \tag{12}$$

together with the plain specimen fatigue life equation may not be applied to other notch types quite well. At the same time, Yang et al. [31] found also that, at a given fatigue life N_f, the products of the critical distances of different notched components determined by two fatigue failure curves of a smooth specimen and notched specimen, and their corresponding linear elastic stress concentration factors, k_t, are almost equal. Thus it could be speculated that the equation between the product of the critical distance and k_t and fatigue life N_f

$$k_t \cdot L(N_f) = A' \cdot N_f^{B'} \tag{13}$$

together with the plain specimen fatigue life equation may be applied to other notch types quite well. Fatigue fracture test data given by Yang et al. [31] proved that prediction results obtained by means of the equation (13) together with the plain specimen fatigue life equation are much better than those by the equation (12) together with the plain specimen fatigue life equation.

From the study of Taylor and his coworkers [25–30], Yang et al. [31] and the *S-N* equation of a low/medium/high cycle fatigue of metallic materials described in Section 2, a Notch *S-N* equation of a low/medium/high cycle fatigue of metallic materials is proposed in this section.

3.3.1 A Brief Description of a Linear Elastic Notch Stress Field

A notch geometry and coordinate system of a V-shaped notch taken by Filippi and Lazzarin [32] are shown in Figure 3.10 (the opening angle 2α).

Under Mode I loading, the most important stress component is the tensile stress $\sigma_y(\theta = 0)$ along the notch bisector, which can be simplified in terms of the maximum tensile stress σ_{max} at the apex of the notch as follows [32]:

$$\sigma_y = \sigma_{max} \cdot \frac{1}{4(q-1)+q\omega_1}\left[4(q-1)(\frac{r}{r_0})^{\lambda_1 - 1} + q\omega_1(\frac{r}{r_0})^{\mu_1 - 1}\right] \tag{14a}$$

where $r_0 = \dfrac{\rho(q-1)}{q}$, $q = \dfrac{2\pi - 2\alpha}{\pi}$, and parameters λ_1, μ_1 and ω_1, which are dependent of the opening angle 2α, were given by Filippi and Lazzarin [32].

The maximum tensile stress σ_{max} at the apex of the notch is commonly expressed in terms of a stress concentration factor k_I and a net section nominal stress σ_{net}; then (14a) can be rewritten as:

$$\sigma_y = k_I \cdot \sigma_{net} \cdot \frac{1}{4(q-1)+q\omega_1}\left[4(q-1)(\frac{r}{r_0})^{\lambda_1 - 1} + q\omega_1(\frac{r}{r_0})^{\mu_1 - 1}\right] \tag{14b}$$

Concerning the stress concentration factor k_I in (14b), many results have been obtained (e.g., [33–36]).

For blunt V-shaped notches in axisymmetric shafts under torsion, the most important stress component is the torsion stress $\tau_{zy}(\theta = 0)$ along the notch bisector, which can be simplified in terms of the maximum shear stress τ_{max} as follows [37]:

$$\tau_{zy} = \tau_{max} \cdot \frac{1}{\omega_3}(\frac{r}{r_0})^{\lambda_3 - 1}\left[1 + (\frac{r}{r_3})^{\mu_3 - \lambda_3}\right] \cdot \left[1 - \frac{r - r_0}{R}\right] \tag{15a}$$

where $r_3 = (1 - \mu_3) \cdot \rho$, and parameters λ_3, μ_3 and ω_3, which are dependent of the opening angle 2α, were given by Filippi and Lazzarin [37]. R in (15a) is a net section radius of shaft.

Formula (15a) can be rewritten as

$$\tau_{zy} = k_{III} \cdot \tau_{net} \cdot \frac{1}{\omega_3}(\frac{r}{r_0})^{\lambda_3 - 1}\left[1 + (\frac{r}{r_3})^{\mu_3 - \lambda_3}\right] \cdot \left[1 - \frac{r - r_0}{R}\right] \tag{15b}$$

where k_{III} is the stress concentration factor, and τ_{net} is a net section nominal stress.

Concerning the stress concentration factor k_{III} in (15b), many results have been obtained (e.g., [38,39]).

3.3.2 Notch *S-N* Equation Under Mode I Loading

It is assumed that, at a pointed V–notch (notch angle α, Williams' eigenvalue, λ), there are two branch notches with the notch radii, ρ_1 and ρ_2, whose stress concentration factors are k_1 and k_2, respectively. Here, the two branch notches are denoted by Notch 1 and Notch 2.

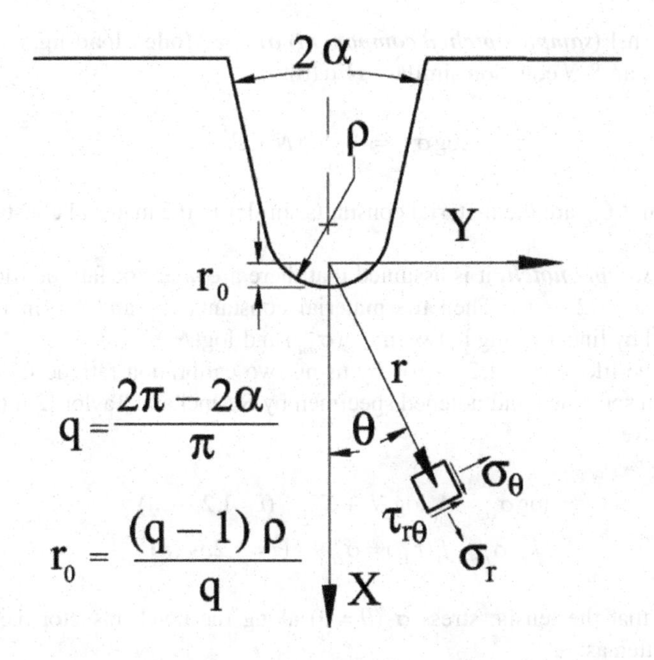

FIGURE 3.10 A notch geometry and coordinate system of a V-shaped notch.

Source: (Data from Filippi and Lazzarin [32])

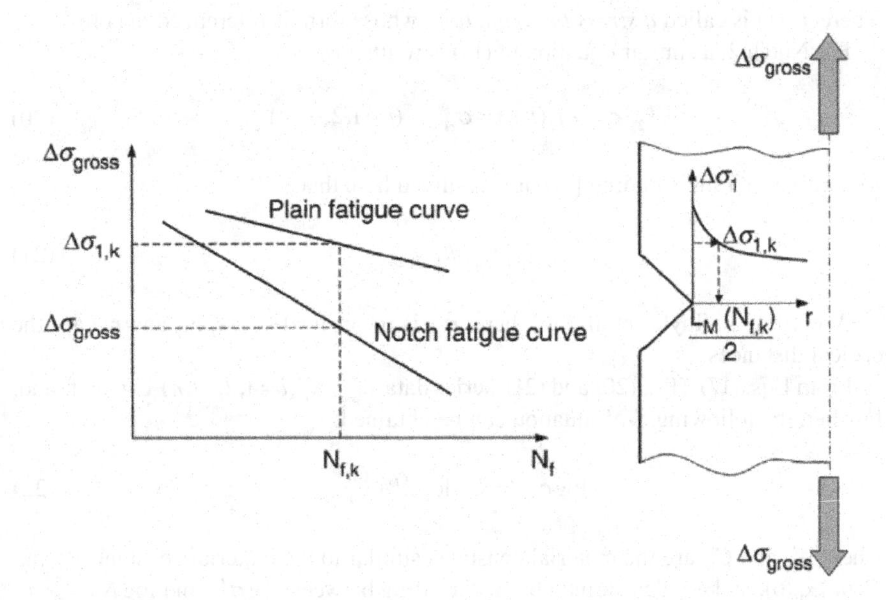

FIGURE 3.11 Plain and notched calibration fatigue curve (a); Linear-elastic stress field in the vicinity of the stress concentrator apex and application of the TCD in terms of the PM to estimate the number of cycles to failure (b).

Source: (Data from Susmel and Taylor [29])

For Notch 1 (***sample notched component***) **under Mode I loading,** it is assumed that there is an *S-N* equation similar to Eq.(6):

$$\log\sigma_{an} = A_{1n}\log N + C_{1n} \tag{16}$$

where A_{1n} and C_{1n} are the material constants similar to the material constants A_1 and C_1 in Eq.(6).

For the ***sample notch,*** it is assumed that there are n sets of fatigue life test data, $\sigma_{ani}, N_i \quad (i = 1, 2, \cdots, n)$. Then the material constants A_{1n} and C_{1n} in Eq.(16) are determined by linear fitting between $\log(\sigma_{ani})$ and $\log(N_i)$.

At fatigue life $N_i\ (i = 1, 2, \cdots, n)$, by using two calibration fatigue failure curves of a smooth specimen and notched specimen by Susmel and Taylor [29] (see Figure 3.11), we have

$$\log\sigma_{ci} = A_1\log N_i + C_1 \qquad (i = 1, 2, \cdots, n) \tag{17}$$

$$k_1\cdot\sigma_{ani}\cdot f_1(r_{c1i}) = \sigma_{ci} \qquad (i = 1, 2, \cdots, n) \tag{18}$$

Note here that the tensile stress $\sigma_y(\theta = 0)$ along the notch bisector, i.e., (14b), is simply written as:

$$\sigma_y = k_I\cdot\sigma_{net}\cdot f(r) \tag{19}$$

where $f_1(r)$ is called *a stress field function*, whose form is referenced to (14b).

For Notch 2, a similar equation to (18) is written as

$$k_2\cdot\sigma'_{ani}\cdot f_2(r_{c2i}) = \sigma_{ci} \qquad (i = 1, 2, \cdots, n) \tag{20}$$

According to Yang's finding [31], it is assumed here that:

$$k_1\cdot r_{c1i} = k_2\cdot r_{c2i} \tag{21}$$

According to Taylor et al. [26], here, σ_{ci} is the critical stress, r_{c1i} and r_{c2i} are the critical distances.

From Eqs. (17), (18), (20) and (21), series data $\sigma'_{ani}, N_i\ (i = 1, 2, \cdots, n)$ can be found. Further, the following *S-N* equation can be obtained:

$$\log\sigma_{an} = A'_{1n}\log N + C'_{1n} \tag{22}$$

where A'_{1n} and C'_{1n} are the material constants similar to the material constants A_1 and C_1 in Eq.(6), and are determined by linear fitting between $\log\sigma'_{ani}$ and $\log N_i$.

It is important to point out here that the *S-N* equation (22) of *Notch* 2 is different from the *S-N* equation (16) of *Notch* 1 (calibration specimen). The latter is called an inherent Notch *S-N* equation, while the former is called a predicted Notch *S-N* equation.

3.3.3 Notch *S-N* Equation Under Mode III Loading

According to the method in Section 3.2, the following equations parallel to equations (16) and (22) can be obtained easily:

$$\log \sqrt{3}\tau_a = A_{0n} \log N + C_{0n} \tag{23}$$

$$\log \sqrt{3}\tau_a = A'_{0n} \log N + C'_{0n} \tag{24}$$

which are an inherent Notch *S-N* equation and a predicted Notch *S-N* equation under Mode III loading, respectively.

3.3.4 Notch *S-N* Equation Under Mode I/III Loading

Following the way of dealing with the multiaxial fatigue life prediction of plain specimens, here, the Notch *S-N* equation under Mode I/III loading for the calibration notch specimen can be written as:

$$\log \sigma_{e,a} = A_{\rho n} \log N + C_{\rho n} \tag{25}$$

where

$$A_{\rho n} = A_{1n} \cdot \rho + A_{0n} \cdot (1-\rho) \tag{26}$$

$$C_{\rho n} = C_{1n} \cdot \rho + C_{0n} \cdot (1-\rho) \tag{27}$$

For *Notch* 2, similarly, the Notch *S-N equation* under Mode I/III loading can be written as:

$$\log \sigma_{e,a} = A'_{\rho n} \log N + C'_{\rho n} \tag{28}$$

where

$$A'_{\rho n} = A'_{1n} \cdot \rho + A'_{0n} \cdot (1-\rho) \tag{29}$$

$$C'_{\rho n} = C'_{1n} \cdot \rho + C'_{0n} \cdot (1-\rho) \tag{30}$$

Here, the Eqs. (25) and (28) are called the inherent multiaxial Notch *S-N* equation and the predicted multiaxial Notch *S-N* equation, respectively.

3.4 EXPERIMENTAL VERIFICATIONS AND DISCUSSIONS

The Notch *S-N* equations of metallic materials described in Section 3 are rewritten as follows:

The inherent Notch *S-N* equations are:

$$\log \sigma_{an} = A_{1n} \log N + C_{1n} \tag{16}$$

and

$$\log \sqrt{3}\tau_a = A_{0n} \log N + C_{0n} \tag{23}$$

which are also called the inherent axial (or bending) Notch *S-N* equation and the inherent torsion Notch *S-N* equation, respectively.

The predicted Notch *S-N* equations are:

$$\log \sigma_{an} = A'_{1n} \log N + C'_{1n} \tag{22}$$

and

$$\log \sqrt{3}\tau_a = A'_{0n} \log N + C'_{0n} \tag{24}$$

which are also called the predicted axial (or bending) Notch *S-N* equation and the predicted torsion Notch *S-N* equation.

In a low/medium/high cycle fatigue regime of metallic materials, in this section, the Notch *S-N* equations will be verified by using the experimental fatigue data of metallic materials from the literature. Here, it is very importantly emphasized that provided that the Notch *S-N* equations are verified to be suitable for fatigue assessment of notched metallic materials under uniaxial loading, a multiaxial loading fatigue assessment can be performed by using stress type multiaxial fatigue equations, including stress type invariant fatigue equations, e.g., Eqs. (25) and (28), and stress type critical plane equations, e.g., the MWCM by Susmel and Lazzarin [1]. In this work, a fatigue life assessment of notched metallic materials under multiaxial loading will be carried out by using the multiaxial Notch *S-N* equations, see Eqs. (25) and (28).

An effect of nonproportional loading on fatigue life is implemented by using Itoh's formula (A-1) and a linear formula (A-2) proposed in Ref.[54] (see Appendix *A*).

An effect of mean stress on fatigue life is considered by using a linear mean stress model proposed here (see Appendix *B*).

In order to quantitatively evaluate the accuracy of the fatigue life prediction, the following error indexes are defined:

$$\text{R.E.} = \frac{(N(cal) - N(\exp)) \times 100}{N(\exp)} \tag{31}$$

$$M.E. = \frac{1}{n}\sum_{i=1}^{n}(\text{R.E.})_i \tag{32}$$

$$E.R. = [\min\{R.E.\}, \max\{R.E.\}] \tag{33}$$

where *n* is the number of experimental cases. The three error indexes defined in (31) to (33) are called a relative error, a mean error and an error range, respectively.

Note here that the material constants, σ_0 and τ_0 in (*B*-2), (*B*-3) and (*B*-4) in Appendix *B*, can be evaluated by using Eqs. (16) and (23) (or Eqs. (22) and (24)), respectively. For example, for Notch 1 (En3B) described in the following, by letting $N = 2000000$ (cycles) and using the inherent Notch *S-N* equations of Notch 1, the calculated σ_0 and τ_0 are:

$$\sigma_0 = 95 \text{ (MPa)}, \ \tau_0 = 160 \text{ (MPa)}$$

While for Notch 2, by letting $N = 2000000$ (cycles) and using the predicted Notch S-N equations of Notch 2, the calculated σ_0 and τ_0 are:

$$\sigma_0 = 181 \text{ (MPa)}, \tau_0 = 225 \text{ (MPa)}$$

In this section, verification examples are divided into three types. Type I examples are used to illustrate the implementation process of notch fatigue life assessment described in the above section, i.e, the process, in which the material constants in the predicted Notch S-N equations (see Eqs. (22) and (24)) are obtained according to two calibration fatigue failure curves of a smooth specimen and notched specimen by Susmel and Taylor [42] and the fatigue life evaluation model of notch metallic materials described in the above section, is illustrated in detail. And further, the predicted Notch S-N equations are used to perform the fatigue life assessment of notched components.

Type II examples, including the notched components made of Cast Iron [43], 16MnR steel [3], SAE 1045 steel [44], SAE 1045 steel [45] and En3B steel [28], are taken to illustrate that the inherent Notch S-N equations (see Eqs. (16) and (23)) are used to perform the fatigue life assessment of notched components, in which the approaches, including how to obtain the material constants in the inherent Notch S-N equations by using experimental data of fatigue lives of notched specimens under bending and torsion loading, and further how to carry out a multiaxial fatigue life assessment of notched components by using the inherent Notch S-N equations, are completely the same as those for plain materials. The details are given in Appendix *C*.

Type III examples, including the notched components made of S690 steel [12], EN3B steel [29], C40 carbon steel [46], AISI416 steel [47], Ti–6Al–4V alloy [48] and 40CrMoV13.9 steel [49], are taken to illustrate that the Notch S-N equations are naturally existing. The details are given in Appendix *D*.

Type I examples given here include: medium/high cycle fatigue of notched components made of low-carbon steel [40], medium/high cycle fatigue of notched components made of En3B steel [41] and low-cycle fatigue of notched components made of IMI 829 Ti alloy [42]. The details are as follows.

Example 1: Medium/High Cycle Fatigue of Notched Components Made of Low-Carbon Steel

Plain and notch fatigue experiments of low-carbon steel reported by Qilafku et al. [40] are taken as an example to illustrate the process of notch fatigue assessment described in Section 3. For the plain material, the material constants in the S-N equations (see Eqs. (6) and (7)) are listed in Table 3.9. Notch geometry is shown in Figure 3.12, in which USC18 and USC19 are No. of specimens used in Ref. [12], which are denoted, respectively, by Notch 1 and Notch 2 in this study. The stress concentration factors, k_I and k_{III}, of the two notches are given in Table 3.10.

Here, Notch 1 specimen is taken as a calibration notched specimen. According to two calibration fatigue failure curves of a smooth specimen and notched specimen by Susmel and Taylor [29] and the fatigue life evaluation model of notch metallic materials described in Section 3, the material constants in the predicted

TABLE 3.9
Material Constants in *S-N* Equations

		Material Constants in Bending *S-N* Equation			Material Constants in Torsion *S-N* Equation			
Materials	**Specimens**	A_1	C_1	**R-square**	A_0	C_0	**R-square**	**Note**
En3B steel	Plain	−0.05088	2.84316	0.93371	−0.03961	2.91508	0.69838	
[41]	Notch1	−0.24764	3.53879	0.85686	−0.11637	3.17584	0.88148	Inherent
	Notch2	−0.16557	3.30173		−0.08704	3.13939		Predicted
Low-carbon	Plain	−0.05002	2.67743	0.85523	−0.04171	2.7756	0.94528	
steel [40]	Notch1	−0.20343	3.17293	0.75598	−0.09243	2.92078	0.9065	Inherent
	Notch2	−0.18463	3.14258	0.75971	−0.08675	2.93467	0.91522	Predicted
16MnRsteel	Notch1	−0.18268	3.1392	0.98329	−0.10159	2.97289	0.93502	Inherent
[3]	Notch2	−0.11446	2.98776	0.97	−0.0887	2.99189	0.97	Inherent
Cast iron [43]	Notch	−0.16067	2.96615	1	−0.07503	2.89289	1	Inherent
SAE1045 [44]	Notch	−0.15284	3.24679	0.96	−0.07881	2.93139	0.98	Inherent
SAE1045 [45]	Notch	−0.13868	2.67583	0.98	−0.12432	2.74659	0.96	Inherent
En3B steel [29]	Notch1	−0.20323	3.34774	0.98				Inherent
	Notch2	−0.15945	3.2195	0.98				Inherent
IMI 829 Ti	Plain	−0.05765	3.10669	0.89				
alloy [42]	Notch1	−0.05765	3.4067	0.98				Inherent
	Notch2	−0.22413	3.6211	0.98				Predicted

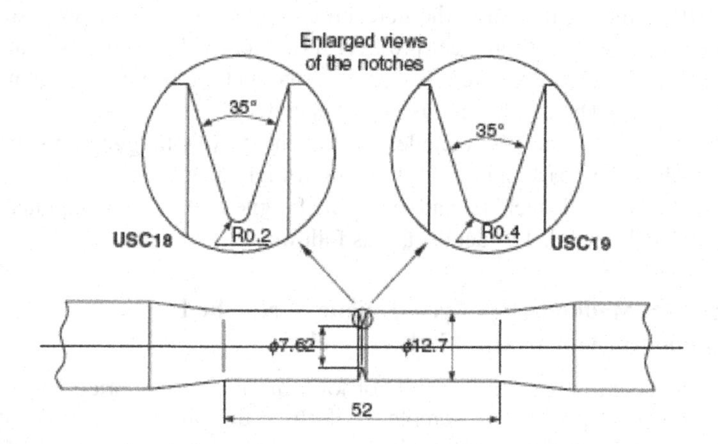

FIGURE 3.12 Notch geometry of low-carbon steel specimens.

TABLE 3.10
Stress Concentration Factors of the Two Notches Shown in Figure 3.12

Notch	k_I	k_{III}
Notch 1	4.41	2.35
Notch 2	3.21	1.87

TABLE 3.11
Experimental Fatigue Data of Notch 1 and Process Data Used to Determine the Material Constants in the S-N Equation of Notch 2 (Tension Loading Fatigue)

$\sigma_{a,n1}$ (exp) (MPa)	N (exp)	$\sigma_{a,e}$ (MPa)	r_{c1} (mm)	r_{c2} (mm)	f_1	f_2	$\sigma_{a,n2}$ (MPa)
180.6	45294	266	0.2405	0.3311	0.3334	0.4037	205.4
145.3	55284	264.1	0.159	0.2188	0.4117	0.4902	168
165.5	71111	261.6	0.2099	0.2889	0.3581	0.4312	189.2
140.3	74158	261.2	0.1513	0.2082	0.4222	0.5014	162.6
120	93409	259	0.1101	0.1516	0.4892	0.5707	141.6
157.1	98439	258.5	0.194	0.2671	0.3728	0.4477	180.1
140.3	121418	256.5	0.157	0.2161	0.4142	0.4932	162.2
159.9	132047	255.7	0.2052	0.2824	0.3625	0.4363	182.9
119	302439	248	0.119	0.1638	0.4722	0.5539	139.6
100	380948	245.9	0.0809	0.1113	0.5569	0.6394	119.9
95.8	598088	241.8	0.0756	0.1041	0.5719	0.6532	115.5
100	700007	240.4	0.0855	0.1176	0.5446	0.6262	119.7

Notch *S-N* equation of Notch 2 under bending loading fatigue can be obtained in the following approach (see Table 3.11):

$\sigma_{a,n1}$ (MPa): experimental load of Notch 1 (stress amplitude);
N (cycles): experimental fatigue life of Notch 1;
$\sigma_{a,e}$ (MPa): stress amplitude subjected to plain specimen, which is calculated by using the bending *S-N* equation of plain specimens (see Eq. (15)) and experimental fatigue life N;
r_{c1} (mm): the critical distance of Notch 1 (see Figure 3.13);
r_{c2} (mm): the critical distance of Notch 2 (see Figure 3.13);
f_1: stress field function value (at $r = r_{c1}$) of Notch 1:

$$f_1 = f_1(r_{c1}),$$

where $f_1(r)$ is a *stress field function* of Notch 1.

f_2: stress field function value (at $r = r_{c2}$) of Notch 2:

$$f_2 = f_2(r_{c2}),$$

where $f_2(r)$ is a *stress field function* of Notch 2.

$\sigma_{a,n2}$: stress amplitude subjected to Notch 2 (see Figure 3.13), which is calculated by using the following equations:

$$\frac{k_{l2}}{k_{l1}} \cdot \frac{\sigma_{a,n2}}{\sigma_{a,n1}} \cdot \frac{f_2}{f_1} = 1$$

$$k_{l1} \cdot r_{c1} = k_{l2} \cdot r_{c2}.$$

where k_{t1} and k_{t2} are the stress concentration factors of Notch 1 and Notch 2 (see Table 3.10), respectively.

Thus, from $\{\sigma_{a,n2}\}_i$ and $\{N\}_i$ $(i = 1,2,...,12)$ in Table 3.11, the material constants in the S-N equation of Notch 2 can be obtained by the linear fitting approach. The obtained material constants are called predicted material constants (see Table 3.9).

Naturally, from $\{\sigma_{a,n1}\}_i$ and $\{N\}_i$ $(i = 1,2,\cdots,12)$ in Table 3.11, the material constants in the S-N equation of Notch 1 can be obtained by the linear fitting approach. The inherent material constants are also listed in Table 3.9.

Similarly, the material constants in the Notch S-N equation of Notch 2 under torsion loading fatigue can be calculated (see Table 3.12). The obtained material constants, including the inherent material constants for Notch 1 and the predicted material constants for Notch 2, are given in Table 3.9.

After determining the material constants in the Notch S-N equations for Notch 1 and Notch 2 (see Table 3.9), the Notch S-N equations will be used to perform multiaxial fatigue life assessment of the notched components.

Under tension and torsion loading fatigue, a comparison of experimental fatigue lives with those calculated by the **inherent *S-N* equation** of Notch 1 are given in Figure 3.14: error ranges and mean errors are [−56,154]% and 12.4% for tension loading fatigue and [−44,261]% and 17.4% for torsion loading fatigue.

For Notch 2, under tension and torsion loading fatigue, a comparison of experimental fatigue lives with those calculated by the **predicted *S-N* equation** of Notch 2 are given in Figure 3.15: error ranges and mean errors are [−22,118]%

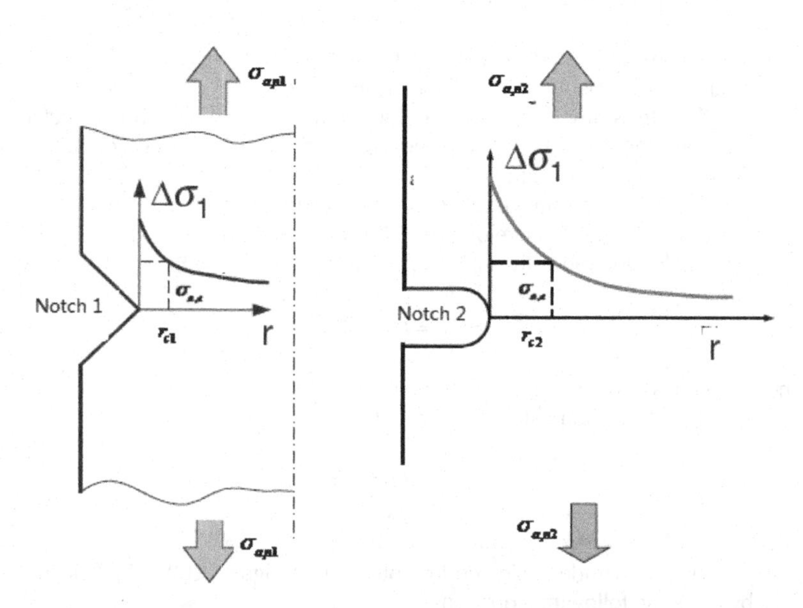

FIGURE 3.13 At equal fatigue life N, Notch 1 has a stress amplitude $\sigma_{a,e}$ at the Critical Distance $r = r_{c1}$. Notch 2 has also a stress amplitude $\sigma_{a,e}$ at the Critical Distance $r = r_{c2}$. The stress amplitude $\sigma_{a,e}$ is the effective stress acted on a plain specimen according to two calibration fatigue failure curves of a smooth specimen and notched specimen by Susmel and Taylor [29]: $\log(\sigma_{a,e}) = A_1 \log(N) + C_1$.

TABLE 3.12
Experimental Fatigue Data of Notch 1 and Process Data Used to Determine the Material Constants in the S-N Equation of Notch 2 (Torsion Loading Fatigue)

$\tau_{a,n1}$ (exp) (MPa)	N (exp)	$\tau_{a,e}$ (MPa)	r_{c1} (mm)	r_{c2} (mm)	f_1	f_2	$\tau_{a,n2}$ (MPa)
200.8	17783	229	0.2483	0.3112	0.4856	0.5584	218.8
200.8	19539	228.1	0.2505	0.314	0.4831	0.5563	218.5
188.1	25119	225.7	0.22	0.2757	0.5108	0.585	205.8
174.5	38580	221.7	0.1906	0.2389	0.5404	0.6164	191.7
174.5	67892	216.5	0.2025	0.2537	0.5287	0.6032	191.7
149.7	84581	214.5	0.1365	0.171	0.6108	0.6856	167.2
168.5	151991	209.4	0.2017	0.2528	0.5287	0.6041	184.8
140.8	586384	197.9	0.1445	0.1811	0.5977	0.6743	156.4
140.2	811131	195.2	0.1485	0.186	0.5935	0.6688	155.9
120.2	3195549	184.4	0.1102	0.1381	0.6534	0.7264	135.5

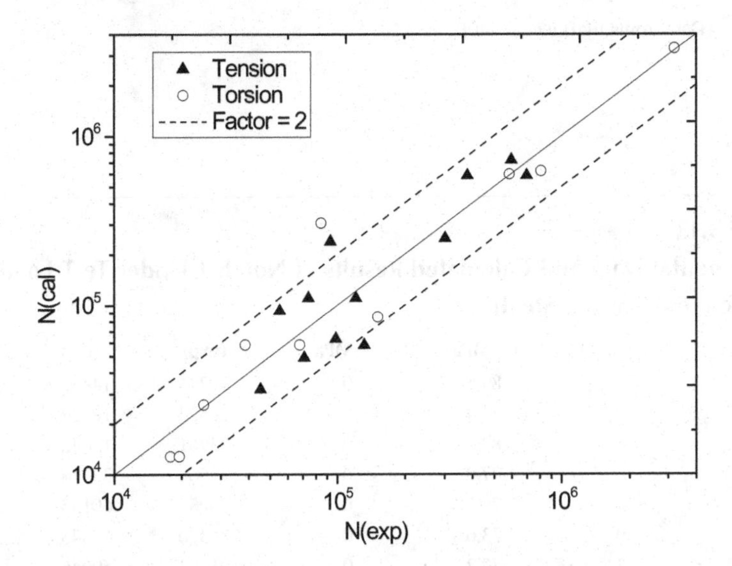

FIGURE 3.14 A comparison of experimental fatigue lives with those calculated by the inherent Notch S-N equation of Notch 1 (tension and torsion loading fatigue).

Source: (Data from Qilafku et al. [40])

and 30% for tension loading fatigue, and [–70,62]% and –17% for torsion loading fatigue.

From Figures 3.14 and 3.15, it can be seen that, under tension and torsion loading fatigue, the predicted fatigue lives of notch specimens by using the Notch S-N equation are in good agreement with those measured experimentally.

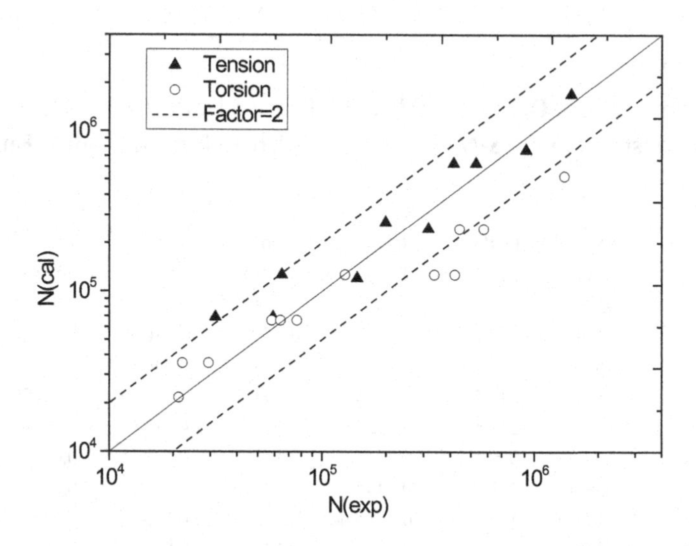

FIGURE 3.15 A comparison of experimental fatigue lives with those calculated by the predicted Notch *S-N* equation of Notch 2 (tension and torsion loading fatigue).

Source: (Data from Qilafku et al. [40])

TABLE 3.13

Experimental Data and Calculated Results of Notch 1 Under Te-T Loading Fatigue (Low-Carbon Steel)

σ_a (Mpa)	σ_m (MPa)	τ_a (MPa)	τ_m (MPa)	N (exp)	N (cal)	R.E. (%)
179.6	0	89.8	0	20244	14836	−26.7
172.9	0	86.4	0	21322	18422	−13.6
177.9	0	89	0	23961	15645	−34.7
155.7	0	77.9	0	33566	33314	−0.7
140.3	0	70.2	0	74989	60135	−19.8
147.2	0	73.6	0	85370	45843	−46.3
140.4	0	70.2	0	101043	59946	−40.7
112.8	0	56.4	0	149081	207409	39.1
113.9	0	56.9	0	214329	196477	−8.3
126.5	0	63.2	0	234690	108370	−53.8
135.2	0	67.6	0	243999	74252	−69.6
106.6	0	53.3	0	442993	285798	−35.5
96.9	0	48.5	0	697378	490519	−29.7
85.6	0	42.8	0	836175	991756	18.6
92.5	0	46.3	0	3018073	638431	−78.8

Note: M.E. = 26.7%, E.R. = [−78.8,39.1]%

Under tension-torsion loading (Te-T) fatigue, experimental data and calculated results by using the intrinsic multiaxial Notch *S-N* equation of Notch 1 are shown in Table 3.13, with error indexes: M.E. = 26.7%, E.R. = [−79, 39]%, which shows the calculated results of fatigue lives are satisfactory.

For Notch 2, under Te-T loading fatigue, experimental data and calculated results by using the predicted multiaxial Notch *S-N* equation of Notch 2 are shown in Table 3.14, with error indexes: M.E. = −14.4%, E.R. = [−65.6,72]%, from which it is seen that the calculated fatigue lives are in good agreement with those measured experimentally.

Example 2: Medium/High Cycle Fatigue of Notched Components Made of En3B

Plain and notch fatigue experiments of En3B (a commercial cold-rolled low-carbon steel) reported by Susmel and Taylor [41] are taken as the second example to illustrate the implementation process of notch fatigue lifetime assessment described in Section 3. For the plain material, the material constants in the *S-N* equations (see Eqs. (6) and (7)) are listed in Table 3.9. Notch geometry is shown in Figure 3.16, in which USC30 and USC31 are no. of specimens used in Ref [12], which are

TABLE 3.14
Experimental Data and Calculated Results of Notch 2 Under Te-T Loading Fatigue (Low-Carbon Steel)

σ_a (MPa)	σ_m (MPa)	τ_a (MPa)	τ_m (MPa)	N(exp)	N(cal)	R.E.(%)
183.1	0	91.6	0	26927	25230	−6.3
183.1	0	91.6	0	35353	25230	−28.6
183.2	0	91.6	0	44070	25167	−42.9
154.3	0	77.2	0	64185	73161	14
128.8	0	64.4	0	130954	225287	72
137.7	0	68.9	0	147161	148526	0.9
140.5	0	70.2	0	355366	131287	−63.1
102.7	0	51.3	0	1028934	922472	−10.3
113	0	56.5	0	1479262	508550	−65.6

Note: M.E. = −14.4%, E.R. = [−65.6,72]%

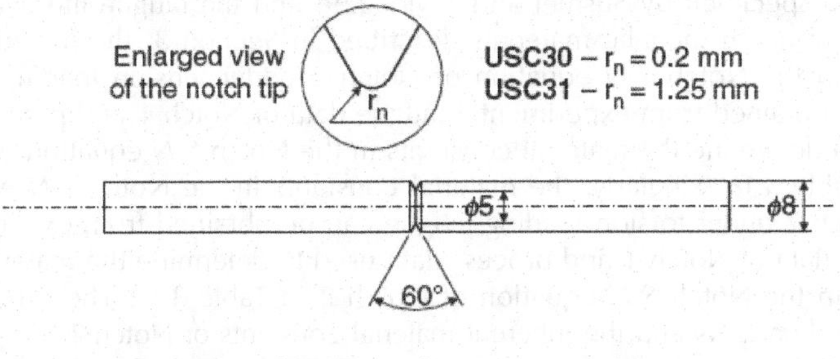

Enlarged view of the notch tip
r_n
USC30 – r_n = 0.2 mm
USC31 – r_n = 1.25 mm
ϕ5
ϕ8
60°

FIGURE 3.16 Notch geometry of En3B specimens.

TABLE 3.15
Stress Concentration Factors of the Two Notches Shown in Figure 3.16

Notch	k_I	k_{III}
Notch 1	3.607	2.040
Notch 2	1.740	1.275

TABLE 3.16
Experimental Fatigue Data of Notch 1 and Process Data Used to Determine the Material Constants in the Notch *S-N* Equation of Notch 2 (Tension Loading Fatigue)

$\sigma_{a,n1}$ (exp) (MPa)	N (exp)	$\sigma_{a,e}$ (MPa)	r_{c1} (mm)	r_{c2} (mm)	f_1	f_2	$\sigma_{a,n2}$ (MPa)
331	10852	434.3	1.96E-01	4.06E-01	3.64E-01	5.93E-01	4.21E+02
203.7	96785	388.6	8.72E-02	1.81E-01	5.29E-01	7.55E-01	2.96E+02
137.5	236230	371.3	3.00E-02	6.23E-02	7.49E-01	8.97E-01	2.38E+02
137.5	314269	366	3.19E-02	6.62E-02	7.38E-01	8.91E-01	2.36E+02
331	18762	422.4	2.07E-01	4.29E-01	3.54E-01	5.81E-01	4.18E+02
331	22505	418.5	2.11E-01	4.37E-01	3.50E-01	5.77E-01	4.17E+02
203.7	148533	380.2	9.19E-02	1.91E-01	5.17E-01	7.46E-01	2.93E+02
300.5	26164	415.3	1.76E-01	3.66E-01	3.84E-01	6.16E-01	3.88E+02
137.5	1313155	340.3	4.18E-02	8.66E-02	6.86E-01	8.63E-01	2.27E+02
142.6	224361	372.3	3.45E-02	7.14E-02	7.24E-01	8.84E-01	2.42E+02
142.6	237312	371.2	3.48E-02	7.22E-02	7.22E-01	8.83E-01	2.42E+02
203.7	52908	400.7	8.09E-02	1.68E-01	5.45E-01	7.68E-01	3.00E+02

denoted, respectively, by Notch 1 and Notch 2 in this study. The stress concentration factors, k_I and k_{III}, are given in Table 3.15.

Here, the Notch 1 specimen is taken as a calibration notched specimen. According to two calibration fatigue failure curves of a smooth specimen and notched specimen by Susmel and Taylor [29] and the fatigue life evaluation model of notch metallic materials described in Section 3, the material constants in the Notch *S-N* equation of Notch 2 under tension loading fatigue can be obtained from experimental fatigue data of Notch 1 and process data used to determine the material constants in the Notch *S-N* equation of Notch 2 in Table 3.16. Similarly, the material constants in the Notch *S-N* equation of Notch 2 under torsion loading fatigue can be obtained from experimental fatigue data of Notch 1 and process data used to determine the material constants in the Notch *S-N* equation of Notch 2 in Table 3.17. These predicted material constants and the inherent material constants of Notch 1 are given in Table 3.9.

TABLE 3.17
Experimental Fatigue Data of Notch 1 and Process Data Used to Determine the Material Constants in Notch *S-N* Equation of Notch 2 (Torsion Loading Fatigue)

$\tau_{a,n1}$ (exp) (Mpa)	N (exp)	$\tau_{a,e}$ (MPa)	r_{c1} (mm)	r_{c2} (mm)	f_1	f_2	$\tau_{a,n2}$ (MPa)
171.1	1025331	274.4	5.06E-02	8.09E-02	7.86E-01	9.05E-01	2.38E+02
264.8	31705	314.9	1.52E-01	2.43E-01	5.83E-01	7.56E-01	3.27E+02
264.8	30028	315.6	1.51E-01	2.41E-01	5.84E-01	7.57E-01	3.27E+02
171.1	306633	287.9	3.87E-02	6.19E-02	8.25E-01	9.26E-01	2.44E+02
171.1	904657	275.8	4.94E-02	7.90E-02	7.90E-01	9.07E-01	2.38E+02
203.7	344380	286.5	8.89E-02	1.42E-01	6.90E-01	8.43E-01	2.67E+02
203.7	614536	280	9.66E-02	1.55E-01	6.74E-01	8.31E-01	2.64E+02
244.5	70232	305.2	1.32E-01	2.11E-01	6.12E-01	7.82E-01	3.06E+02
297.4	7700	333.1	1.78E-01	2.85E-01	5.50E-01	7.24E-01	3.61E+02

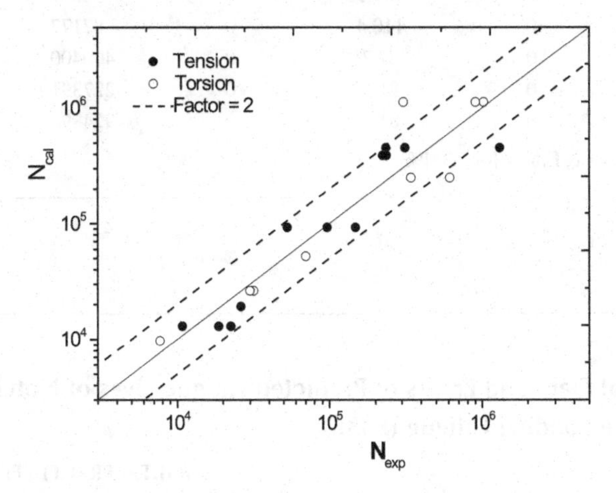

FIGURE 3.17 Comparison of experimental fatigue lives with those calculated by the inherent Notch *S-N* equation of Notch 1 (En3B).

Source: (Data from Susmel and Taylor [41])

After determining the material constants in the Notch *S-N* equations for Notch 1 and Notch 2 (see Table 3.9), the Notch *S-N* equations will be used to perform fatigue life assessment of the notched components.

For Notch 1, under tension and torsion loading fatigue, a comparison of experimental fatigue lives with those calculated by the inherent Notch *S-N* equation of Notch 1 are given in Figure 3.17, with error ranges and mean errors: [−65,75]% and 13.5% for tension loading fatigue and [−59,265]% and 20.4% for torsion loading fatigue, which shows that the calculated fatigue lives are in good agreement with experimental ones.

For Notch 1, under Te-T loading fatigue, experimental data and calculated results by using the intrinsic multiaxial Notch *S-N* equation are shown in Table 3.18, with error indexes: M.E. = −19%, E.R. = [−67,34]%, from which it is seen that the calculated fatigue lives are satisfactory.

Under **out-of-phase loading fatigue**, experimental data of Notch 1 and calculated results are listed in Table 3.19. Error ranges and mean errors of fatigue lives predicted by using Itoh's formula (A-1) and the linear formula (A-2) are, respectively, [−76,71]% and 11.7%, and [−69,117]% and 11.7%, which illustrates that the two formulas have almost the same accuracy.

TABLE 3.18
Experimental Data of Notch 1 and Calculated Results Under Te-T Loading Fatigue (En3B)

σ_a (MPa)	σ_m (MPa)	τ_a (MPa)	τ_m (MPa)	*N* (exp)	R.E. (%)
259.6	0	155.9	0	14743	−11
216.3	0	129.9	0	30837	2.6
183.9	0	110.4	0	87177	−20.6
146	0	87.7	0	460400	−54.3
135.2	0	81.2	0	227391	34.1
135.2	0	81.2	0	924890	−67

Note: M.E. = −19%, E.R. = [−67,34]%

TABLE 3.19
Experimental Data and Errors of Predicted Fatigue Lives of Notch 1 Under Out-of-Phase Loading Fatigue (En3B)

σ_a(MPa)	τ_a (MPa)	*N* (exp)	*F*	*F′*	$\bar{\sigma}_{ea} / \sigma_{ea}$	ER (I.T) (%)	ER (L.F) (%)	ER (I.T) (%)	ER (L.F) (%)
183.9	110.4	67416	0.9618	0.6813	1.0055	9.1	5.9	−14.3	−14.2
216.3	129.9	30764	0.9614	0.8012	1.0058	9.3	−10.7	−14.1	−14.3
146	87.7	210914	0.9612	0.5407	0.9996	6.1	27.1	−16.6	−16.4
259.6	155.9	9202	0.9614	0.9616	1.0763	51.5	−0.8	19.1	18.6
135.2	81.2	189952	0.9613	0.5008	1.1032	70.6	117.4	34.1	34.5
129.8	77.9	1646841	0.962	0.4809	**0.7344**	−76	−68.5	Deleting an unusual data	
Error range (%)						[−76,71]	[−69,117]	[−17,34]	**[−16,35]**
Mean error (%)						11.7	11.7	1.6	1.6

Note: All out-of-phase angles = 90°; all $\sigma_m = 0$; all $\tau_m = 0$. $\alpha' = -0.013$, $\lambda = 0.78797$, $\gamma = 0.30178$. After deleting the unusual data, $\alpha' = 0.0396$, $\lambda = 1.03567$, $\gamma = 0.00348$.

Note here that the sixth data $\bar{\sigma}_{ea}/\sigma_{ea} = 0.7344$ is an **unusual data** because it is obviously deviated from the mean value (0.9874). After deleting the unusual data, error ranges and mean errors of fatigue lives predicted by using Itoh's formula (A-1) and the linear formula (A-2) are [−17,34]% and 1.6%, and [−16,35]% and 1.6%. Similarly, the two formulas have almost the same accuracy.

Under Te-T loading fatigue with **mean stress effect**, experimental data and process data of fatigue life analysis of Notch 1 are listed in Table 3.20. Due to lack of experimental fatigue data of uniaxial mean stress effect, a multiaxial fatigue life evaluation equation taken here is

$$\log\sigma_{ea}\left(\alpha+\beta\frac{\sigma_{em}}{\sigma_0{}^{\rho}(\sqrt{3}\tau_0)^{1-\rho}}\right) = A_{pn}\log N + C_{pn} \tag{34}$$

where α and β are fitting constants between $\{\bar{\sigma}_{ea}/\sigma_{ea}\}_i$ and $\left\{\frac{\sigma_{em}}{\sigma_0{}^{\rho}(\sqrt{3}\tau_0)^{1-\rho}}\right\}_i$, $(i = 1,2,\ldots,6)$ according to the following relation:

$$\frac{\bar{\sigma}_{ea}}{\sigma_{ea}} = \left(\alpha+\beta\frac{\sigma_{em}}{\sigma_0{}^{\rho}(\sqrt{3}\tau_0)^{1-\rho}}\right) \tag{35}$$

where $\{\bar{\sigma}_{ea}\}_i$ is calculated by using the following relation:

$$\log\bar{\sigma}_{ea} = A_{pn}\log N + C_{pn} \tag{36}$$

Note here that the material parameters A_{pn} and C_{pn} in (34) and (36) are calculated by using the material constants in the Notch S-N equation of Notch 1 (see Table 3.9 and the interpolation formulas (8) and (9)).

By using Eq.(34), error range and mean error of the predicted fatigue lives under Te-T loading fatigue with mean stress effect are [−18,69]% and 3%, which illustrates that the predicted fatigue lives are in good agreement with experimental ones.

TABLE 3.20
Experimental Data and Process Data of Fatigue Life Analysis of Notch 1 Under Te-T Loading Fatigue with Mean Stress Effect (En3B)

σ_a (MPa)	σ_m (MPa)	τ_a (MPa)	τ_m (MPa)	N (exp)	$\bar{\sigma}_{ea}/\sigma_{ea}$	$\frac{\sigma_{em}}{\sigma_0{}^{\rho}(\sqrt{3}\tau_0)^{1-\rho}}$	ER (%)
194.7	194.7	202.5	202.5	6364	1.158	2.4332	−13.8
194.7	194.7	202.5	202.5	6680	1.1479	2.4332	−17.9
162.2	162.2	168.7	168.7	11899	1.2419	2.0271	66.8
108.2	108.2	112.5	112.5	396137	0.9904	1.352	−22.9
97.3	97.3	101.2	101.2	626504	1.0139	1.216	−2.6
129.8	129.8	135	135	83907	1.0917	1.6221	8.5

Note: $\sigma_0 = 95$(MPa), $\tau_0 = 160$(MPa), $\alpha = 0.84848$, $\beta = 0.14011$

Under **out-of-phase loading fatigue with mean stress effect**, experimental data and process data of fatigue life analysis of Notch 1 are listed in Tables 3.21 and 3.22. F_{Mean} in Table 3.21 is defined as

$$F_{Mean} = \left(\alpha + \beta \frac{\sigma_{em}}{\sigma_0{}^{\rho}(\sqrt{3}\tau_0)^{1-\rho}} \right) \tag{37}$$

where α and β are fitting constants between $\{\sigma_{ena} / \sigma_{ea}\}_i$ and $\left\{ \frac{\sigma_{em}}{\sigma_0{}^{\rho}(\sqrt{3}\tau_0)^{1-\rho}} \right\}_i$, $(i = 1, 2, \ldots, 6)$, and σ_{ena} is calculated by

$$\log \sigma_{ena} = A_{\rho n} \log N + C_{\rho n} \tag{38}$$

Note also here that the material parameters $A_{\rho n}$ and $C_{\rho n}$ in (38) are calculated by using the material constants in the Notch *S-N* equation of Notch 1 and the interpolation formulas (8) and (9).

TABLE 3.21
Experimental Data and Process Data of Fatigue Life Analysis of Notch 1 Under Out-of-Phase Loading Fatigue with Mean Stress Effect (En3B)

σ_a (MPa)	σ_m (MPa)	τ_a (MPa)	τ_m (MPa)	N (exp)	$\sigma_{ena} / \sigma_{ea}$	$\frac{\sigma_{em}}{\sigma_0{}^{\rho}(\sqrt{3}\tau_0)^{1-\rho}}$	F_{Mean}	$\frac{\sigma_{ena}}{\sigma_{ea}} / F_{Mean}$
129.8	129.8	135	135	62951	1.1496	1.6221	1.0601	1.0845
91.9	91.9	95.6	95.6	685098	1.0563	1.1486	0.9427	1.1205
146	146	151.9	151.9	45580	1.0831	1.8248	1.1103	0.9755
108.2	108.2	112.5	112.5	1397923	**0.7892**	1.352	0.9932	0.7946
119	119	123.7	123.7	203792	1.0151	1.4869	1.0266	0.9889
135.2	135.2	140.6	140.6	59130	1.1163	1.6896	1.0768	1.0367

Note: $\sigma_0 = 95$(MPa), $\tau_0 = 160$(MPa), $\alpha = 0.65813$, $\beta = -0.24779$

TABLE 3.22
Process Data of Fatigue Life Analysis of Notch 1 Under Out-of-Phase Loading Fatigue with Mean Stress Effect (En3B)

No.	F	F'	F_{Mean}	$\frac{\sigma_{ena}}{\sigma_{ea}} / F_{Mean}$	$(1 + \alpha' {*} F)$	$ZF1$	ER (I.F) (%)	$ZF2$	ER (L.F) (%)
1	0.5551	0.4932	1.0601	1.0845	1.0001	1.0602	56.8	1.0593	57.6
2	0.555	0.3492	0.9427	1.1205	1.0001	0.9428	88	0.9458	84.7
3	0.5549	0.5547	1.1103	0.9755	1.0001	1.1104	−12.9	1.1075	−11.7
4	0.5553	0.4112	0.9932	**0.7946**	1.0001	0.9933	−72.1	0.9947	−72.3
5	0.5554	0.4522	1.0266	0.9889	1.0001	1.0267	−6.1	1.027	−6.2
6	0.5552	0.5138	1.0768	1.0367	1.0001	1.0769	22.1	1.0753	23.1

Note: all out-of-phase angles = 90°; $\alpha' = 0.00021$; $\lambda = 1.01315$, $\gamma = -0.02821$

$ZF1$ and $ZF2$ in Table 3.22 are defined as

$$ZF1 = (\alpha + \beta \frac{\sigma_{em}}{\sigma_0{}^{\rho}(\sqrt{3}\tau_0)^{1-\rho}}) \cdot (1 + \alpha' \cdot F) \tag{39}$$

$$ZF2 = (\alpha + \beta \frac{\sigma_{em}}{\sigma_0{}^{\rho}(\sqrt{3}\tau_0)^{1-\rho}}) \cdot (\lambda + \gamma \cdot F') \tag{40}$$

Under **out-of-phase loading fatigue with mean stress effect**, fatigue life evaluation equations corresponding to Itoh's formula (A-1) and the linear formula (A-2) are respectively:

$$\log[\sigma_{ea} \cdot (\alpha + \beta \frac{\sigma_{em}}{\sigma_0{}^{\rho}(\sqrt{3}\tau_0)^{1-\rho}}) \cdot (1 + \alpha' \cdot F)] = A_{\rho n} \log N + C_{\rho n} \tag{41}$$

and

$$\log[\sigma_{ea} \cdot (\alpha + \beta \frac{\sigma_{em}}{\sigma_0{}^{\rho}(\sqrt{3}\tau_0)^{1-\rho}}) \cdot (\lambda + \gamma \cdot F')] = A_{\rho n} \log N + C_{\rho n} \tag{42}$$

Note here that the materials constant α' in Itoh's formula (A-1) is such a fitting constant between $\left\{\frac{\sigma_{ena}}{\sigma_{ea}} / F_{Mean}\right\}_i$ and $\{F\}_i$, $(i = 1, 2, \ldots, 6)$, and that material constants, λ and γ, in the linear formula (A-2), are fitting constants between $\left\{\frac{\sigma_{ena}}{\sigma_{ea}} / F_{Mean}\right\}_i$ and $\{F'\}_i$, $(i = 1, 2, \ldots, 6)$.

Under out-of-phase loading fatigue with mean stress effect, for Notch 1, error ranges and mean errors predicted by using Eqs. (41) and (42) are [−72,88]% and 12.5%, and [−72,85]% and 12.5%, with almost the same accuracy.

Note here that the fourth data $\sigma_{ena} / \sigma_{ea}$ = **0.7892** is an unusual data because it is obviously deviated from the mean value. After deleting the unusual data, error range and mean error of fatigue lives, [−28,34]% and 1.5%, predicted by Eq. (41) are completely the same as those by Eq. (42) (see Table 3.23).

TABLE 3.23
Process Data of Fatigue Life Analysis of Notch 1 Under Out-of-Phase Loading Fatigue with Mean Stress Effect (Deleting an Unusual Data) (En3B)

No.	F	F'	F_{Mean}	$\frac{\sigma_{ena}}{\sigma_{ea}} / F_{Mean}$	(1 + α'*F)	$ZF1$	ER (I.F) (%)	$ZF2$	ER (L.F) (%)
1	0.5551	0.4932	1.0903	1.0544	1.0000	1.0903	34.2	1.0903	34.2
2	0.555	0.3492	1.0467	1.0092	1.0000	1.0467	5.2	1.0467	5.2
3	0.5549	0.5547	1.109	0.9766	1.0000	1.109	−12.3	1.109	−12.3
5	0.5554	0.4522	1.0779	0.9418	1.0000	1.0779	−28.3	1.0779	−28.3
6	0.5552	0.5138	1.0966	1.018	1.0000	1.0966	10.4	1.0965	10.4

Note: all out-of-phase angles = 90°; α' = 1.44E-05; λ = 1.00017, γ = −0.00035

Under Te-T loading fatigue, experimental data and calculated results by using the predicted multiaxial Notch *S-N* equation of Notch 2 are listed in Table 3.24, with error indexes: M.E. = −26.5%, E.R. = [−59,−1]%, from which it is seen that the calculated fatigue lives are satisfactory.

Under **out-of-phase loading fatigue**, experimental data and process data of fatigue life analysis of Notch 2 are shown in Table 3.25. An accuracy of fatigue lives predicted by using the linear formula (A-2), with error range and mean error, [−73,46]% and 9.4%, is almost the same as that by using Itoh's formula (A-1), with error range and mean error, [−73,43]% and 9.5%.

Note here that the first data $\bar{\sigma}_{ea} / \sigma_{ea}$ = **0.8033** is an unusual data because it is obviously deviated from the mean value. After deleting the unusual data, error range and mean error of predicted fatigue lives, [−26,22]% and 1.4%, by using the

TABLE 3.24

Experimental Data and Calculated Results by Using the Predicted Multiaxial Notch *S-N* Equation of Notch 2 Under Te-T Loading Fatigue (En3B)

σ_a (MPa)	σ_m (MPa)	τ_a (MPa)	τ_m (MPa)	*N* (exp)	R.E. (%)	R.E. (%)
259.6	0	155.9	0	82952	62380	−24.8
200	0	115.5	0	437907	426084	−2.7
180	0	103.9	0	2174897	893883	−58.9
275	0	158.8	0	46254	45653	−1.3
230	0	132.8	0	188480	160019	−15.1
190	0	109.7	0	1400006	610403	−56.4

Note: M.E. = −26.5%, E.R. = [−59,−1]%

TABLE 3.25

Experimental Data and Calculated Results of Notch 2 Under Out-of-Phase Loading Fatigue (En3B)

σ_a (MPa)	τ_a (MPa)	N (exp)	*F*	*F'*	$\bar{\sigma}_{ea} / \sigma_{ea}$	ER (I.T) (%)	ER (L.F) (%)	ER (I.T) (%)	ER (L.F) (%)
260	150.1	314817	0.9999	0.9122	**0.8033**	−72.6	−72.7	Deleting unusual data	
285	164.5	31700	0.9997	0.9997	1.0166	43.1	34.3	17.9	9
285	164.5	36976	0.9997	0.9997	0.9945	22.6	15.2	1.1	−6.5
230	132.8	150125	0.9999	0.8071	1.0091	35.8	45.7	11.9	21.8
270	155.9	59622	0.9999	0.9474	0.9805	11	8.1	−8.5	−11.6
230	132.8	245935	0.9999	0.8071	0.9405	−17.1	−11.1	−31.7	−25.7
250	144.3	79328	0.9997	0.8769	1.0169	43.3	46.4	18.1	21.1
Error range (%)						[−73,43]	[−73,46]	[−32,18]	[−26,22]
Mean error (%)						9.5	9.4	1.5	1.4

Note: All out-of-phase angles = 90°; all $\sigma_m = 0$; all $\tau_m = 0$. $\alpha' = -0.034$, $\lambda = 0.87957$, $\gamma = 0.09519$. After deleting the unusual data, $\alpha' = -0.007$, $\lambda = 0.88465$, $\gamma = 0.11958$.

linear formula (A-2), are almost same as those by using Itoh's formula (A-1), with error range and mean error, [−32,18]% and 1.5%.

Under Te-T loading fatigue with **mean stress effect**, experimental data and process data of fatigue life analysis of Notch 2 are given in Table 3.26. The fatigue lives calculated by using the predicted multiaxial Notch *S-N* equation of Notch 2 are in good agreement with experimental ones, with error indexes: M.E. = 5.2%, M.R. = [−34,85]%.

Under **out-of-phase loading fatigue with mean stress effect**, experimental data and fatigue life prediction results of Notch 2 are listed in Table 3.27. An accuracy of fatigue lives predicted by Eq. (42), with error indexes: M.E. = 9.7%, E.R. = [−37,137]%, is almost the same as that by Eq. (41), with error indexes: M.E. = 10.6%, M.R. = [−40,130]%.

TABLE 3.26

Experimental Data and Process Data of Fatigue Life Analysis of Notch 2 Under Te-T Loading Fatigue with Mean Stress Effect (En3B)

σ_a (MPa)	σ_m (MPa)	τ_a (MPa)	(MPa)	*N* (exp)	$\bar{\sigma}_{ea} / \sigma_{ea}$	$\frac{\sigma_{em}}{\sigma_0^{\rho}(\sqrt{3}\tau_0)^{1-\rho}}$	R.E. (%)
150	150	150	150	844615	0.988262	1.12821	1.3
180	180	180	180	28108	1.26575	1.353852	84.8
160	160	160	160	370618	1.02808	1.203424	−14.2
165	165	165	165	249286	1.048132	1.241031	−20.5
190	190	190	190	34298	1.169363	1.429066	−34.2
170	170	170	170	110056	1.127977	1.278638	13.8

Note: $\sigma_0 = 181.3$ (MPa), $\tau_0 = 225$(MPa), $\alpha = 0.06374$, $\beta = 0.81805$

TABLE 3.27

Experimental Data and Fatigue Life Prediction Results of Notch 2 Under Out-of-Phase Loading Fatigue with Mean Stress Effect (En3B)

σ_m (MPa)	σ_m (MPa)	τ_a (MPa)	τ_m (MPa)	Out-of-Phase Angle	N (exp)	ER (I.F) (%)	ER (L.F) (%)
170	170	170	170	90	112944	4.6	6.2
155	155	155	155	90	367445	−39.6	−37.2
180	180	180	180	90	49200	63.4	63.4
190	190	190	190	90	52000	7.9	6.2
200	200	200	200	90	67873	−41	−42.8
160	160	160	160	90	304439	−41.4	−39.5
235	235	135.7	135.7	90	59243	24.2	6.1
160	160	160	160	90	77755	129.6	136.9
175	175	175	175	90	111250	−12.7	−12
Error range (%)						[−40,130]	[−37,137]
Mean error (%)						10.6	9.7

Example 3: Low Cycle Fatigue of Notched Components Made of IMI 829 Ti Alloy

Low-cycle fatigue experiments (with constant-amplitude load- and strain-controlled tests being conducted using fully reversed sinusoidal loading) of plain and notched specimens made of IMI 829 Ti alloy reported by Putchkov et al.[42] are taken as the third example to illustrate the process of notch fatigue lifetime assessment described in Section 3. For the plain material, the material constants in the *S-N* equations are listed in Table 3.9. Notch geometry details studied by Putchkov et al.[42] are shown in Figure 3.18. In this study, notches with notch radii 0.25 mm and 0.35 mm are denoted by Notch 1 and Notch 2, respectively. Their stress concentration factors are 3.2 and 2.8, respectively (see Figure 3.18).

Here, the Notch 1 specimen is taken as a calibration notched specimen. According to two calibration fatigue failure curves of a smooth specimen and notched specimen by Susmel and Taylor [29] and the fatigue life evaluation model of notch metallic materials described in Section 3, the material constants in the Notch *S-N* equation of Notch 2 under tension loading fatigue can be obtained from experimental fatigue data of Notch 1 and process data used to determine the material constants in the Notch *S-N* equation of Notch 2 in Table 3.28. These predicted material constants and the inherent material constants of Notch 1 are given in Table 3.9.

For plain specimens, a comparison of experimental fatigue lives with those calculated by the plain *S-N* equation is shown in Figure 3.19, with error indexes: M.E. = 2 6.4%, E.R. = [−49.8,93.6]%.

For Notch 1, a comparison of experimental fatigue lives with those calculated by the intrinsic Notch *S-N* equation is also shown in Figure 3.19, with error indexes: M.E. = 0.8%, E.R. = [−20.7,14.1]%.

For Notch 2, a comparison of experimental fatigue lives with those calculated by the predicted Notch *S-N* equation is also shown in Figure 3.19, where there is

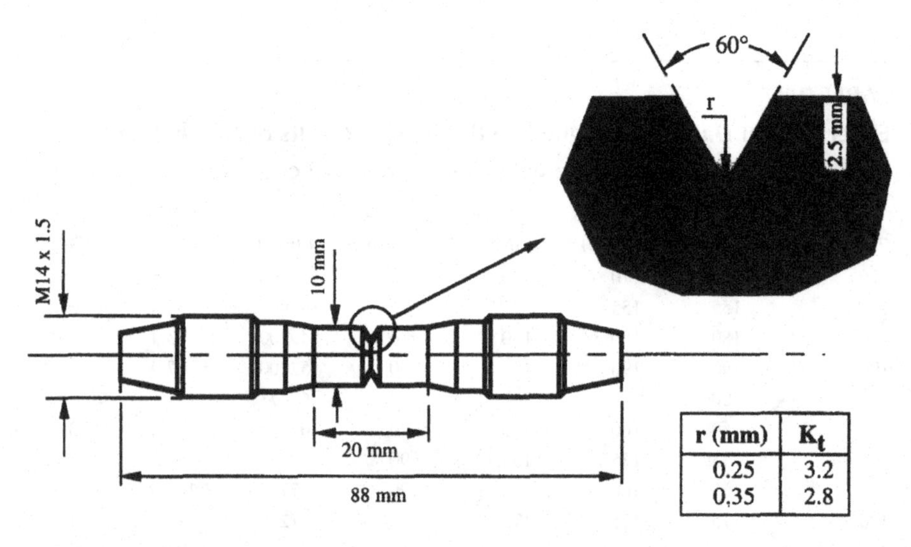

FIGURE 3.18 The geometrical details of notched components.

Source: (Data from Putchkov *et al.*[42])

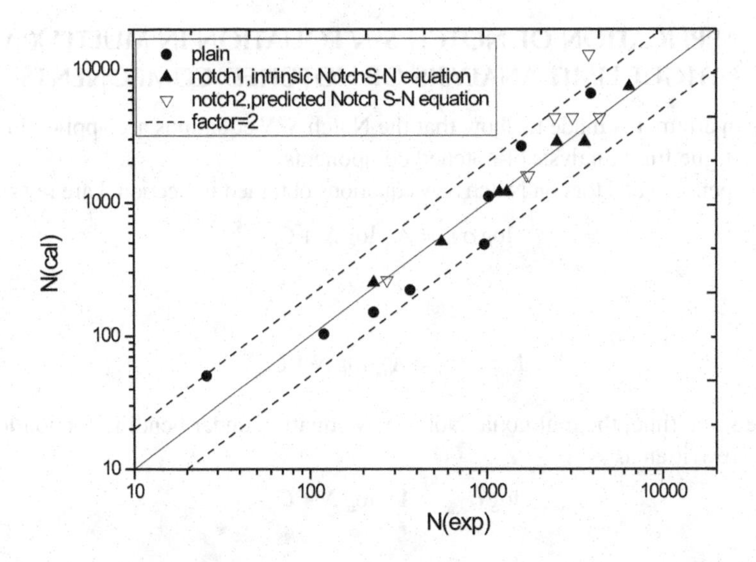

FIGURE 3.19 Comparison of experimental fatigue lives with those calculated for plain and notched components made of IMI 829 Ti alloy.

Source: (Data from Putchkov et al.[42])

TABLE 3.28
Experimental Fatigue Data of Notch 1 and Process Data Used to Determine the Material Constants in the Notch S-N Equation of Notch 2 (Tension Loading Fatigue, IMI 829 Ti Alloy)

$\sigma_{a,n1}$ (exp) (MPa)	N (exp)	$\sigma_{a,e}$ (MPa)	r_{e1} (mm)	r_{e2} (mm)	$\sigma_{a,n2}$ (cal) (MPa)
640	3613	1590.7	0.024347	0.027825	697.7
800	1295	1687.6	0.048969	0.055965	846.7
1200	230	1864.4	0.114844	0.13125	1220.8
640	2512	1624.3	0.021591	0.024675	700.7
800	1174	1697.2	0.048051	0.054915	847.6
1000	555	1772.1	0.082688	0.0945	1032.5
500	6467	1538.2	9.19E-06	1.05E-05	573

an unusual data with relative error 240.3%. After deleting the unusual data, the calculated error indexes are: M.E. = 16.4%, E.R. = [−8,77.0]%.

From the comparison of experimental fatigue lives with those calculated for plain and notched components made of IMI 829 Ti alloy shown in Figure 3.19, thus, it is seen that the fatigue lives calculated by using the plain *S-N* equation for the plain specimens, the intrinsic Notch *S-N* equation for Notch 1, and the predicted Notch *S-N* equation for Notch 2 are in good agreement with experimental ones.

3.5 APPLICATION OF NOTCH *S-N* EQUATION IN MULTIAXIAL FATIGUE LIMIT ANALYSIS OF NOTCHED COMPONENTS

Here, an attempt is made to show that the Notch *S-N* equations are applied in multiaxial fatigue limit analysis of notched components.

The bending and torsion Notch *S-N* equations obtained in Section 3 are rewritten as

$$\log \sigma_{an} = A_{1n} \log N + C_{1n} \tag{43}$$

and

$$\log \sqrt{3}\tau_a = A_{0n} \log N + C_{0n} \tag{44}$$

At the same time, the multiaxial Notch *S-N* equation under bending-torsion loading is also rewritten as

$$\log \sigma_{e,a} = A_{\rho n} \log N + C_{\rho n} \tag{45}$$

where

$$A_{\rho n} = A_{1n} \cdot \rho + A_{0n} \cdot (1-\rho) \tag{46}$$

$$C_{\rho n} = C_{1n} \cdot \rho + C_{0n} \cdot (1-\rho) \tag{47}$$

It is assumed here that, for a notch, its bending and torsion fatigue limits, σ_{on} and τ_{on} are known, its multiaxial fatigue limits (denoted by σ_{eo}) can be obtained from Eqs. (43) to (47):

$$\sigma_{eo} = \sigma_{on}{}^{\rho} \left(\sqrt{3}\tau_{on}\right)^{(1-\rho)} \tag{48}$$

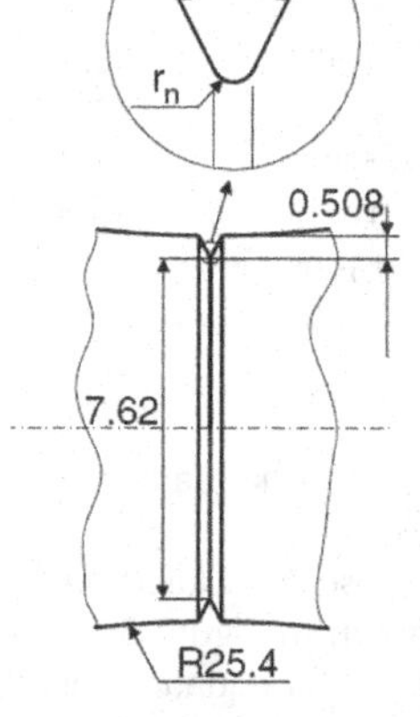

Code	Material	r_n (mm)	$K_{t,b}$	$K_{t,t}$
N-HSF1	0.4% C steel (Normalised)	0.005	18.0	8.6
N-HSF2	3% Ni steel	0.005	18.0	8.6
N-HSF3	3/3.5% Ni steel	0.010	13.3	6.7
N-HSF4	Cr-Va steel	0.011	12.1	6.2
N-HSF5	3.5% NiCr steel (n. impact)	0.022	8.7	4.7
N-HSF6	3.5% NiCr steel (l. impact)	0.022	8.7	4.7
N-HSF7	NiCrMo steel (75–80 tons)	0.031	7.5	4.2

FIGURE 3.20 Notch geometrical sizes and materials studied for studying multiaxial fatigue limits of notched components.

Source: (Data from Susmel [12])

TABLE 3.29
Fatigue Limit Data of Notched Components Under Bending-Torsion Loading and the Calculated Results

Materials	σ_{an} (exp) (MPa)	τ_{an} (exp) (MPa)	$\sigma_{e,a}$ (Mpa)	ρ	$\sigma_{e,o}$ (MPa)	R.E. (%)
0.4% C steel	76.7	140	254.3	0.3016	259.7	2.1
(Normalized)	120.1	107.3	221.3	0.5428	228.4	3.2
σ_{on} =	148.4	74.7	196.9	0.7537	204.2	3.7
179.1(MPa)	166.1	51.7	188.7	0.8802	190.9	1.2
τ_{on} =	180.5	27.6	186.7	0.9667	182.3	−2.4
176(MPa)	M.E. = 1.6%, E.R. = [−2.4,3.7]%					
3% Ni steel,	72.6	132.8	241.2	0.301	245.1	1.6
σ_{on} =	126.7	106.4	223.6	0.5665	231.1	3.3
209.9(MPa)	171.5	83.5	224.3	0.7645	221.2	−1.4
τ_{on} =	195.6	56.8	218.9	0.8934	214.9	−1.8
151.3(MPa)	209.2	25.3	213.7	0.9788	210.9	−1.3
	M.E. = 0.1%, E.R. = [−1.8,1.6]%					
3/3.5% Ni	97.6	172.6	314.5	0.3104	313.3	−0.4
steel	175.5	153.5	318.6	0.5509	309.5	−2.8
σ_{on} =	214.4	108.4	285	0.7523	306.4	7.5
302.6(MPa)	265.4	78.7	298.4	0.8895	304.3	2
τ_{on} =	286.2	40.1	294.5	0.9718	303	2.9
183.7(MPa)	M.E. = 1.8%, E.R. = [−2.8,7.5]%					
Cr-Va steel	77.3	147.3	266.6	0.29	258.5	−3
σ_{on} =	137.6	117	245	0.5617	241.4	−1.5
216.1(MPa)	172	88.6	230.5	0.7462	230.4	0
τ_{on} =	210.3	58	233.1	0.9023	221.5	−5
160.6(MPa)	210.9	27.9	216.4	0.9747	217.5	0.5
	M.E. = −1.8%, E.R. = [−5.0,0.5]%					
3.5% NiCr steel	102.8	188.2	341.8	0.3008	360.5	5.5
(n. impact)	170	145.3	303.7	0.5598	323.3	6.4
σ_{on} =	228.2	113.8	301.5	0.7568	297.5	−1.3
268.6(MPa)	261	77	293.1	0.8905	281.3	−4
τ_{on} =	290.8	36.9	297.7	0.9767	271.2	−8.9
236.2(MPa)	M.E. = −0.5%, E.R. = [−8.90,6.4]%					
3.5% NiCr	157.1	160.6	319.5	0.4918	279.8	−12.4
steel (l.	168.1	147.7	306.1	0.5491	275.8	−9.9
impact)	198.7	99.4	262.9	0.7558	262.2	−0.3
σ_{on} =	223.8	67.2	252.3	0.8872	253.9	0.7
247(MPa)	235.4	31.8	241.8	0.9737	248.6	2.8
τ_{on} =	M.E. = −3.8%, E.R. = [−12.4,2.8]%					
182.2(MPa)						
NiCrMo steel	104.2	194.8	353.1	0.2951	367.5	4.1
(75–80 tons)	169.8	147.6	306.9	0.5533	329	7.2
σ_{on} =	237.3	123.8	319.8	0.742	303.5	−5.1
271.7(MPa)	252.9	75.5	284.7	0.8883	285	0.1
τ_{on} =	265.7	36.6	273.2	0.9727	274.9	0.6
240.8(MPa)	M.E. = 1.4%, E.R. = [−5.1,7.2]%					

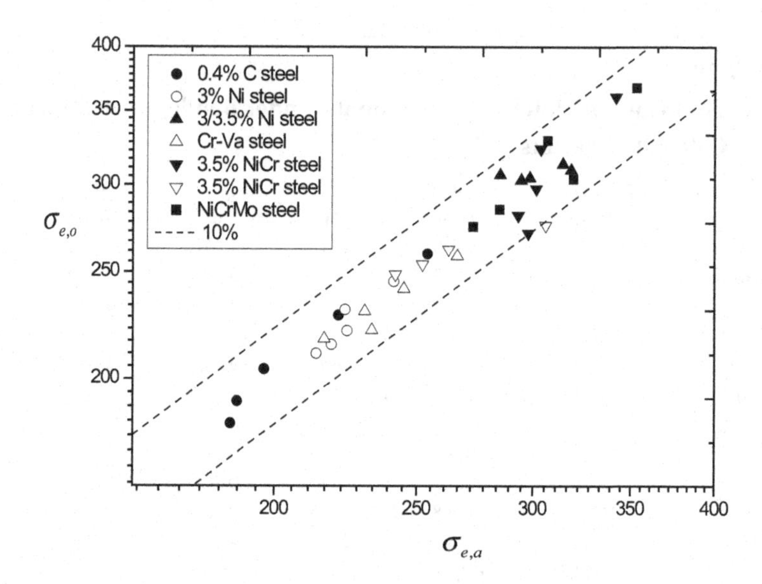

FIGURE 3.21 Comparison of experimental results of multiaxial fatigue limits with predicted ones under bending-torsion loading.

Source: (Data from Gough [50]).

The experimental verifications (see Table 3.29 and Figure 3.21; notch geometric sizes and materials studied for studying multiaxial fatigue limits of notched components gotten by Susmel [12] are shown in Figure 3.20) given below show that the multiaxial fatigue limit prediction equation (48) is not only simple in computation but also high in accuracy. In Figure 3.21 and Table 3.29, $\sigma_{e,a}$ and $\sigma_{e,o}$ are experimental and predicted results, respectively. $\sigma_{e,a}$ is calculated from experimental data σ_{an} and τ_{an} with the von-Mises stress definition, and the calculation of $\sigma_{e,o}$ are carried out according to the right of Eq. (48) by using the bending and torsion fatigue limits, σ_{on} and τ_{on}, and multiaxial parameter ρ.

Note here that, for 3.5% NiCr steel (l. impact), experimental data with a relative error = −12.4% (see Table 3.29), appear to be unreliable due to the fact that ρ = 0.4918 of the experimental data is obviously different from the corresponding one of the other materials. So the experimental data is not shown in Figure 3.21.

3.6 CONCLUSIONS AND FINAL COMMENTS

From this study, the following conclusions are made:

(1) The Notch *S-N* equations are naturally existing and a different notch has its different *S-N* equation.
(2) In a low/medium/high cycle fatigue regime of metallic materials, the Notch *S-N* equation proposed in this study—by using concepts of the critical stress, the critical distance and the local approach in dealing with the fracture failure of notches—has been verified to have a high accuracy and

a high efficiency in performing the fatigue failure assessment of notched metallic materials.

(3) The multiaxial fatigue limit equation for notched components proposed in this study by using the Notch *S-N* equations has been verified to have a high accuracy.

(4) The linear model for dealing with mean stress effect proposed in Appendix *B* and the linear formula dealing with nonproportional loading fatigue proposed recently in Ref.[54] (see Appendix *A*) have been verified to have a high accuracy.

A few comments given here are as follows:

(1) In recent years, the author was devoted to a unified lifetime estimation equation for a low/medium/high cycle fatigue of metallic materials. This originated from the fact that "on one day of the beginning of the year of 2019, the author fixed his eyes on the well-known Rambrg–Osgood relationship (see eqns (3) and (4)) on Pages 215, 218 in Ref [12] and, after thinking for a moment, felt deeply that, for the LCF of metallic materials, such a way that a '**stress quantity**', which is calculated based on the *linear elastic analysis* of the studied component, instead of a '**strain quantity**', which is calculated based on the elastic-plastic analysis of the studied component, is taken to be a *mechanical quantity, S*, to establish a relation of the *mechanical quantity, S,* to the fatigue life, *N*, is practicable", which occurred in a paper (A unified lifetime estimation equation for a low/medium/high cycle fatigue of metallic materials under uniaxial and multiaxial loading—Part 1:plain materials, date:2019/07/01). In order to roughly understand the author's viewpoint, the reader can see Section 2 of this study, or see the Appendix in Ref.[11]. Moreover, the continued paper ("A unified lifetime estimation equation for a low/medium/high cycle fatigue of metallic materials under uniaxial and multiaxial loading—Part II: notched materials", date:2019/07/09) was also submitted for publication. But the two papers mentioned previously were not accepted for publication.

The core of this work is roughly same as that of the continued paper mentioned previously.

(2) The author considers that although many experimental examples given here have shown that the Notch *S-N* equations proposed in this study have a high accuracy in performing the fatigue failure assessment of a low/medium/high cycle fatigue of metallic materials, including mean stress effect and nonproportional loading effect, more experimental verifications are needed.

(3) It is worthy to mention that the Notch *S-N* curve method, which is on the basis of the linear elastic stress analysis of the notched components, is very efficient because there is no need for elastic plastic analysis even for low-cycle fatigue of notched metallic materials.

REFERENCES

1 Susmel, L., Lazzarin, P. A bi-parametric modified Wöhler curve for high cycle multiaxial fatigue assessment. *Fatigue and Fracture of Engineering Materials and Structures* 25, 63–78 (2002).

2 Susmel, L., Meneghetti, G., Atzori, B. A simple and efficient reformulation of the classical Manson–Coffin curve to predict lifetime under multiaxial fatigue loading. Part I: plain materials. *Transactions of the ASME, Journal of Engineering Materials and Technology* 131, 021009-1 to 9 (2009 April).

3 Gao Z,, Qiu, B, Wang, X, Jiang Y., An investigation of fatigue of a notched member, *International Journal of Fatigue* 32, 1960–1969 (2010).

4 Cristofori, A., Susmel, L., Tovo, R. A stress invariant based criterion to estimate fatigue damage under multiaxial loading. *International Journal of Fatigue* 30, 1646–1658 (2008).

5 Papadopoulos, I.V., Davoli, P., Gorla, C., Filippini, M., Bernasconi, A. A comparative study of multiaxial highcycle fatigue criteria for metals. *Int J Fatigue* 19, 219–235 (1997).

6 Sines, G. Behaviour of metals under complex static and alternating stresses. In: *Metal fatigue*. New York: McGraw Hill; 1959. p. 145–169.

7 Crossland, B. Effect of large hydroscopic pressures on the torsional fatigue strength of an alloy steel. In: Proceedings of international conference on fatigue of metals, London, New York; 1956. p. 138–149.

8 Li, B., De Freitas, M. A procedure for fast evaluation of high cycle fatigue under multiaxial random loading. *J Mech Design* 124, 558–563 (2002).

9 Wang, C.H., Brown, M.W. Multiaxial random load fatigue: Life prediction techniques and experiments. In: Fourth international conference on biaxial/multiaxial fatigue, Paris; 1994 May 31–June 3.

10 Bishop, N.W.M., Sherratt, F. *Finite element based fatigue calculations*. UK: NAFEMS Ltd; 2000.

11 Yan, X.Q. An empirical fracture equation of mixed mode cracks. *Theoretical and Applied Fracture Mechanics* 116, 103146 (2021).

12 Susmel, L. *Multiaxial notch fatigue, from nominal to stress/strain quantities*. Cambridge, UK: Woodhead Publishing Limited, CRC Press; 2009.

13 Gao, Z., Zhao, T., Wang, X., Jiang, Y. Multiaxial fatigue of 16MnR steel. *ASME J Press Vess Technol* 131(2), 021403 (2009).

14 Szusta, A., Seweryn, J. Fatigue damage accumulation modelling in the range of complex low-cycle loadings: The strain approach and its experimental verification on the basis of EN AW-2007 aluminum alloy. *Int J Fatigue* 33, 255–264 (2011).

15 Szusta, A. Seweryn, J. Experimental study of the low-cycle fatigue life under multiaxial loading of aluminum alloy EN AW-2024-T3 at elevated temperatures. *Int J Fatigue* 33, 255–264 (2011).

16 Liu, B.W., Yan, X.Q. A new model of multiaxial fatigue life prediction with influence of different mean stresses. *International Journal of Damage Mechanics* 28(9), 1323–1343 (2019).

17 Neuber, H. Zur Theorie der technischen Formzahl. *Forschg Ing-Wes* 7, 271–281 (1936).

18 Neuber, H. *Theory of notch stresses: Principles for exact calculation of strength with reference to structural form and material*. 2nd ed. Berlin: Springer Verlag; 1958.

19 Peterson, R.E. Notch sensitivity. In: Sines, G., Waisman, J.L., editors. *Metal fatigue*. New York, USA: McGraw Hill; 1959. p. 293–306.

20 Taylor, D. The theory of critical distances: A new perspective in fracture mechanics. Oxford, UK: Elsevier; 2007.
21 Susmel, L., Taylor, D. The theory of critical distances to estimate the static strength of notched samples of Al6082 loaded in combined tension and torsion. Part II: Multiaxial static assessment. *Engineering Fracture Mechanics* 77, 470–478 (2010).
22 Berto, F., Lazzarin, P. Recent developments in brittle and quasi-brittle failure assessment of engineering materials by means of local approaches. *Materials Science and Engineering R* 75, 1–48 (2014).
23 Lazzarin, P., Zambardi, R. A finite-volume-energy based approach to predict the static and fatigue behaviour of components with sharp V-shaped notches. *International Journal of Fracture* 112, 275–298 (2001).
24 Lazzarin, P., Berto, F. Some expressions for the strain energy in a finite volume surrounding the root of blunt V-notches. *International Journal of Fracture* 135, 161–185 (2005).
25 Taylor, D. Predicting the fracture strength of ceramic materials using the theory of critical distances. *Engng Frac Mech* 71, 2407–2416 (2004).
26 Taylor, D., Merlo, M., Pegley, R., Cavatorta, M.P. The effect of stress concentrations on the fracture strength of polymethylmethacrylate. *Material Science and Engineering* A382, 288–294 (2004).
27 Taylor, D., Cornetti, P., Pugno, N. The fracture mechanics of finite crack extension. *Engng Frac Mech* 72, 1021–1038 (2005).
28 Susmel, L., Taylor, D. An elasto-plastic reformulation of the theory of critical distances to estimate lifetime of notched components failing in the low/medium-cycle fatigue regime. *ASME J Eng Technol* 132, 210021–210028 (2010).
29 Susmel, L., Taylor, D. A novel formulation of the theory of critical distances to estimate lifetime of notched components in the medium-cycle fatigue regime. *Fatigue Fract Eng Mater Struct* 30(7), 567–581 (2007).
30 Susmel, L., Taylor, D. On the use of the theory of critical distances to estimate fatigue strength of notched components in the medium-cycle fatigue regime. In: Proceedings of FATIGUE 2006 Atlanta, USA; 2006.
31 Yang, X.G., Wang, J.K. Liu, J.L. High temperature LCF life prediction of notched DS Ni-based superalloy using critical distance concept. *International Journal of Fatigue* 33, 1470–1476 (2011).
32 Filippi, S., Lazzarin, P. Distributions of the elastic principal stress due to notches in finite size plates and rounded bars uniaxially loaded. *International Journal of Fatigue* 26, 377–391 (2004).
33 Noda, N.A., Sera, M., Takase, Y. Stress concentration factors for round and flat test specimens with notches. *Int. J. Fatigue* 17(3), 163–178 (1995).
34 Noda, N.A., Takase, Y. Stress concentration formula useful for all notch shape in a round bar (comparison between torsion, tension and bending). *International Journal of Fatigue* 28, 151–163 (2006).
35 Noda, N.A., Takase, Y. Stress concentration formulas useful for any shape of notch in a round test specimen under tension and under bending. *Fatigue Fract Eng Mater Struct* 22, 1071–1082 (1999).
36 Mykhaylo, P. Savruk and Andrzej Kazberuk, two-dimensional fracture mechanics problems for solids with sharp and rounded V-notches. *Int J Fract* 161, 79–95 (2010).
37 Zappalorto, M., Lazzarin, P., Filippi, S. Stress field equations for U and blunt V-shaped notches in axisymmetric shafts under torsion. *Int J Fract* 164, 253–269 (2010).
38 Zappalorto, M., Berto, F., Lazzarin, P. Practical expressions for the notch stress concentration factors of round bars under torsion. *International Journal of Fatigue* 33, 382–395 (2011).

39 Zappalorto, M., Lazzarin, P., Berto, F. Elastic notch stress intensity factors for sharply V-notched rounded bars under torsion. *Engineering Fracture Mechanics* 76, 439–453 (2009).
40 Qilafku, G., Kadi, N., Dobranski, J., Azari, Z., Gjonaj, M., Pluvinage, G. Fatigue specimens subjected to combined loading: Role of hydrostatic pressure. *International Journal of Fatigue* 23, 689–701 (2001).
41 Susmel, L., Taylor, D., The odified Wöhler Curve Method applied along with the theory of critical distances to estimate finite life of notched components subjected to complex multiaxial loading paths. *Fatigue and Fracture of Engineering Materials and Structures* 31(12), 1047–1064. doi: 10.1111/j.1460-2695.2008.01296.x
42 Putchkov, I.V., Temis, Y.M., Dowson, A.L., Damrit, D. Development of a finite element based strain accumulation model for the prediction of fatigue lives in highly stressed Ti components. *Int. J. Fatigue* 17(6), 385–398 (1995).
43 Berto, F., Lazzarin, P., Tovo, R. Multiaxial fatigue strength of severely notched cast iron specimens. *International Journal of Fatigue* 67, 15–27 (2014).
44 Kurath, P., Downing, S.D., Galliart, D.R. Summary of non-hardened notched shaft-round robin program. In: Leese, G.E., Socie, D.F., editors. *Multiaxial fatigue: Analysis and experiments*, SAE AE-14. Warrendale, PA: Society of Automotive Engineers; 1989. p. 13–32.
45 Yip, M.-C., Jen, Y.-M. (1996) Biaxial fatigue crack initiation life prediction of solid cylindrical specimens with transverse circular hole. *International Journal of Fatigue* 18, 111–117. doi: 10.1016/0142-1123(95)00057-7
46 Atzori, B., Berto, F., Lazzarin, P., Quaresimin, M. Multi-axial fatigue behaviour of a severely notched carbon steel. *International Journal of Fatigue* 28, 485–493 (2006).
47 Berto, F., Lazzarin, P. Fatigue strength of structural components under multi-axial loading in terms of local energy density averaged on a control volume. *International Journal of Fatigue* 33, 1055–1065 (2011).
48 Berto, F., Campagnolo, A., Lazzarin, P. Fatigue strength of severely notched specimens made of Ti–6Al–4V under multiaxial loading. *Fatigue Fract Engng Mater Struct* 38, 503–517 (2015).
49 Berto, F., Lazzarin, P., Marangon, C. Fatigue strength of notched specimens made of 40CrMoV13.9 under multiaxial loading. *Materials and Design* 54, 57–66 (2014).
50 Gough, H.J. Engineering steels under combined cyclic and static stresses. *Proceedings of the Institution of Mechanical Engineers* 160, 417–440 (1949).
51 Itoh, T., Sakane, M., Ohnami, M., Socie, D.F. Nonproportional low cycle fatigue criterion for type 304 stainless steel. *ASME J. Eng. Mater. Technol.* 117, 285–292 (1995).
52 Freitas, M., Li, B., Santos, J.L.T. Multiaxial fatigue and deformation: Testing and prediction. ASTM STP 1387; 2000.
53 Liu, B.W., Yan, X.Q. A multi-axial fatigue limit prediction equation for metallic materials. *ASME Journal of Pressure Vessel Technology* 142, 034501-6 (2020).
54 Yan, X.Q. Applicability of the Wöhler Curve Method for a low/medium/high cycle fatigue of metallic materials. Submitted to *Applications in Science and Engineering* for publication.
55 Gasiak, G., Pawliczek, R. Application of an energy model for fatigue life prediction of construction steels under bending, torsion and synchronous bending and torsion. *Int J Fatigue* 25(12), 1339–1346 (2003).
56 Yan, X.Q. A mean stress model of fatigue life of metal materials under multiaxial loading. *Material Science & Engineering International Journal* 5(2), 60–69 (2021).

APPENDIX A: ON DEALING WITH NONPROPORTIONAL LOADING FATIGUE

Regarding nonproportional loading fatigue, Itoh's formula [51] is usually used to reveal the relationship of the amplitude of the equivalent nonproportional stress, σ_{ean}, and the amplitude of the effective proportional stress, σ_{ea}, i.e.,

$$\sigma_{ean} = \sigma_{ea}(1+\beta F) \tag{A-1}$$

where β is material sensitivity parameter to load-path nonproportionality defined on $\sigma - \sqrt{3}\tau$ stress plane; F is usually called a nonproportional loading factor which expresses the severity of nonproportional loading.

Recently, it was found in Ref.[54] that the prediction results of fatigue life obtained by using Itoh's formula sometimes are not satisfactory. Thus the following attempt was made:

$$\sigma_{ean} / \sigma_{ea} = \lambda + \gamma F \tag{A-2}$$

where λ, γ are constants to be determined by linear fitting of $\sigma_{ean} / \sigma_{ea}$ and F. When λ in formula (A-2) is equal to 1, the linear fitting relation (A-2) becomes Itoh's formula. In Ref.[54], the formulas (A-1) and (A-2) were called Itoh's formula (I.F) and the linear formula (L.F), respectively.

Under out-of-phase loading fatigue, the nonproportional factor F (NPF) is usually decided as the ratio of the two principal axes of elliptical loading paths [52]. According to this definition of NPF, two similar elliptical loading paths with different radii, in which one is, for example, much larger than the other, have a uniform value, F, which appears to be unreasonable. In order to reveal the effect of the relative size of fatigue loading path on the nonproportional factor F, a new definition of NPF is proposed here as follows:

It is assumed that there are m out-of-phase loading cases, in which nonproportional factor, F, and effective proportional stress, σ_{ea}, are denoted by F_i, and $(\sigma_{ea})_i$, $(i=1,2,..,m)$, respectively, and the maximum of $(\sigma_{ea})_i$ is denoted by $(\sigma_{ea})_{\max}$. Then a new definition of nonproportional factor (NPF') is

$$F_i' = F_i \cdot \frac{(\sigma_{ea})_i}{(\sigma_{ea})_{\max}}, \quad (i=1,2,..,m) \tag{A-3}$$

APPENDIX B: MULTIAXIAL FATIGUE LIFE PREDICTION EQUATION WITH NONZERO MEAN STRESS

In this Appendix, a linear mean stress model proposed in Ref.[56] is described briefly.

$$\log \sigma_{ea}\left(\beta_\rho + \gamma_\rho \frac{\sigma_{em}}{\sigma_{e0}}\right) = A_\rho \log N + C_\rho \tag{B-1}$$

where

$$\sigma_{eo} = \sigma_o^{\rho}\left(\sqrt{3}\tau_o\right)^{(1-\rho)} \tag{B-2}$$

and σ_o and τ_o are, respectively, a tensile fatigue limit and a torsion fatigue limit.

Under the axial and torsion fatigue loading, Eq. (B-1) can be written as

$$\log \sigma_a (\beta_1 + \gamma_1 \frac{\sigma_m}{\sigma_0}) = A_1 \log N + C_1 \tag{B-3}$$

and

$$\log \sqrt{3}\tau_a (\beta_0 + \gamma_0 \frac{\tau_m}{\tau_0}) = A_0 \log N + C_0 \tag{B-4}$$

in which β_1, γ_1, β_0 and γ_0 are material constants determined by fitting experimental data of fatigue life.

By the way, the material parameters β_ρ and γ_ρ in Eq. (B-1) can be determined by using the interpolation formulas similar to formulas (8) to (9). Eq. (B-2) is originated from the multiaxial fatigue limit prediction equation recently proposed by Liu and Yan [53].

Concerning the linear mean stress model proposed here, the following two comments are given:

1. For some metal materials, for example, cast iron reported by Berto, Lazzarin and Tovo [43], an accuracy of fatigue lives predicted by means of the linear mean stress model and Marin's mean stress model is very high and the two models have almost the same accuracy.
2. For S355J0 alloy steel reported by Gasiak and Pawliczek [55], under the axial and pure torsion fatigue loading, an accuracy of fatigue lives predicted by means of the linear mean stress model and Marin's mean stress model is very high. But under multiaxial fatigue loading, an accuracy of fatigue lives predicted by using Marin's mean stress model is very poor, while on the contrary, fatigue lives predicted by means of the linear mean stress model are very satisfactory (see Ref.[56]).

APPENDIX C: EXPERIMENTAL INVESTIGATIONS: THE INHERENT NOTCH *S-N* EQUATIONS ARE USED TO PERFORM THE FATIGUE LIFE ASSESSMENT OF NOTCHED COMPONENTS

In this Appendix, some examples are taken to illustrate that the inherent Notch *S-N* equations (see Eqs. (16) and (23)) are used to perform the fatigue life assessment of notched components, in which the approaches, including how to obtain the material constants in the inherent Notch *S-N* equations by using experimental data of fatigue lives of notched specimens under bending and torsion loading, and further how to carry out a multiaxial fatigue life assessment of notched components by using the inherent Notch *S-N* equations, are completely the same as those for plain materials.

Example 1: High-Cycle Fatigue of Notched Components Made of Cast Iron

Experimental data (see Table 3.C-1) of high-cycle fatigue of notched cast iron specimens reported by Berto et al. [43] are employed to illustrate that the inherent Notch *S-N* equations (see Eqs. (16) and (23)) are used to perform the fatigue life assessment of notched materials. The notch geometry is shown in Figure 3.C-1, with theoretical stress concentration factors, k_I = 7.64 for tension, and k_{III} = 3.23 for torsion, both referred to as the net transverse sectional area. The material constants in the inherent Notch *S-N* equations, which are determined by using experimental data of fatigue lives of the notch specimens under tension and torsion loading fatigue (see Table 3.C-1) are listed in Table 3.9. In the following, the inherent Notch *S-N* equations will be used to perform a fatigue life assessment of the notched specimens.

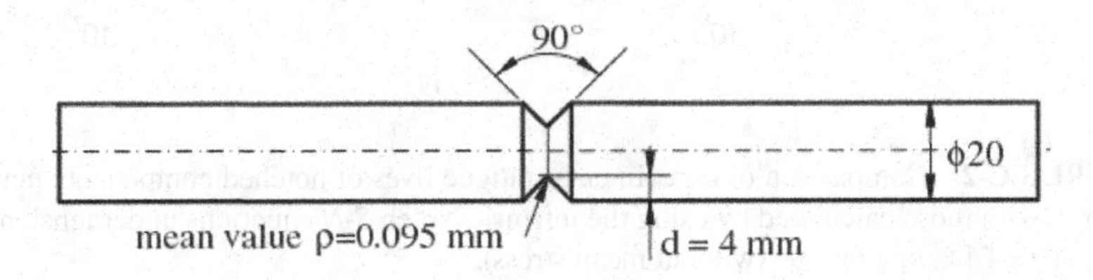

FIGURE 3.C-1 Geometry of the V-notched specimens.

Source: (Data from Berto et al. [43])

TABLE 3.C-1
Experimental Data of the V-Notched Specimens

Series code	Load	No spec.		σ_a or τ_a (MPa) 10^6	2×10^6	5×10^6
1	Tension, $R = -1$	7	σ	100.5	89.9	77.6
2	Torsion, $R = -1$	9	τ	160.0	151.9	141.8
3	Tension, $R = 0$	6	σ	63.5	57.6	50.5
4	Torsion, $R = 0$	9	τ	115.2	109.6	102.5
5	$R = -1, \Phi = 0, \lambda = 1.0$	12	σ	81.0	74.0	65.6
6	$R = -1, \Phi = 90°, \lambda = 1.0$	13	σ	86.4	82.6	77.8
7	$R = 0, \Phi, = 0, \lambda = 1.0$	7	σ	60.3	56.4	51.6
8	$R = 0, \Phi, = 90°, \lambda = 1.0$	6	σ	57.1	53.3	48.6
9	$R = -1, \Phi, = 0, \lambda = 0.6$	5	σ	105.9	99.7	92.1
			τ	63.5	59.2	55.3
10	$R = -1, \Phi = 90°, \lambda = 0.6$	4	σ	92.2	85.9	78.2
			τ	55.3	51.5	46.9

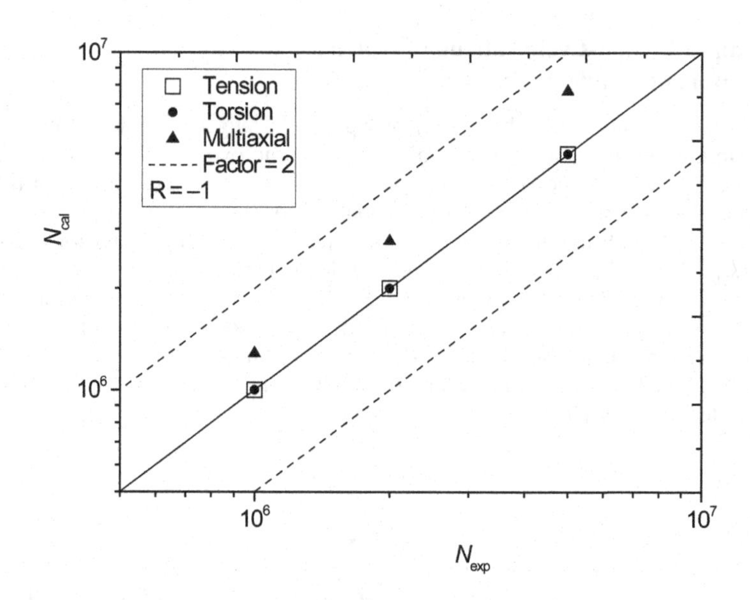

FIGURE 3.C-2 Comparison of experimental fatigue lives of notched components made of cast iron with those calculated by using the intrinsic Notch *S-N* equations under tension, torsion and Te-T loading fatigue (without mean stress).

Source: (Data from Berto et al. [43])

Under tension and torsion loading fatigue, the inherent Notch *S-N* equations of notched cast-iron specimens are completely linear. Under **Te-T loading fatigue**, fatigue lives predicted by using the intrinsic multiaxial Notch *S-N* equation are in good agreement with experimental ones (see Figure 3.C-2).

Under tension and torsion loading fatigue with **mean stress effect**, a comparison of experimental fatigue lives with those calculated by using the linear mean stress model (see (B-1)-(B-4)) are given in Table 3.C-2, from which it can be seen that the predicted results are in excellent agreement with experimental ones (also see Figure 3.C-3).

Under **out-of-phase loading fatigue**, experimental data and calculated results are shown in Table 3.C-3, with error indexes: E.R. = [–33,47]%, M.E. = 6.1% for Itoh's formula (A-1) and E.R. = [–0.07,0.03]%, M.E. = –0.003% for the linear formula (A-2), which shows that the predicted results by the linear formula (A-2) are much better than those by Itoh's formula (A-1).

In the process of fatigue life analysis performed by using experimental data of No. 7, No. 8, No. 9 and No. 10 shown in Table 3.C-1, it is found that some data shown in Table 3.C-1 appear wrong. By comparing data listed in No. 5 and No. 6, it is found that, under equal fatigue life, the Mises stress of **out-of-phase loading fatigue** is larger than one of **in-phase loading fatigue**. At fatigue life N = 10E6, for example, the Mises stress of out-of-phase loading fatigue, 172.8(MPA), is larger than one of in-phase loading fatigue, 162(MPA). But comparisons between data in No. 7 and data in No. 8 and between data in No. 9 and data in No. 10 show that the Mises stress of out-of-phase loading fatigue is smaller than one of in-phase loading fatigue, contrary to the relation mentioned

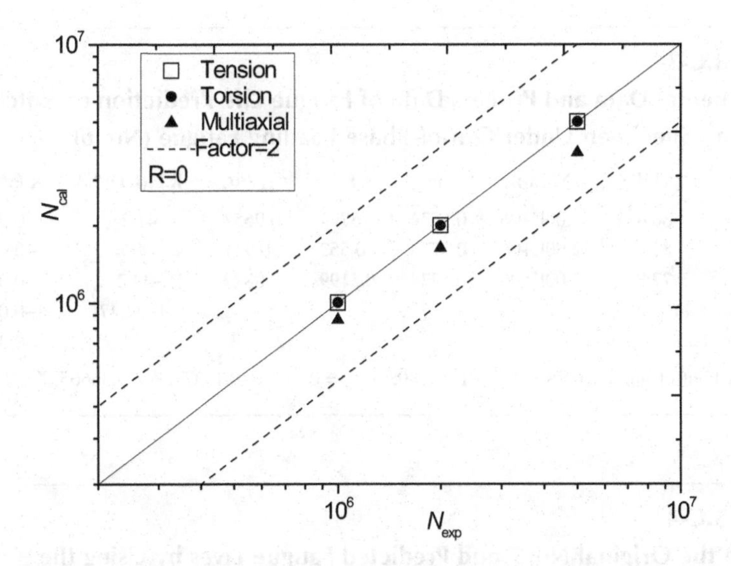

FIGURE 3.C-3 Comparison of experimental fatigue lives of notched components made of cast iron with those calculated by using the intrinsic Notch *S-N* equations under tension, torsion and Te-T loading fatigue with mean stress effect.

Source: (Data from Berto et al. [43])

TABLE 3.C-2

Experimental Data and Calculated Results Under Tension and Torsion Loading Fatigue with Mean Stress Effect (Notched Cast-Iron Specimens)

No. 3				No. 4			
σ_a (MPa)	σ_m (MPa)	N (exp)	R.E. (%)	τ_a (MPa)	τ_m (MPa)	N (exp)	R.E. (%)
63.5	63.5	1.00E+06	0.2	115.2	115.2	1.00E+06	0.3
57.6	57.6	2.00E+06	−0.2	109.6	109.6	2.00E+06	−0.3
50.5	50.5	5.00E+06	0.1	102.5	102.5	5.00E+06	0.3
Error range (%)			[−0.2,0.2]	Error range (%)			[−0.3,0.3]
Mean error (%)			0.03	Mean error (%)			0.1

Note: $\alpha_1 = 1.3576$, $\beta_1 = 0.2744$ $\alpha_0 = 1.34$, $\beta_0 = 0.0605$

previously. Thus a change is done in this study: $\Phi = 0$ in No. 7 is changed into $\Phi = 90^\circ$; $\Phi = 90^\circ$ in No. 8 is changed into $\Phi = 0$. For data in No. 9 and No. 10, a similar change is done.

Table 3.C-4 shows the experimental data in the original No. 7 and a comparison of experimental fatigue lives with those calculated by the inherent Notch *S-N* equation, with error indexes: M.E. = −52%, E.R. = [−48,−56]%.

TABLE 3.C-3
Experimental Data and Process Data of Fatigue Life Prediction of Notched Cast-Iron Specimen Under Out-of-Phase Loading Fatigue (No. 6)

σ_a (MPa)	τ_a (MPa)	N(exp)	F	F′	$\bar{\sigma}_{ea}/\sigma_{ea}$	R.E.(I.T)(%)	R.E.(L.F)(%)
86.4	86.4	1.00E+06	0.5774	0.5774	0.9658	47.2	0.0304
82.6	82.6	2.00E+06	0.5774	0.552	0.931	4.5	−0.0668
77.8	77.8	5.00E+06	0.5774	0.5199	0.8872	−33.2	0.0256
Error range (%)						[−33,47]	[−0.07,0.03]
Mean error (%)						6.1	−0.003

Note: All out-of-phase angles = 90°; all $\sigma_m = 0$; all $\tau_m = 0$. $\alpha' = -0.13373$, $\lambda = 1.36665$, $\gamma = 0.17671$

TABLE 3.C-4
Data in the Original No. 7 and Predicted Fatigue Lives by Using the Multiaxial Notch *S-N* Equation (Notched Cast-Iron Specimen)

σ_a (MPa)	σ_m (MPa)	τ_a (Mpa)	τ_m (MPa)	N(exp)	N(cal)	R.E (%)
60.3	60.3	60.3	60.3	1.00E+06	5.19E+05	−48.1
56.4	56.4	56.4	56.4	2.00E+06	9.66E+05	−51.7
51.6	51.6	51.6	51.6	5.00E+06	2.20E+06	−56

TABLE 3.C-5
Data Obtained by Changing $\Phi = 90^\circ$ in No. 8 into $\Phi = 0$ and Predicted Fatigue Lives by the Notch *S-N* Equation (Notched Cast-Iron Specimen)

σ_a (MPa)	σ_m (MPa)	τ_a (MPa)	τ_m (MPa)	N(exp)	N(cal)	R.E. (%)
57.1	57.1	57.1	57.1	1.00E+06	8.62E+05	−13.8
53.3	53.3	53.3	53.3	2.00E+06	1.63E+06	−18.5
48.6	48.6	48.6	48.6	5.00E+06	3.81E+06	−23.8

Table 3.C-5 shows experimental data obtained by changing $\Phi = 90^\circ$ in No. 8 into $\Phi = 0$ and a comparison of experimental fatigue lives with those calculated by the Notch *S-N* equation, with error indexes: M.E. = −19%, E.R. = [−14,−24]%.

Table 3.C-6 lists experimental data obtained by changing $\Phi = 0$ in No. 7 into $\Phi = 90^\circ$ and a comparison of experimental fatigue lives with those calculated by the Notch *S-N* equation and Itoh's formula (A-1), and by the Notch *S-N* equation and the linear formula (A-2), with error ranges, [−13,16]% and [−0.1,0.2]%, which illustrates that the fatigue lives predicted by the two models are in excellent agreement with experimental ones, and that an accuracy of fatigue lives predicted by the linear formula (A-2) is higher than that by Itoh's formula (A-1).

TABLE 3.C-6
Data Obtained by Changing $\Phi = 0$ in No. 7 into $\Phi = 90^\circ$ and Predicted Fatigue Lives (Notched Cast-Iron Specimen)

σ_a (MPa)	σ_m (MPa)	τ_a (MPa)	τ_m (MPa)	Φ	N(exp)	R.E.(I.F) (%)	ER(L.F) (%)
60.3	60.3	60.3	60.3	90	1.00E+06	15.6	−0.1
56.4	56.4	56.4	56.4	90	2.00E+06	1.9	0.2
51.6	51.6	51.6	51.6	90	5.00E+06	−13.2	−0.1

TABLE 3.C-7
Data in the Original No. 9 and Predicted Fatigue Lives by the Notch *S-N* Equation (Notched Cast-Iron Specimen)

σ_a (MPa)	σ_m (MPa)	τ_a (MPa)	τ_m (MPa)	N(exp)	N(cal)	R.E. (%)
105.9	0	63.54	0	1.00E+06	4.49E+05	−55.1
99.7	0	59.82	0	2.00E+06	7.04E+05	−64.8
92.1	0	55.26	0	5.00E+06	1.27E+06	−74.6

TABLE 3.C-8
Data Obtained by Changing $\Phi = 90^\circ$ in No. 10 into $\Phi = 0$ and Predicted Fatigue Lives by the Notch *S-N* Equation (Notched Cast-Iron Specimen)

σ_a (MPa)	σ_m (MPa)	τ_a (MPa)	τ_m (MPa)	N(exp)	N(cal)	R.E. (%)
92.2	0	55.3	0	1.00E+06	1.26E+06	26.0
85.9	0	51.5	0	2.00E+06	2.13E+06	6.7
78.2	0	46.9	0	5.00E+06	4.29E+06	−14.2

Table 3.C-7 gives experimental data in the original No. 9 and a comparison of experimental fatigue lives with those calculated by the Notch *S-N* equation, with error indexes: M.E. = −65%, E.R. = [−55,−75]%.

Table 3.C-8 lists experimental data obtained by changing $\Phi = 90^\circ$ in No. 10 into $\Phi = 0$ and a comparison of experimental fatigue lives with those calculated by the Notch *S-N* equation, with error indexes: M.E. = 6.2%, E.R. = [−14,26]%.

Table 3.C-9 shows experimental data obtained by changing $\Phi = 0$ in No. 9 into $\Phi = 90^\circ$ and a comparison of experimental fatigue lives with those calculated by the Notch *S-N* equation and Itoh's formula (A-1), and by the Notch *S-N* equation and the linear formula (A-2), with error ranges, [−31, 44]% and [−0.2,0.4]%, which shows that an accuracy of fatigue lives predicted by the linear formula (A-2) is higher than that by Itoh's formula (A-1).

TABLE 3.C-9
Data Obtained by Changing $\Phi = 0$ in No. 9 into $\Phi = 90^\circ$ and Predicted Fatigue Lives by the Notch S-N Equation (Notched Cast-Iron Specimen)

σ_a (MPa)	σ_m (MPa)	τ_a (MPa)	τ_m (MPa)	Φ	N(exp)	R.E.(I.F) (%)	R.E.(L.F) (%)
105.9	0	63.54	0	90	1.00E+06	43.9	−0.2
99.7	0	59.82	0	90	2.00E+06	4.7	0.4
92.1	0	55.26	0	90	5.00E+06	−30.9	−0.2

Example 2 Low/Medium Cycle Fatigue of Notched Components Made of 16MnR Steel

Notch fatigue experiments of 16MnR reported by Gao et al.[3] are taken as an example to illustrate that the inherent Notch *S-N* equations proposed in this study are suitable for performing the **low/medium cycle** fatigue life assessment of notched materials. Notch geometry is shown in Figure 3.C-4. The stress concentration factors are given in Table 3.C-10. For the notched material, the material constants in the inherent Notch *S-N* equations are listed in Table 3.9.

In the following, the inherent Notch *S-N* equations (see Eqs. (16) and (23)) are used to perform the fatigue life assessment of the two notches shown in Figure 3.C-4.

A comparison of experimental fatigue lives with those calculated by using the inherent Notch *S-N* equations of Notch 1 is given in Figure 3.C-5, with error indexes: E.R. = [−39,28]%, M.E. = 3.3% for tension loading fatigue, and E.R. = [-53,74]%, M.E. = 13.2% for torsion loading fatigue. A similar comparison for Notch 2 is shown in Figure 3.C-6, with error indexes: E.R. = [−60,82]% and M.E. = 7.4% for tension loading fatigue, and E.R. = [−27,47]%, M.E. = 3.9% for torsion loading fatigue. From Figure 3.C-5 and Figure 3.C-6, it is seen that fatigue lives of notched 16MnR specimens predicted by using the inherent Notch *S-N* equations are in good agreement with experimental ones.

Under **Te-T loading fatigue**, a comparison of experimental fatigue lives with those calculated by using the multiaxial intrinsic *S-N* equation of Notch 1 is shown in Table 3.C-11, with error indexes: M.E. = −21.1%, E.R. = [−44,3]%, from which it is seen that the predicted multiaxial fatigue lives are in good agreement with experimental ones.

Under **out-of-phase loading fatigue**, for Notch 1, a comparison of experimental fatigue lives with those calculated by the multiaxial intrinsic Notch *S-N* equation with **out-of-phase loading effect** is shown in Table 3.C-12, with error indexes: E.R. = [−66,446]%, M.E. = 57.7% for Itoh's formula (A-1), and E.R. = [−41,74]%, M.E. = 6.7% for the linear formula (A-2), which illustrates that the predicted fatigue lives by the linear formula (A-2) are much more accurate than those by Itoh's formula (A-1).

Note here that the sixth data is an unusual data because of R.E. (I.F) = 446%. After deleting the unusual data, the calculated error indexes are: E.R. = [-49,185]%, M.E. = 19.2% for Itoh's formula (A-1), and E.R. = [−21,19]%, M.E. = 1.0% for the linear formula (A-2). Obviously, the predicted fatigue lives by the linear formula (A-2) are much more accurate than those by Itoh's formula (A-1).

TABLE 3.C-10
Stress Concentration Factors of the Two Notches Shown in Figure 3.C-4

Notch	k_I	k_{III}
Notch 1	2.1634	1.435
Notch 2	1.3745	1.131

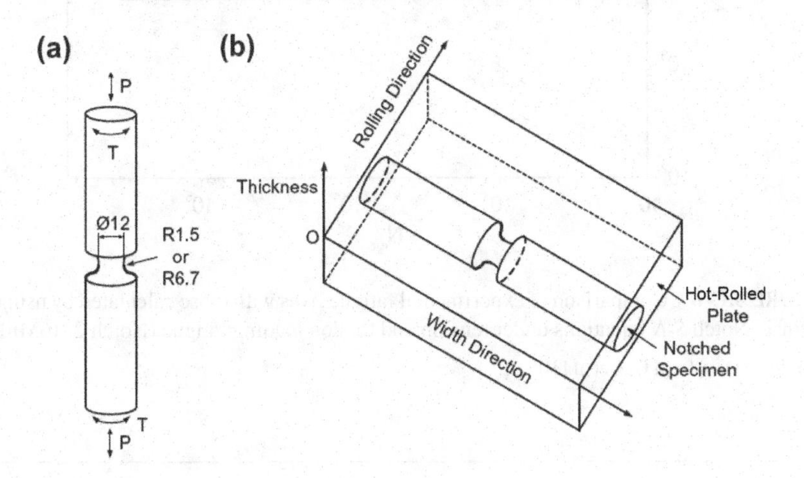

FIGURE 3.C-4 Notch geometry of 16MnR.
Source: (Data from Gao et al.[3])

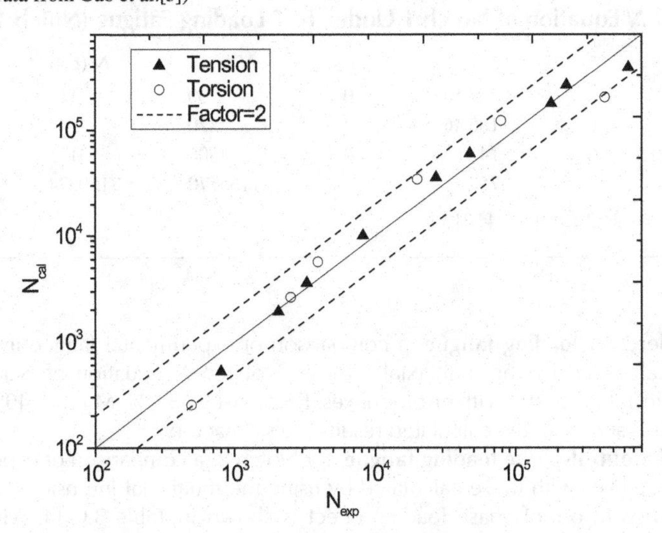

FIGURE 3.C-5 Comparison of experimental fatigue lives with those calculated by using the intrinsic Notch S-N equations under tension and torsion loading fatigue (Notch 1, 16MnR).
Source: (Data from Gao et al.[3])

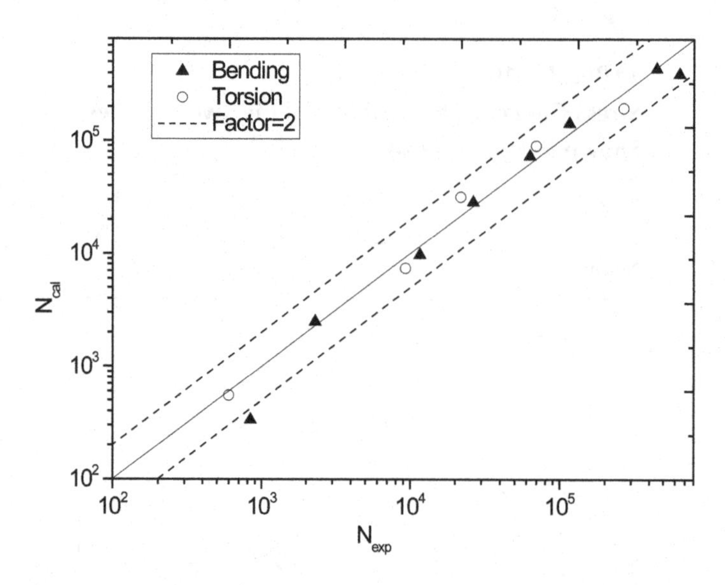

FIGURE 3.C-6 Comparison of experimental fatigue lives with those calculated by using the intrinsic Notch *S-N* equations under tension and torsion loading fatigue (Notch 2, 16MnR).

Source: (Data from Gao et al.[3])

TABLE 3.C-11

Experimental Fatigue Data and Calculated Results by Using the Multiaxial Intrinsic *S-N* Equation of Notch 1 Under Te-T Loading Fatigue (Notch 1, 16MnR)

σ_a (MPa)	σ_m	τ_a	τ_m	*N* (exp)	*N* (cal)	R.E. (%)
410.19	0	237.56	0	200	113	−43.5
317.11	0	186.46	0	820	547	−33.3
240.19	0	141.64	0	3500	3135	−10.4
129.26	0	75.37	0	155770	160374	3

Note: M.E. = −21.2%, E.R. [−44,3]%

Under **Te-T loading fatigue**, a comparison of experimental fatigue lives with those calculated by the multiaxial intrinsic Notch *S-N* equation of Notch 2 is shown in Table 3.C-13, with error indexes: E.R. = [−69,−34]%, M.E. = −49%, from which it is seen that the calculated results are satisfactory.

Under **out-of-phase loading fatigue**, for Notch 2, a comparison of experimental fatigue lives with those calculated by using the multiaxial intrinsic Notch *S-N* equation with out-of-phase loading effect is shown in Table 3.C-14, with error indexes: E.R. = [−49,65]%, M.E. = 9.0% for Itoh's formula (A-1), E.R. = [−36,43]%, M.E. = 6.0% for the linear formula (A-2), which shows that the predicted fatigue lives by the linear formula (A-2) are much more accurate than those by Itoh's formula (A-1).

TABLE 3.C-12
Experimental Fatigue Data and Calculated Results by Using the Multiaxial Intrinsic Notch *S-N* Equation with Out-of-Phase Loading Effect (Notch 1, 16MnR)

σ_a	σ_m	τ_a	τ_m	N(exp)	R.E.(I.F) (%)	R.E.(L.F) (%)	R.E.(I.F) (%)	R.E.(L.F) (%)
422.81	0	248.61	0	400	−65.8	42	−49.2	13.1
373.21	0	220.33	0	660	−55.1	11.2	−33.3	5.1
319.27	0	188.37	0	1460	−45.5	−19	−19	−10.3
244.75	0	145.69	0	5480	−24.2	−41.3	12.3	−21.3
154.25	0	89.53	0	43000	90.8	−26.3	185.2	18.7
127.54	0	73.87	0	50000	**446**	73.6	Deleting an unusual data	
Error range (%)					[−66,446]	[−41,74]	[−49,185]	[−21,19]
Mean error (%)					57.7	6.7	19.2	1.0

Note: (a) All out-of-phase angles = 90°; (b) $\alpha' = -0.067$; $\lambda = 1.2828$, $\gamma = -0.54719$ (c) After deleting an unusual data, $\alpha' = -0.125$; $\lambda = 1.1419$, $\gamma = -0.37584$

TABLE 3.C-13
Experimental Data and Calculated Results Under Te-T Loading Fatigue (Notch 2, 16MnR)

σ_a	σ_m	τ_a	τ_m	N(exp)	N(cal)	R.E. (%)
416.43	0	243	0	340	107	−68.6
360.81	0	209.44	0	720	417	−42.1
322.56	0	191.28	0	2200	1090	−50.5
262.47	0	153.81	0	12000	7892	−34.2
209.45	0	122.43	0	108000	65935	−38.9

Note: M.E. = −49%, E.R. = [−69,−34]%

TABLE 3.C-14
Experimental Fatigue Data and Calculated Results by Using the Multiaxial Intrinsic Notch *S-N* Equation with Out-of-Phase Loading Effect (Notch 2, 16MnR)

σ_a	σ_m	τ_a	τ_m	N(exp)	ER(I.F) (%)	ER(L.F) (%)
323.66	0	192.27	0	9300	−49.3	−25
259.83	0	151.49	0	30000	36.5	43
217.08	0	129.17	0	120000	64.7	42.4
189.18	0	112.45	0	855500	−15.6	−36.4
Error range (%)					[−49,65]	[−36,43]
Mean error (%)					9.0	6

Note: (a) All out-of-phase angles = 90°; (b) $\alpha' = -0.1524$; $\lambda = 0.96385$, $\gamma = -0.15109$

Example 3 Medium/High Cycle Fatigue of Notched Components Made of SAE 1045

Experimental data of medium/high cycle fatigue of notched components (the notch geometry is shown in Figure 3.B.8 in Ref.[12]) made of SAE 1045 reported by Kurath et al. [44] are used to check the validation of the Notch *S-N* equation proposed in this study. A comparison of experimental fatigue lives with those calculated by using the Notch *S-N* equations is shown in Figure 3.C-7, from which it is seen that, for tension, torsion and Te-T loading fatigue, the calculated fatigue lives are in good agreement with experimental ones.

Under **out-of-phase loading fatigue**, a comparison of experimental fatigue lives with those calculated by using the multiaxial intrinsic Notch *S-N* equation with out-of-phase loading effect is shown in Table 3.C-15, with error indexes: E.R. = [−33,238]%, M.E. = 20.0% for Itoh's formula (A-1), and E.R. = [−37,210]%, M.E. = 14.0% for the linear formula (A-2), from which it is seen that the predicted fatigue lives by the linear formula (A-2) and Itoh's formula (A-1) have roughly the same accuracy.

Example 4 Medium Cycle Fatigue of Notched Components Made of SAE 1045

Experimental data of medium-cycle fatigue of notch components (the notch geometry is shown in Figure 3.B.9 in Ref.[12]) made of SAE 1045 reported by Yip

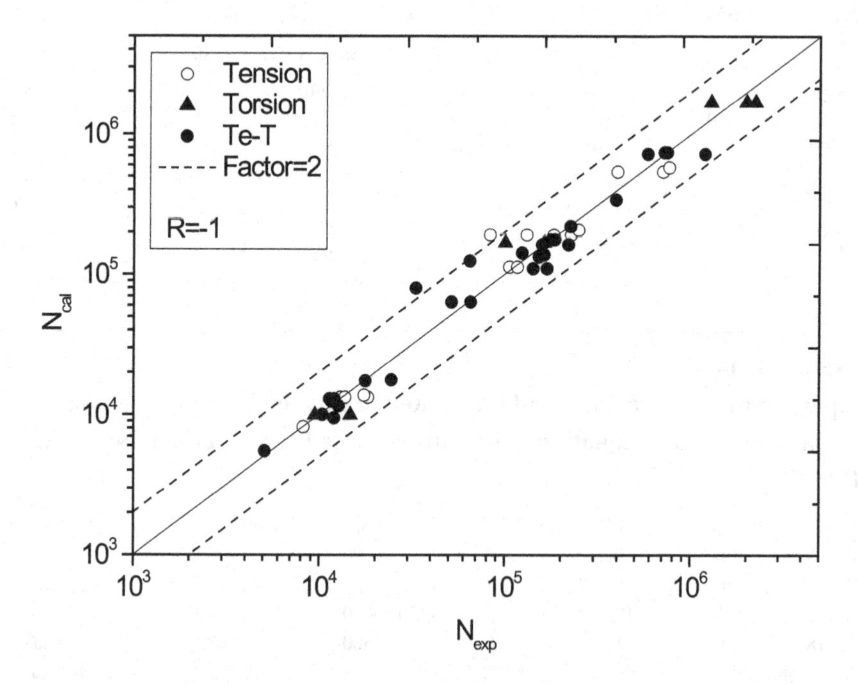

FIGURE 3.C-7 Comparison of experimental fatigue lives with those calculated by using the Notch *S-N* equations (SAE 1045 steel).

Source: (Data from Kurath, et al. [44])

TABLE 3.C-15
Experimental Data and Calculated Results of Notched Specimens Made of SAE 1045 Under Out-of-Phase Loading Fatigue

σ_a	τ_a	N(exp)	F	F′	$\bar{\sigma}_{ea} / \sigma_{ea}$	ER(I.T) (%)	ER(L.F) (%)
183	214.9	13110	0.4917	0.4917	0.9857	8.5	11.3
294.4	167.1	27470	0.9831	0.9785	0.9044	−33.1	−36.9
286.5	167.1	24620	0.9899	0.972	0.9304	−16.9	−22
270.2	178.4	10840	0.8745	0.8654	1.0259	70.2	63
366.1	105.4	23980	0.4986	0.4918	0.9257	−30.8	−29.6
122.5	173.5	157100	0.4077	0.3189	0.9649	−14.1	−12.4
206.1	136.1	45580	0.8743	0.6601	1.1197	238	209.5
194.2	136.1	213800	0.8238	0.6067	0.9464	−11.4	−18.4
194.2	136.1	266200	0.8238	0.6067	0.9207	−28.9	−34.4
Error range (%)						[−33,238]	[−37,210]
Mean error (%)						20	14

Note: (a) All out-of-phase angles = 90°; all $\sigma_m = 0$; all $\tau_m = 0$; (b) $\alpha' = -0.0473$; $\lambda = 0.98729$, $\gamma = -0.027$

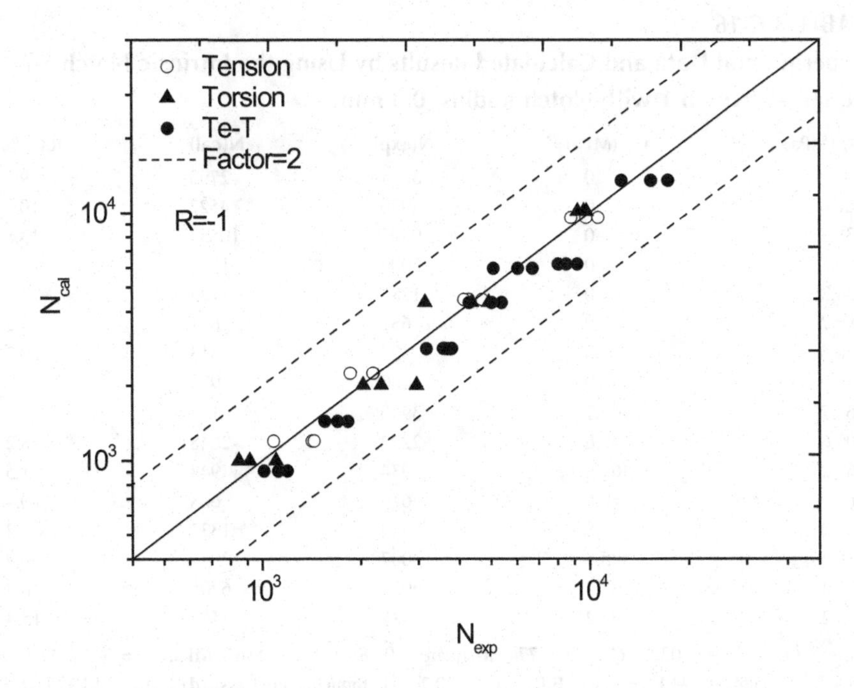

FIGURE 3.C-8 Comparison of experimental fatigue lives with those calculated by using the Notch *S-N* equations (SAE 1045 steel).

Source: (Data from Yip and Jen [45])

and Jen [45] are used to check the validation of the Notch *S-N* equation proposed in this study. A comparison of experimental fatigue lives with those calculated by using the Notch *S-N* equations is shown in Figure 3.C-8, from which it is seen that, for tension, torsion and Te-T loading fatigue, the calculated fatigue lives are in excellent agreement with experimental ones.

Example 5: Low/Medium Cycle Fatigue of Notched Components Made of En3B

The V-notched samples of En3B reported by Susmel and Taylor [28] had a gross diameter equal to 10 mm, a net diameter equal to 5 mm, and a notch opening angle equal to 60 degrees. Three different values of the notch root radius were investigated: 0.3 mm, 1.25 mm, and 4.0 mm. The notched cylindrical samples were tested in force control, using two different load ratios: R = –1 and R = 0. Experimental results and the calculated results by using the intrinsic Notch *S-N* equation are shown in Tables 3.C-16, 3.C-17 and 3.C-18 (see also Figure 3.C-9), from which it is seen that the calculated fatigue lives are in good agreement with those experimental ones.

TABLE 3.C-16
Experimental Data and Calculated Results by Using the Intrinsic Notch *S-N* Equation (Notch 1 with Notch Radius, 0.3 mm)

σ_a (MPa)	σ_m (MPa)	N(exp)	N(cal)	R.E. (%)
444	0	3074	2793	–9.1
402.6	0	4090	4522	10.6
338.7	0	9301	10583	13.8
489.5	0	2009	1728	–14
534.3	0	1283	1123	–12.4
589.7	0	657	691	5.2
626.5	0	427	513	20.2
549.9	0	1060	975	–8
365.7	365.7	3940	4114	4.4
409.6	409.6	2232	2048	–8.2
463.2	463.2	934	948	1.5
489.5	489.5	670	668	–0.4
438.8	438.8	1649	1333	–19.2
385.5	385.5	3007	2979	–0.9
337.4	337.4	6054	6705	10.8
503.2	503.2	485	560	15.4

Note: (a) A_{1n} = –0.20323, C_{1n} = 3.34774, R-square = 0.98; (b) σ_o = 116.7 MPa, α_0 = 0.85331, β_0 = 0.08584; (c) M.E. = 0.8%, E.R. = [–14,20.2]% without mean stress; (d) M.E. = 0.4%, E.R. = [–19.2,15.4]% with mean stress

TABLE 3.C-17
Experimental Data and Calculated Results by Using the Intrinsic Notch *S-N* Equation (Notch 2 with Notch Radius, 1.25 mm)

σ_a (MPa)	σ_m (MPa)	N(exp)	N(cal)	R.E. (%)
423.8	0	5028	5187	3.2
489.5	0	2372	2101	−11.4
536.4	0	1405	1183	−15.8
592.1	0	659	637	−3.4
611.9	0	509	518	1.8
631.4	0	363	426	17.2
514	0	1661	1546	−6.9
412.8	0	5058	6117	20.9
358.7	358.7	6027	7986	32.5
385.5	385.5	5210	4068	−21.9
440.6	440.6	1408	1140	−19
439.9	439.9	749	1158	54.6
406.5	406.5	2867	2463	−14.1
417.6	417.6	2087	1906	−8.7
392.1	392.1	3634	3466	−4.6

Note: (a) $A_{1n} = -0.15945$, $C_{1n} = 3.2195$, R-square = 0.98; (b) $\sigma_o = 164$ MPa, $\alpha_0 = 0.5696$, $\beta_0 = 0.24383$; (c) M.E. = 0.7%, E.R. = [−15.8,20.9]% without mean stress; (d) M.E. = 2.7%, E.R. = [−21.9,54.6]% with mean stress.

TABLE 3.C-18
Experimental Data and Calculated Results by Using the Intrinsic Notch S-N Equation (Notch 3 with Notch Radius, 4 mm)

σ_a (MPa)	σ_m (MPa)	N(exp)	N(cal)	R.E. (%)
488.1	0	7916	8429	6.5
623.7	0	654	654	0
542.3	0	1644	2811	71
569.4	0	1058	1690	59.8
515.2	0	6253	4798	−23.3
596.6	0	1456	1039	−28.6
650.8	0	670	420	−37.4
426.1	426.1	5134	4500	−12.4
482.9	482.9	725	652	−10.1
454.5	454.5	1780	1670	−6.2
443.2	443.2	1560	2462	57.8
448.8	448.8	3946	2029	−48.6
460.2	460.2	928	1377	48.4
437.5	437.5	2785	3003	7.8

Note: (a) $A_{1n} = -0.09589$, $C_{1n} = 3.06495$, R-square = 0.84; (b) $\sigma_o = 289$ MPa, $\alpha_0 = 0.65103$, $\beta_0 = 0.38357$; (c) M.E. = 6.8%, E.R. = [−37.4,59.8]% without mean stress; (d) M.E. = 5.3%, E.R. = [−48.6,57.8]% with mean stress

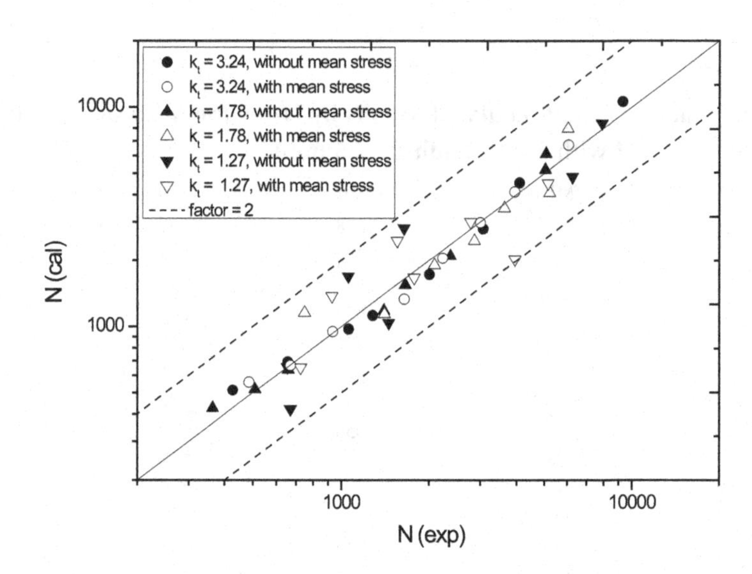

FIGURE 3.C-9 Comparison of experimental fatigue lives with those calculated by using the intrinsic Notch *S-N* equation.

Source: (Data from Susmel and Taylor [28])

APPENDIX D: EXPERIMENTAL INVESTIGATIONS: THE NOTCH *S-N* EQUATION IS VERIFIED TO BE NATURALLY EXISTING BY SOME FATIGUE TEST DATA OF VARIOUS NOTCH SPECIMENS FROM THE LITERATURE

In this Appendix, some fatigue test data of various notch specimens reported in the literature (see Figures 3.D-1 to 3.D-17), are shown to illustrate the following issues:

(1) The Notch *S-N* equations are naturally existing.
(2) A different notch has its different *S-N* equation.

Descriptions of fatigue test data of various notch specimens reported in the literature, shown in Figures 3.D-1 to 3.D-17, are as follows:

Figure 3.D-1 shows fatigue results generated by testing plain and V-notched cylindrical specimens of S690 steel under rotating bending (Data from Susmel [12]).

Figures 3.D-2 and 3.D-3 show geometries of the tested specimens and fatigue results generated under fully reversed loading (data from Susmel and Taylor [29]) (EN3B, a commercial cold-rolled low-carbon steel), respectively.

Figures 3.D-4 to 3.D-9 show geometries of the tested specimens and fatigue test results (C40 carbon steel) (data form Atzori et al. [46]).

Figures 3.D-10 to 3.D-13 show geometries of the tested specimens and fatigue test results (AISI 416 steel) (data form Berto and Lazzarin [47]).

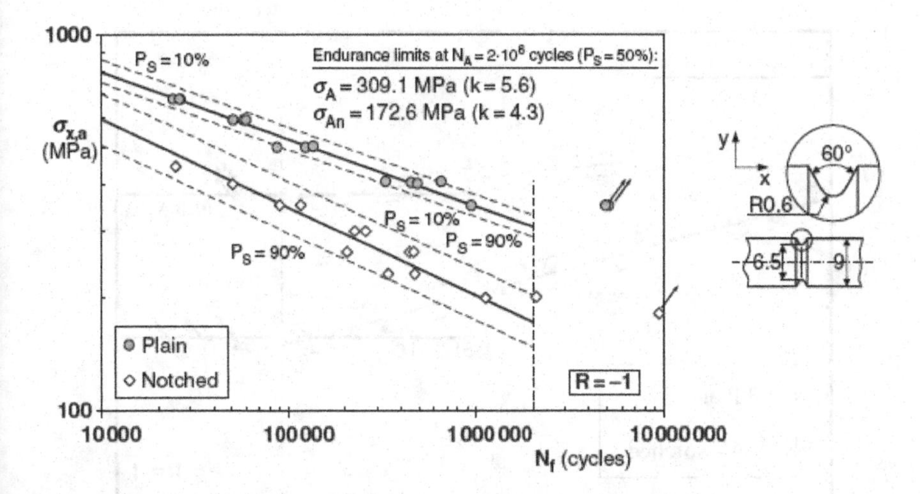

FIGURE 3.D-1 Fatigue results generated by testing plain and V-notched cylindrical specimens of S690 steel under rotating bending.

Source: (Data from Susmel [12]).

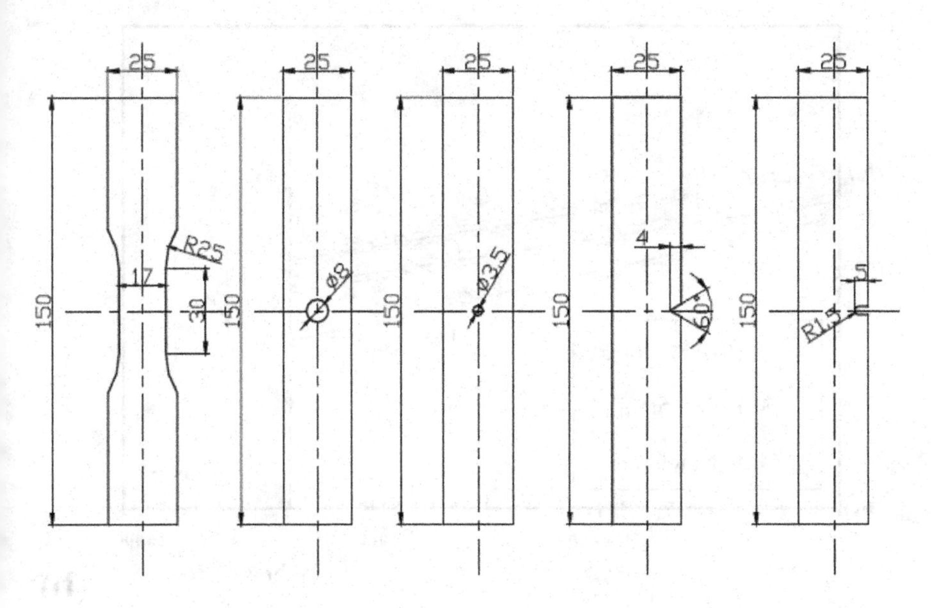

FIGURE 3.D-2 Geometries of the tested specimens (dimensions in millimetres) (En3B).

Source: (Data from Susmel and Taylor [29]).

Figures 3.D-14 and 3.D-15 show geometries of the tested specimens and fatigue test results (Ti–6Al–4V alloy) (data from Berto et al.[48]).

Figures 3.D-16 and 3.D-17 show geometries of the tested specimens and fatigue test results (40CrMoV13.9 steel) (data from Berto et al.[49]).

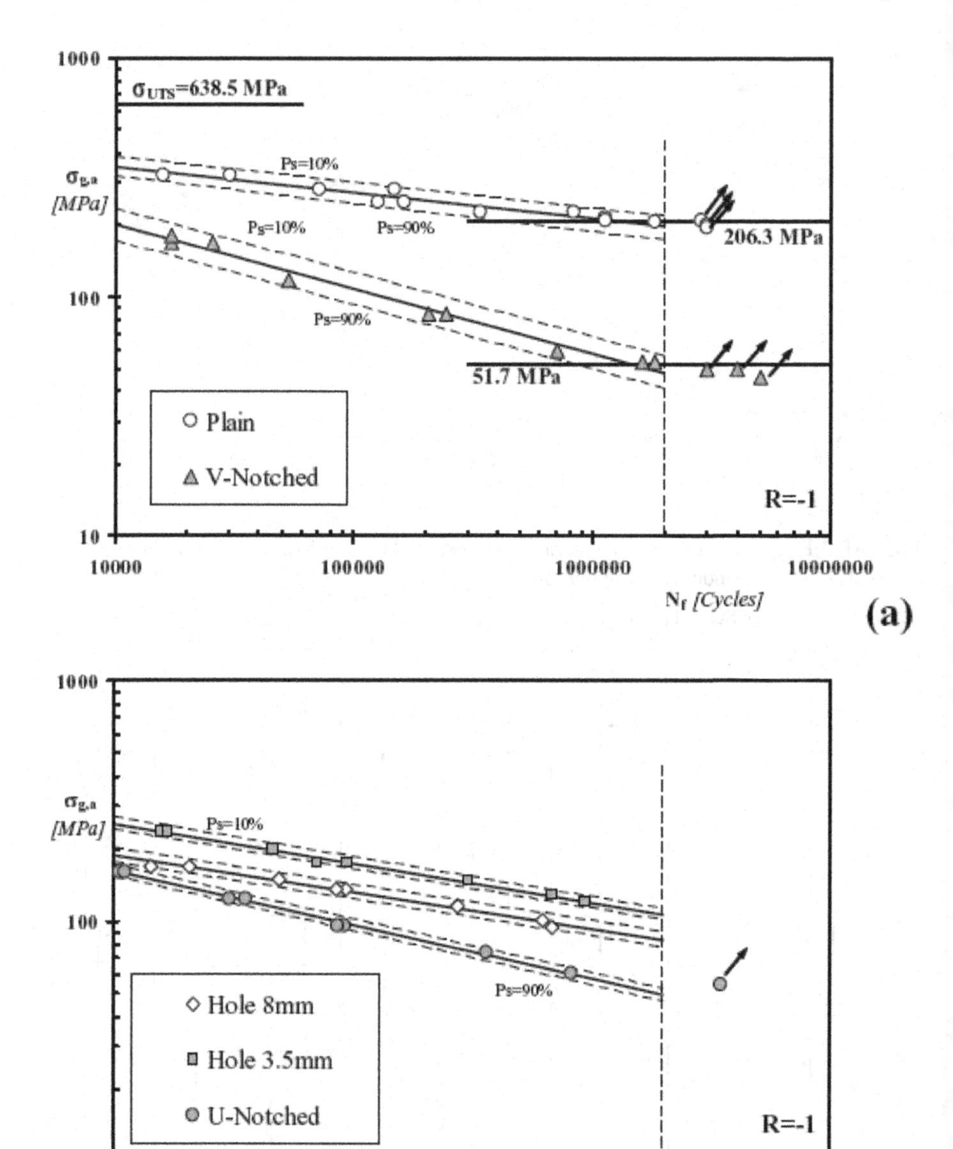

FIGURE 3.D-3 Fatigue results generated under fully reversed loading (En3B).

Source: (Data from Susmel and Taylor [29]).

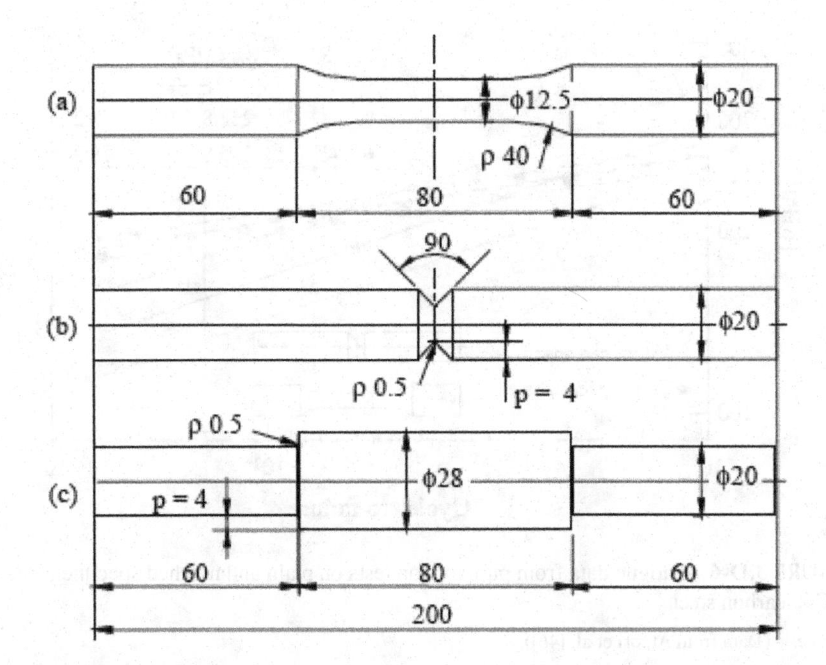

FIGURE 3.D-4 Geometry of plain and notched specimens made of C40 carbon steel.

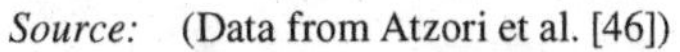
Source: (Data from Atzori et al. [46])

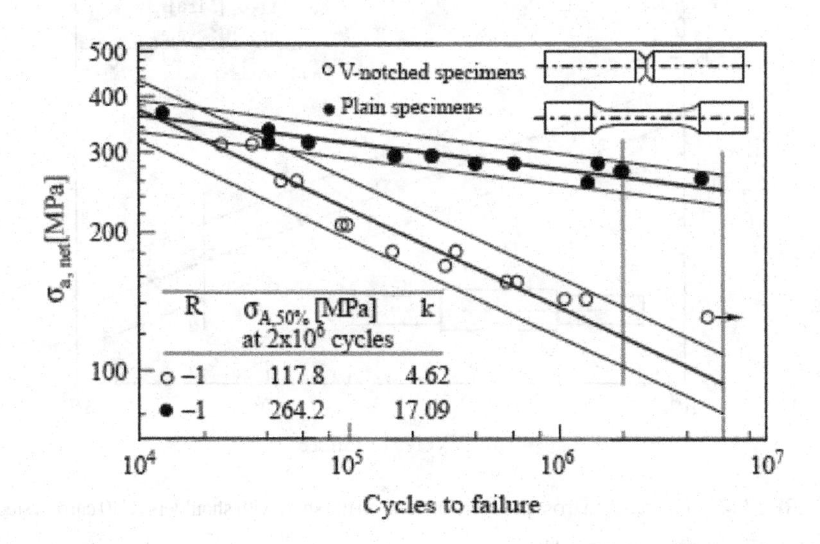

FIGURE 3.D-5 Fatigue data from pure axial tests on plain and notched specimens made of C40 carbon steel.

Source: (Data from Atzori et al. [46]).

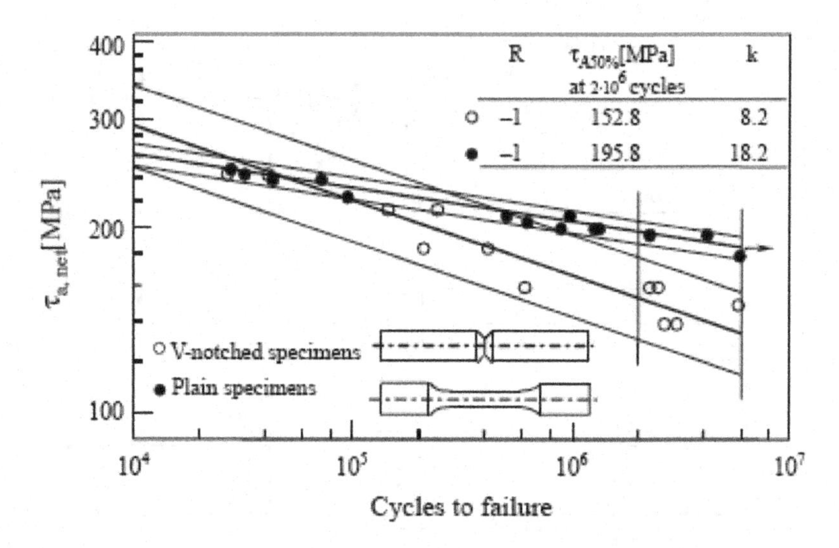

FIGURE 3.D-6 Fatigue data from pure torsion tests on plain and notched specimens made of C40 carbon steel.

Source: (Data from Atzori et al. [46]).

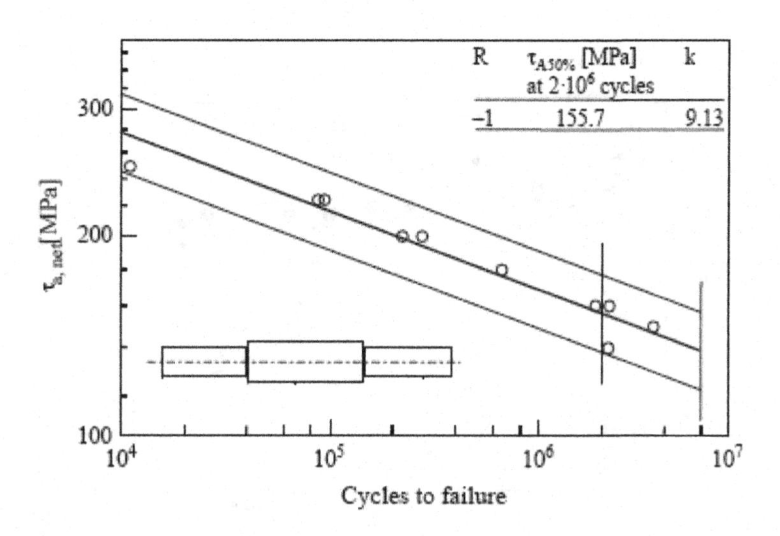

FIGURE 3.D-7 Fatigue data from pure torsion tests on the shaft with shoulders (C40 carbon steel).

Source: (Data from Atzori et al. [46]).

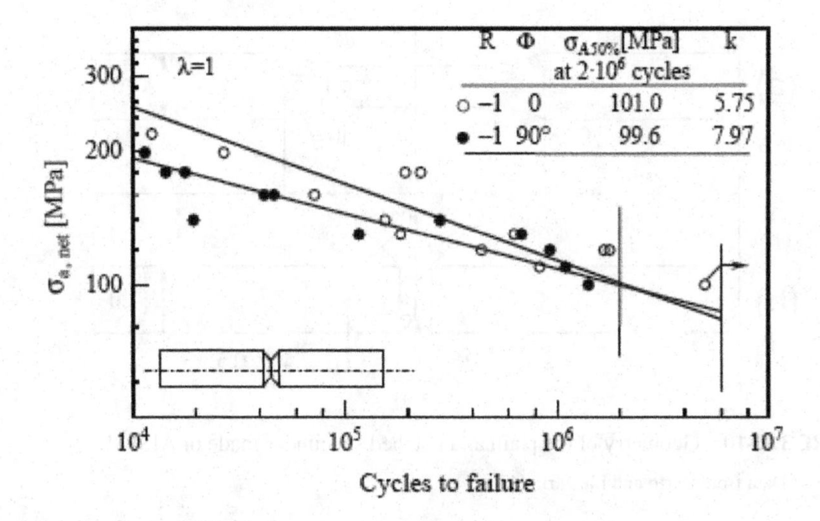

FIGURE 3.D-8 Multiaxial fatigue data obtained under a nominal load ratio R = –1 (ϕ = 0 and 90°) (C40 carbon steel).

Source: (Data from Atzori et al. [46]).

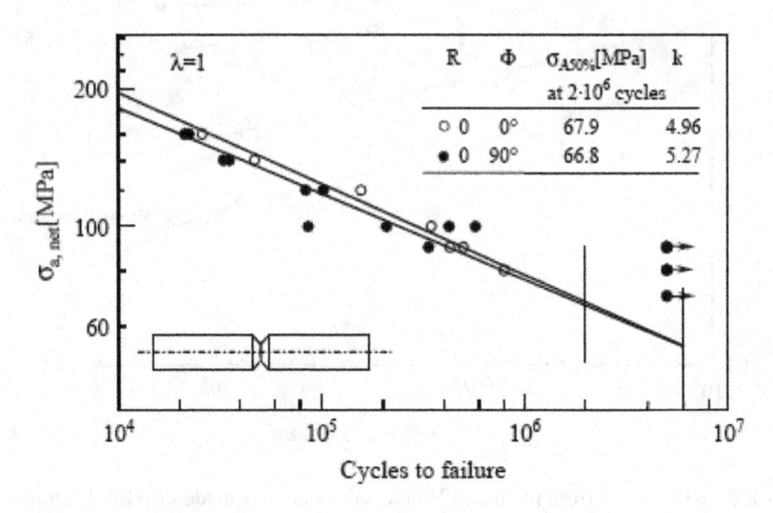

FIGURE 3.D-9 Multiaxial fatigue data obtained under a nominal load ratio R = 0 (ϕ = 0 and 90°) (C40 carbon steel).

Source: (Data from Atzori et al. [46]).

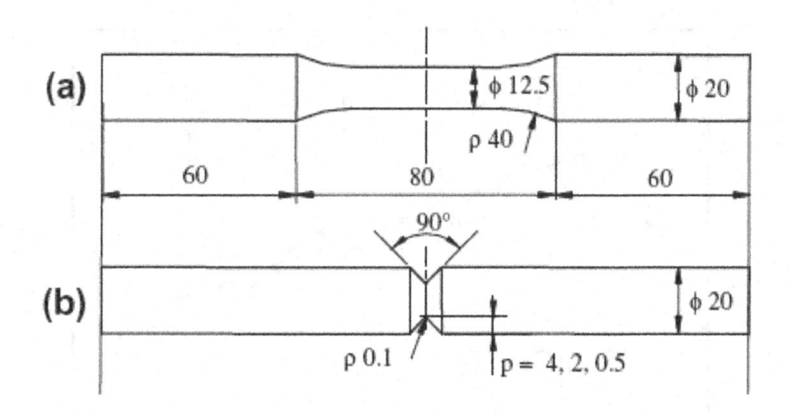

FIGURE 3.D-10 Geometry of the plain and notched specimens made of AISI 41.

Source: (Data from Berto and Lazzarin [47])

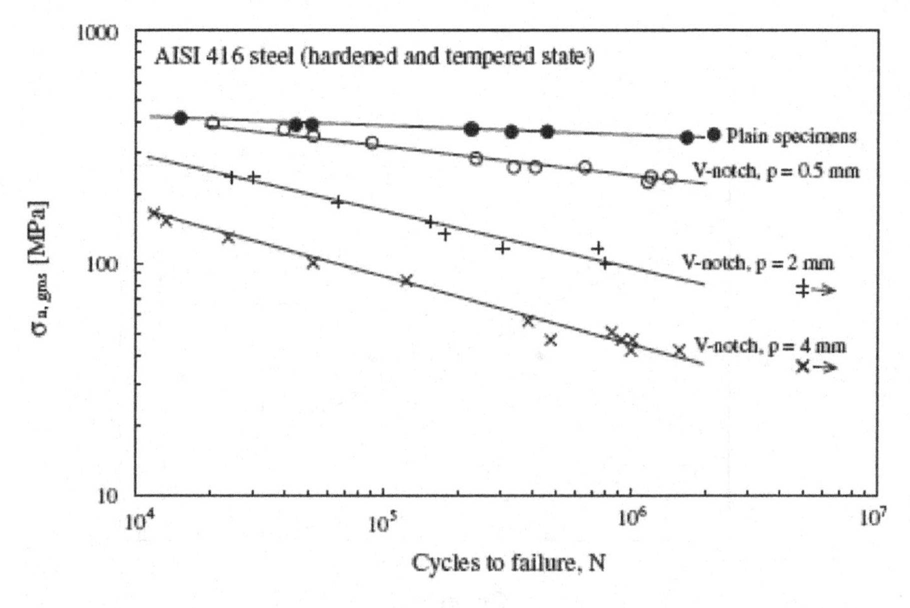

FIGURE 3.D-11 Data from plain and V-notched specimens made of AISI 416 under pure tension loading (stress amplitudes referred to as the gross sectional area).

Source: (Data from Berto and Lazzarin [47]).

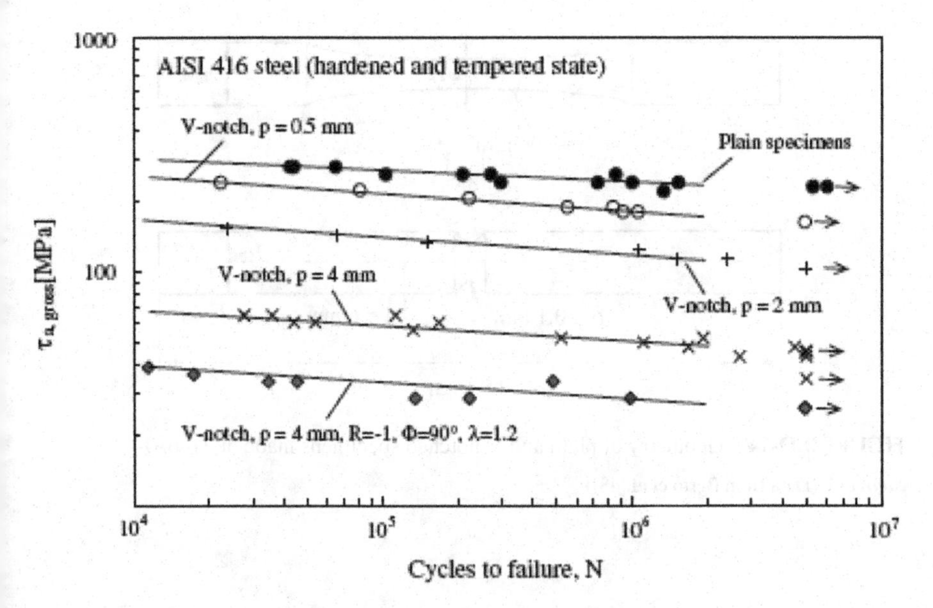

FIGURE 3.D-12 Data from V-notched specimens made of AISI 416 under pure torsion (stress amplitude referred to the gross section).

Source: (Data from Berto and Lazzarin [47]).

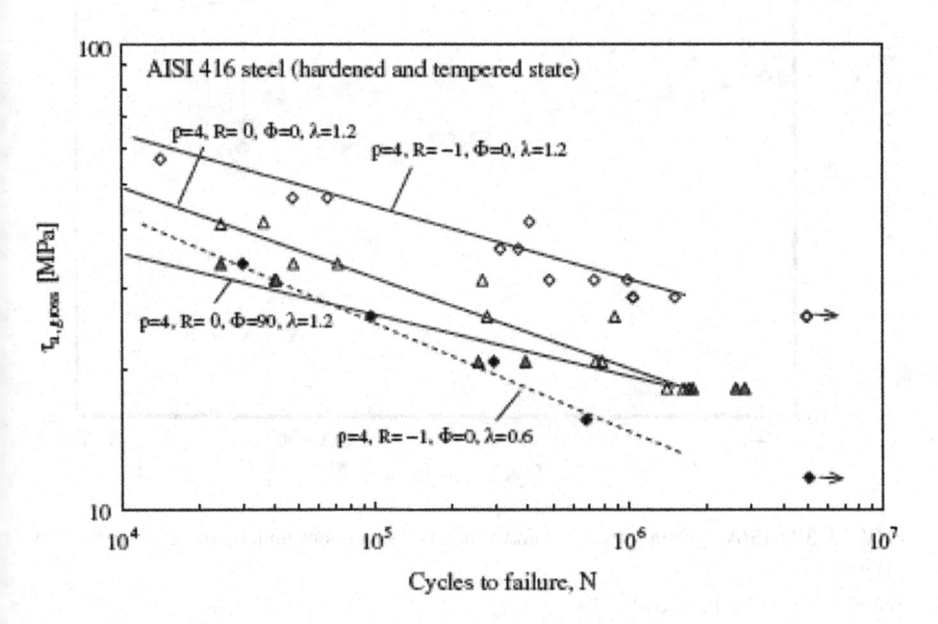

FIGURE 3.D-13 Data from plain and V-notched specimens made of AISI 416 under multiaxial loading (stress amplitude referred to the gross section).

Source: (Data from Berto and Lazzarin [47]).

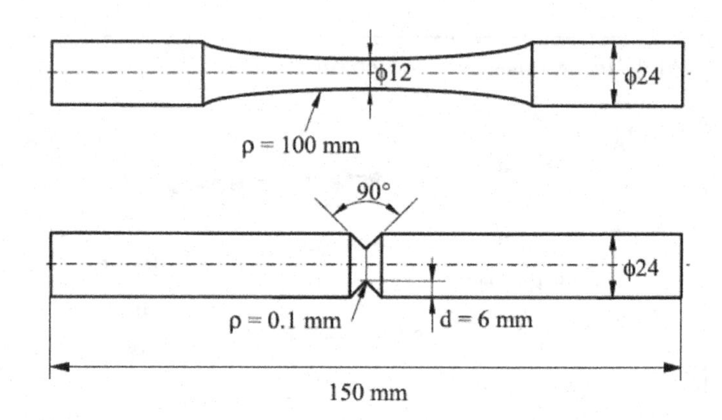

FIGURE 3.D-14 Geometry of plain and V-notched specimens made of Ti–6Al–4V.

Source: (Data from Berto et al.[48])

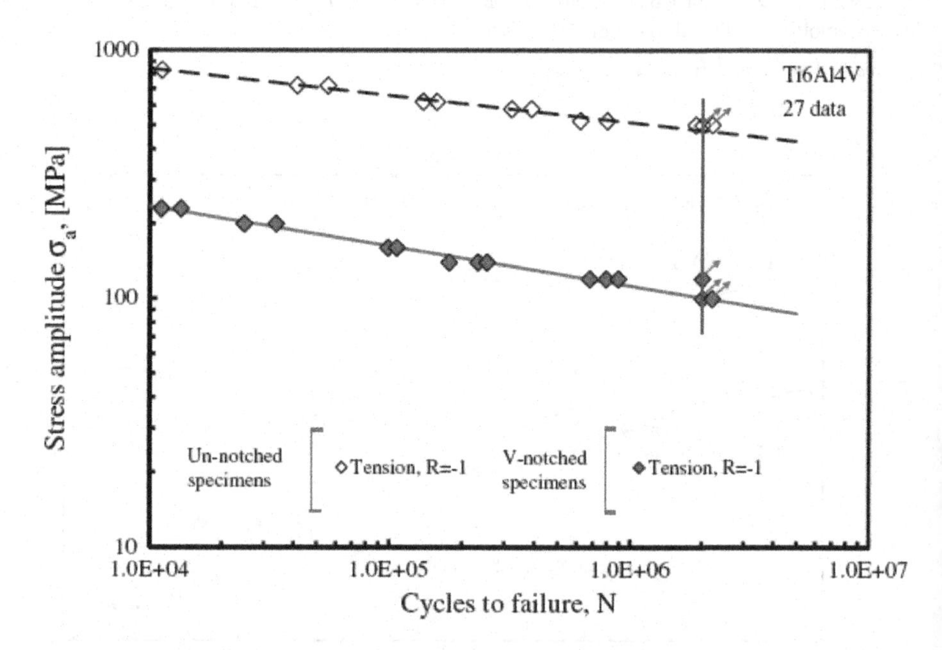

FIGURE 3.D-15(A) Data from plain and V-notched specimens under pure tension (R = –1) (Ti–6Al–4V).

Source: (Data from Berto et al.[48])

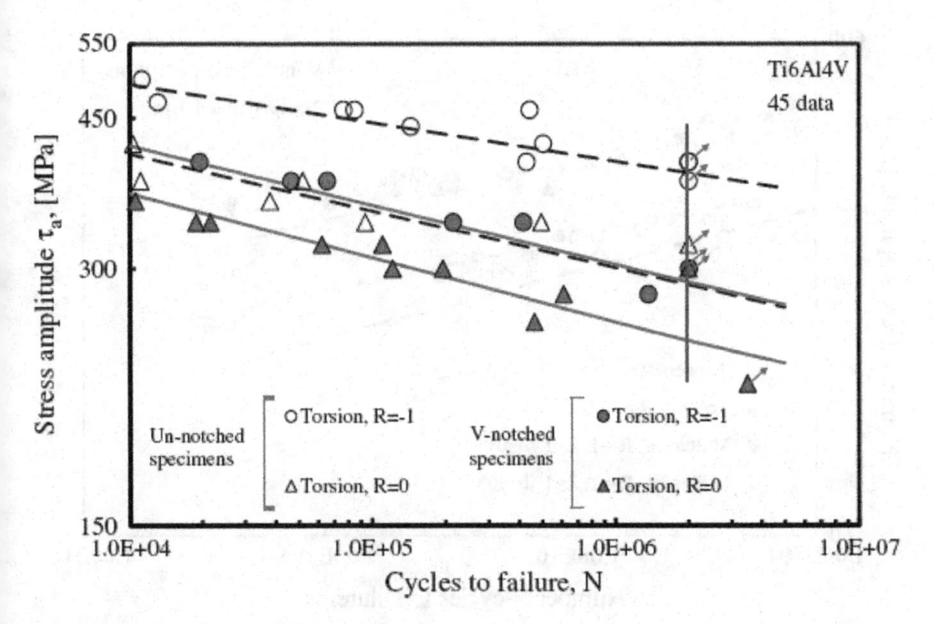

FIGURE 3.D-15(B) Data from plain and V-notched specimens under pure torsion (R = –1 and R = 0) (Ti–6Al–4V).

Source: (Data from Berto et al.[48])

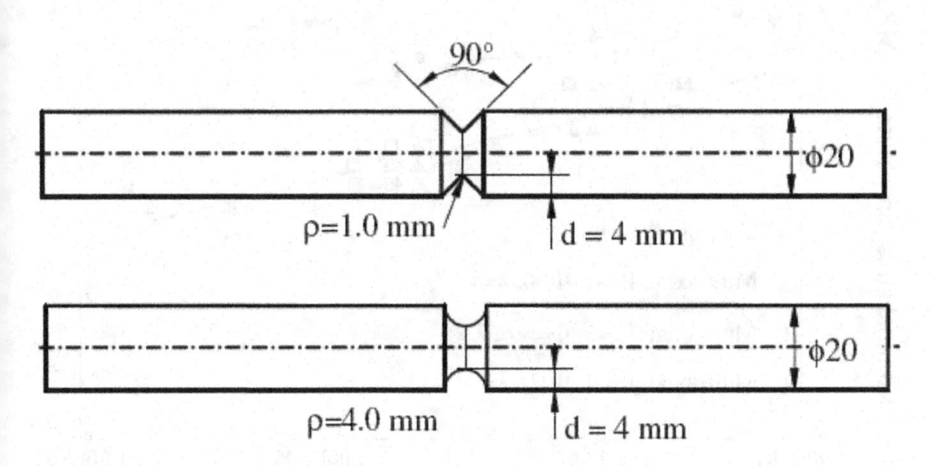

FIGURE 3.D-16 Geometry of V-notches and semicircular notches (40CrMoV13.9 steel).

Source: (Data from Berto et al.[49])

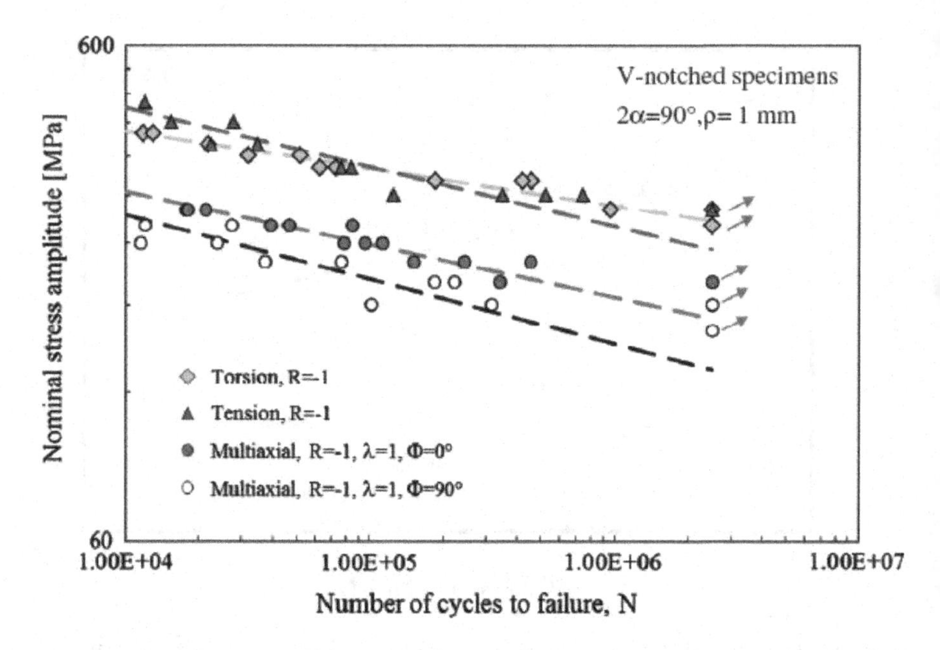

FIGURE 3.D-17(A) Data from V-notches under pure tension, torsion and multiaxial loading (40CrMoV13.9 steel).

Source: (Data from Berto et al.[49]))

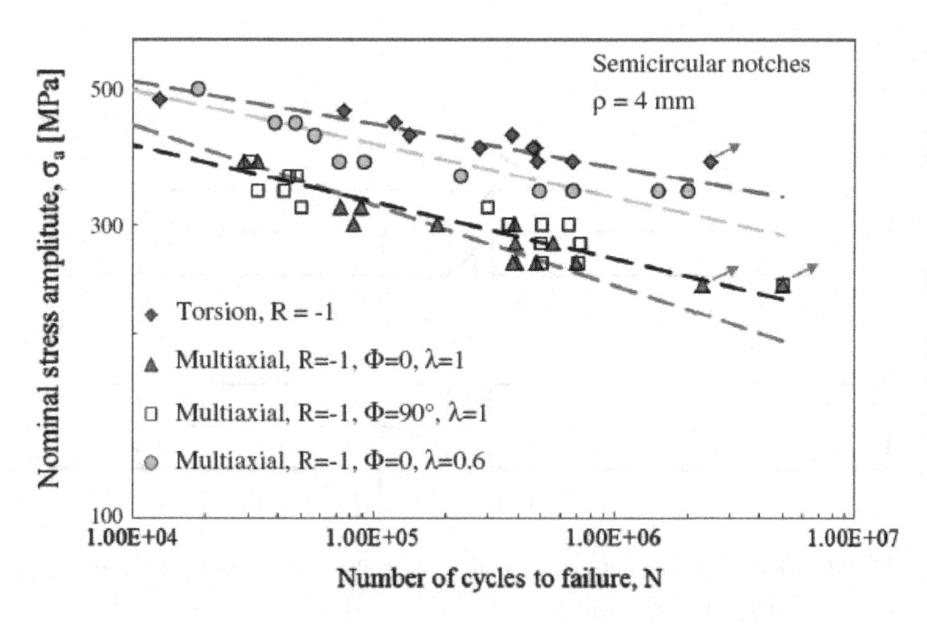

FIGURE 3.D-17(B) Data from semicircular notches under pure torsion and multiaxial loading (40CrMoV13.9 steel).

Source: (Data from Berto *et al.*[49])

4 A Local Approach for Fracture Analysis of V-Notch Specimens Under Mode I Loading

4.1 INTRODUCTION

The prediction of brittle fracture in cracked solids based on the stress intensity factor K of linear elastic fracture mechanics is widely accepted [1]. The suitability of using the single parameter K to correlate fracture is a result of the universal nature of the singular stress field near a crack tip as shown by Williams [2]. Williams further showed that a universal singular stress field of the form also exists in the region surrounding a sharp notch. In fact, the case of a crack is really an extreme case of a reentrant notch where the included angle is zero. In the expression for the universal stress field, $\lambda - 1$ is the order of the stress singularity. It, along with the angular function, $f_{ij}(\theta)$, can be completely determined by an asymptotic analysis of the stress state. Thus, they are not functions of the particular geometry of the solid containing the notch or the specific far-field loading [1]. The stress intensity K is a function of the geometry of the solid and the far-field loading, and thus for a given geometry and loading, K completely characterizes the stress state in a region near the notch tip [1].

The V-notch problem for linear elastic materials was first analyzed by Williams [2] by means of an eigenfunction series expansion and later by England [3] using complex potentials. More recently, solutions valid for cracked and notched components have been published by Seweryn and Molski [4], by Lazzarin and Tovo [5] and by Filippi et al. [6]. These last approaches were based on complex stress functions, according to Muskhelishvili's method.

With respect to the study on stress intensity K for a specific geometry, it is worthy to mention the works of Gross and Mendelson [7], Knesl [8], and Dunn *el al.* [1].For three-point bending (TPB) specimens, values were taken from Gross and Mendelson [7]. For four-point bending (FPB) specimens, the values were from Grenestedt *et al.*[9]. The values of single-edge notched tension (SENT) specimens were from Strandberg [10] and those of double-edge notched tension (DENT) specimens were from Seweryn [11].

When dealing with brittle, or quasi brittle materials—where linear elasticity can be applied—the stress and displacement fields near the tips of V-notches can be characterized by the notch stress intensity factors (NSIFs), K_v, which is a function of the V-notch angle [12]. Using cracked specimens as an example, a fracture criterion

DOI: 10.1201/9781003356721-4

based on critical values of the notch stress intensity factors can be stated, i.e., a crack will propagate from the tip of a notch when the actual value of the notch stress intensity factor reaches a critical value (Carpinteri [13]; Knesl [8]; Seweryn [11]; Dunn et al.[1]). The critical value of the notch stress intensity factor, K_{vc}, of each V-notch angle, has to be obtained through experimental measurements on very sharp notched samples, frequently an involved procedure [12].

As already mentioned, critical values of the notch stress intensity factors are obtained experimentally. Fracture tests are performed on sharp V-notched specimens and the critical loads are measured. From these results the notch stress intensity factors are computed. These notches have to be very sharp (usually the tip notch radius is less than 10 microns). Different angles, in the range from 0° to 180°, have to be tested and quite often different geometries are advisable when scatter appears. All these procedures may become cumbersome [12].

Using the *cohesive crack* concept [12], computation of critical values of the notch stress intensity factors was performed by Gómez and Elices by means of the parameters characterizing the softening function measured experimentally. The results were checked successfully against experimental ones in different materials: steel, aluminum, PMMA and PVC.

Since the beginning of the last century, as has been researched by Neuber [14,15] and by Peterson [16], a lot of advances of local approaches to deal with brittle failure of notched components have been made, in particular, the CD approach [17,18] (often called TCD, Theory of Critical Distance) and the volume-based SED approach [19–21]. In the two well-known local approaches, the size of Neuber's elementary volume is taken to be a material constant.

In this study, a local approach for fracture analysis of V-notch specimens is proposed on the basis of the linear elastic stress field ahead of V-notches and by using the concepts of the critical stress and the critical distance by Taylor [17]. In the local approach, the size of the critical distance is not taken to be a material constant, whereas it is assumed that the critical distances of different notches are different and are connected closely with the variation of notched geometry. By using fracture test data of V-notches from the literature, the local approach has been verified to be both simple in computation and high in accuracy.

4.2 A BRIEF DESCRIPTION OF LINEAR ELASTIC STRESS FIELD AND STRESS INTENSITY OF THE V-NOTCHES

It was shown from previous studies (for example [1, 2]) that under Mode I loading, the stress distribution ahead of the notches is expressed to be:

$$\sigma_{ij} = \frac{K_v}{\sqrt{2\pi r^{1-\lambda}}} f_{ij}(\theta, \lambda) \tag{1}$$

where r and θ are the polar coordinates (as shown in Figure 4.1), K_v is the notch stress intensity factor, λ is a function of the notch angle α, defined below, and f_{ij} is a function of the polar angle θ and, implicitly, of the notch angle α.

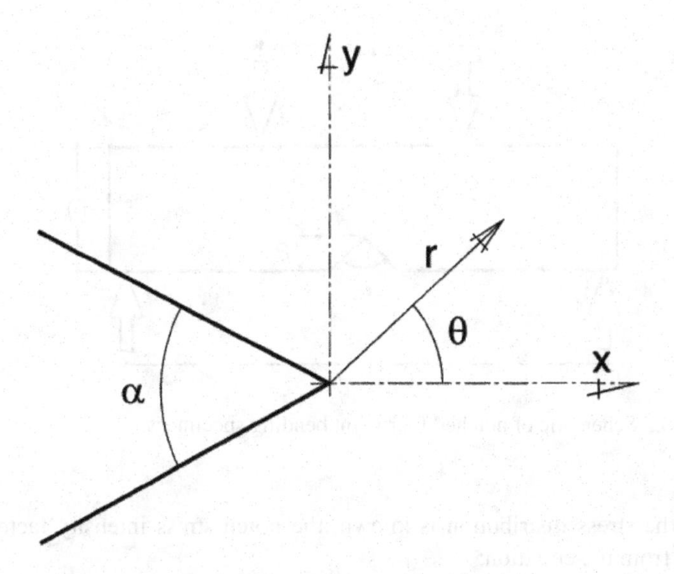

FIGURE 4.1 Coordinate system and symbols used for V-notches.

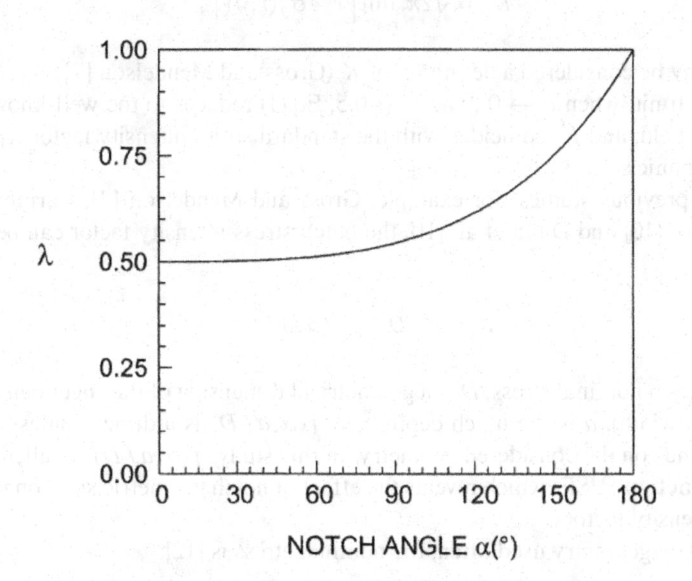

FIGURE 4.2 Values of λ as a function of the V-notch angle.

Source: (Data from GÓMEZ and ELICES [12]).

The parameter λ characterizes the strength of the notch tip stress singularity and is the root of the equation:

$$\sin(\lambda\beta) + \lambda \sin(\beta) = 0 \tag{2}$$

where $\beta = 2\pi - \alpha$. Figure 4.2 shows the values of λ as a function of the notch angle.

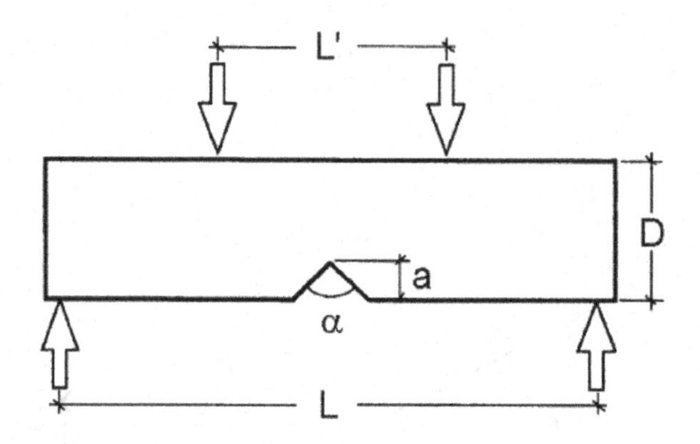

FIGURE 4.3 Schematic of notched four-point bending specimens.

When the stress distribution is known, the notch stress intensity factor can be computed from the equation:

$$K_I^V = \sqrt{2\pi} \lim_{r \to 0^+} \left[r^{1-\lambda} \sigma_\theta (r,0) \right] \tag{3}$$

which may be considered a definition of K (Gross and Mendelson [7]).

In the limit, when $\alpha \to 0^\circ$, and $\lambda \to 0.5$, Eq.(1) reduces to the well-known crack tip stress field, and K_v coincides with the standard stress intensity factor K_I of fracture mechanics.

From previous studies (for example, Gross and Mendelson [7], Carpinteri [13], Strandberg [10] and Dunn et al. [1]), the notch stress intensity factor can be written to be:

$$K_v = \sigma_n \cdot D^{1-\lambda} \cdot f(\alpha, a/D) \tag{4}$$

where σ_n is a nominal stress, D is a geometrical dimension of the specimen (usually, specimen width), a is the notch depth, and $f(\alpha, a/D)$ is a dimensionless function that depends on the considered geometry. In this study, $f(\alpha, a/D)$ is called a **notch shape function** (NSF) which reveals the effect of notch geometric sizes on the notch stress intensity factors.

For some geometry used often, the nominal stress is [12]:

$$\sigma_n = \frac{P}{BD} \text{ for tension specimens,} \tag{5a}$$

$$\sigma_n = \frac{3PL}{2BD^2} \text{ for three-point bending specimens,} \tag{5b}$$

$$\sigma_n = \frac{3P(L-L')}{2BD^2} \text{ for four-point bending specimens,} \tag{5c}$$

where P is the load and L, D and B are respectively the length, depth and thickness of the specimen, and L' is shown in Figure 4.3.

4.3 A LOCAL STRESS FIELD FAILURE MODEL

It is assumed that there are two V notches (denoted by notch 1 and notch 2) which have the notch angles, α_1 and α_2 (with Williams's eigenvalues λ_1 and λ_2), the critical notch stress intensity factors, K_{vc1} and K_{vc2}, and the critical distances, L_{c1} and L_{c2}.

For the notch 1, a failure condition according to the Point Method (PM) by Taylor [17] is assumed to be

$$F(K_{vc1},\lambda_1,L_{c1},\sigma_0)=0 \tag{6}$$

where σ_0 is the critical stress.

For notch 2, similarly, a failure condition is assumed to be

$$F(K_{vc2},\lambda_2,L_{c2},\sigma_0)=0 \tag{7}$$

Referring to the notch tip stress field (1), function F in Eqs. (6) and (7) can be written as:

$$F(K_v,\lambda,r,\sigma_0)=\frac{K_v}{\sqrt{2\pi}r^{1-\lambda}}-\sigma_0 \tag{8}$$

According to the previous investigations [1,7,12], the critical notch stress intensity factors, K_{vc1} and K_{vc2}, can be written as:

$$K_{vc1}=\sigma_{nc1}\cdot D^{1-\lambda_1}\cdot f(\alpha_1,a/D) \tag{9}$$

$$K_{vc1}=\sigma_{nc2}\cdot D^{1-\lambda_2}\cdot f(\alpha_2,a/D) \tag{10}$$

where σ_{nc1} and σ_{nc2} is the critical failure stresses of notch 1 and notch 2, respectively.

Here, it is assumed that L_{c1} and $f(\alpha_1,a/D)$ of Notch 1 are connected with L_{c2} and $f(\alpha_2,a/D)$ of Notch 2,

$$\frac{f(\alpha_1,a/D)}{L_{c1}}=\frac{f(\alpha_2,a/D)}{L_{c2}} \tag{11}$$

From (6), (7) and (11), we have

$$L_{c1}=\left[\frac{K_{vc2}}{K_{vc1}}\left(\frac{f(\alpha_2,a/D)}{f(\alpha_1,a/D)}\right)^{\lambda_2-1}\right]^{\frac{1}{\lambda_1-\lambda_2}} \tag{12}$$

and

$$\sigma_0=\frac{K_{vc1}}{\sqrt{2\pi}}\cdot\left[\frac{K_{vc2}}{K_{vc1}}\left(\frac{f(\alpha_2,a/D)}{f(\alpha_1,a/D)}\right)^{\lambda_2-1}\right]^{\frac{\lambda_1-1}{\lambda_1-\lambda_2}} \tag{13}$$

For any V-notch with notch angle α_3, from the following equation similar to Eq.(11):

$$\frac{f(\alpha_1,a/D)}{L_{c1}}=\frac{f(\alpha_3,a/D)}{L_{c3}} \tag{14}$$

we have

$$\frac{K_{vc3}}{K_{vc1}} = \left(\frac{K_{vc2}}{K_{vc1}}\right)^{\frac{\lambda_1-\lambda_3}{\lambda_1-\lambda_2}} \left(\frac{f(\alpha_2, a/D)}{f(\alpha_1, a/D)}\right)^{\frac{(\lambda_1-\lambda_3)(\lambda_2-1)}{\lambda_1-\lambda_2}} \left(\frac{f(\alpha_3, a/D)}{f(\alpha_1, a/D)}\right)^{1-\lambda_3} \tag{15}$$

By letting notch angle $\alpha_3 = 0$, then from (15) we have

$$\frac{K_{Ic}}{K_{vc1}} = \left(\frac{K_{vc2}}{K_{vc1}}\right)^{\frac{\lambda_1-0.5}{\lambda_1-\lambda_2}} \left(\frac{f(\alpha_2, a/D)}{f(\alpha_1, a/D)}\right)^{\frac{(\lambda_1-0.5)(\lambda_2-1)}{\lambda_1-\lambda_2}} \left(\frac{f(0, a/D)}{f(\alpha_1, a/D)}\right)^{0.5} \tag{16}$$

For brittle materials, the critical stress σ_0 can be taken to be the ultimate tensile stress σ_t [17]. Under the assumption that the fracture toughness of notched material studied, K_{Ic}, is known, by taking Notch 1 to be a crack, the critical stress intensity factor of any V-notch 2, K_{vc2}, is calculated by using the following equation:

$$K_{vc2} = \sqrt{2\pi}\sigma_t \cdot \left\{\frac{f(\alpha_2, a/D)}{f(0, a/D}\right\}^{1-\lambda_2} \cdot \left(\frac{K_{Ic}}{\sqrt{2\pi}\sigma_t}\right)^{2(1-\lambda_2)} \tag{17}$$

Here, the approach that the fracture toughness of any V-notch (pointed V-notch) is calculated by using formulas (17) and (15) is called a **notch shape function (NSF) approach**.

The analysis on the notch shape function $f(\alpha, a/D)$ reported in the literature (for example, numerical results of notch shape function of three-point bending specimens reported by Gross and Mendelson [7]) showed that the variation of the ratio $f(\alpha_2, a/D)/f(\alpha_1, a/D)$ with a/D is negligible. Thus the calculation of the ratio of notch shape function of two different notch angles can be carried out by taking $a/D = 0.5$.

4.4 EXPERIMENTAL VERIFICATIONS

Based on the test data of fracture failure of V-notched components from the literature, in this section, the local approach for fracture analysis of notched components proposed in this study will be verified. In order to quantitatively evaluate the accuracy of the failure analysis, the following error indexes are defined:

$$R.E. = \frac{(\mathrm{K_{vc}}(cal) - \mathrm{K_{vc}}(\exp)) \times 100}{\mathrm{K_{vc}}(\exp)} \tag{18}$$

$$M.E. = \frac{1}{n}\sum_{i=1}^{n}(R.E.)_i \tag{19}$$

$$M.A.E. = \frac{1}{n}\sum_{i=1}^{n} abs(R.E.)_i \tag{20}$$

$$E.R. = [\min(R.E.)_i, \max(R.E.)_i] \tag{21}$$

where n is the number of test cases, $K_{vc}(exp)$ is the experimentally measured fracture toughness; and $K_{vc}(cal)$ are the calculated fracture toughness by means of the local approach proposed in this study.

For the sake of clear discussions, the experimental verifications given here are described in the forms of examples, respectively.

Example 1: Fracture Analysis of V-Notch TPB Specimens Made of PMMA

In 1997, fracture initiation at sharp notches was studied by Dunn *et al.* [1] by using a V-notch TPB specimen (Figure 4.4). For the V-notch TPB specimen shown in Figure 4.4, the notch stress intensity factor was calculated by the following equation given by Dunn *et al.* [1]

$$K_v = \sigma_n h^{1-\lambda} f\left(\frac{a}{h}\right) = \left(\frac{3PL}{2bh^2}\right) h^{1-\lambda} f\left(\frac{a}{h}\right) \tag{22}$$

where $f(a/h)$ was

$$f\left(\frac{a}{h}\right) = c_1\left(\frac{a}{h}\right) + c_2\left(\frac{a}{h}\right)^2 + c_3\left(\frac{a}{h}\right)^3 + c_4\left(\frac{a}{h}\right)^4 + c_5\left(\frac{a}{h}\right)^5 \tag{23}$$

where constants c_1, c_2, c_3, c_4 and c_5 were shown in Table 4.1.

A material studied experimentally in Ref.[1] was PMMA that has the fracture toughness K_{Ic} = 1.02 MPam$^{0.5}$ and the ultimate tensile stress σ_t = 124 Mpa. In this study, fracture analysis for V-notch TPB specimens made of PMMA is performed by using Eq.(17). The obtained results and fracture test data reported in Ref.[1] are shown in Table 4.2, from which it is seen that the predicted results of notch fracture toughness are in good agreement with those measured experimentally, with the error indexes: M.E. = −9.6%, M.A.E. = 9.6%, E.R. = [−17.1,0]%.

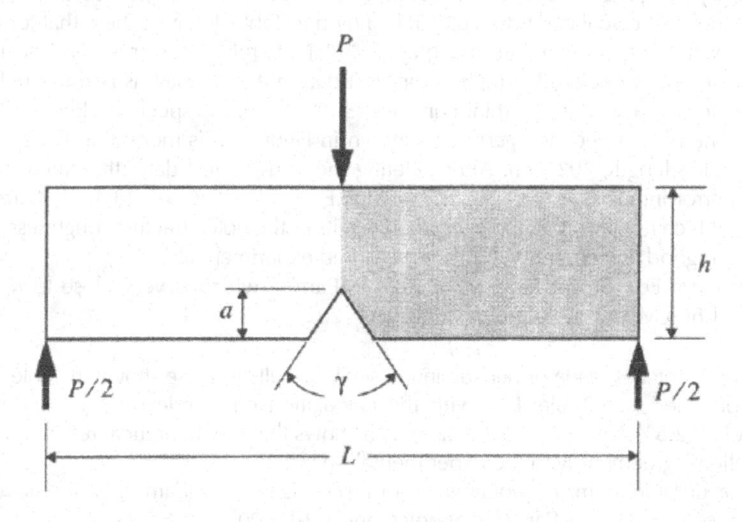

FIGURE 4.4 Geometry of notched three-point flexure specimen.

Source: (Data from Dunn *et al.* [1])

TABLE 4.1
Order of the Corner Stress Singularity and Coefficients of the Nondimensional Function $f(a/h)$ in Eq. (23) for the Notched Three-Point Flexure Specimens (Data from Dunn et al. [1])

γ (degrees)	λ	c_1	c_2	c_2	c_4	c_5
0	0.5000	3.99812	−24.5978	84.3671	−129.951	77.8949
60	0.5122	4.45310	−27.9941	96.0504	−147.987	88.3623
90	0.5445	5.45245	−35.6525	122.909	−189.630	112.681
120	0.6157	7.89944	−55.2765	193.464	−300.322	177.660

Example 2: Fracture Analysis of V-Notch DENT Specimens Made of Plexiglas and Duraluminum

In 1994, fracture behavior of V-notch DENT specimens (Figure 4.5) made of Plexiglas and duraluminum was investigated by Seweryn [11]. Here, fracture analysis for the V-notch DENT specimens reported in Ref.[11] is performed by using the local stress field failure model proposed in this study. Two V-notches with notch angles $2\beta=20^{\circ}$, 100° are taken to be Notch 1 and Notch 2, respectively. For V-notches made of Plexiglas, details of fracture analysis are as follows:

(1) By using the fracture test data of the two V-notches, L_{c1} = 3.3 mm, L_{c2} = 4.4 mm and σ_0 = 13.0 Mpa are obtained from Eqs. (6), (7) and (11) (see Figure 4.6). Here, the values of the notch shape function of V-notch DENT specimens are taken from Ref.[12] (see also Table 4.3).
(2) By using the fracture test data of the two V-notches, the notch fracture toughness K_{vc} calculated from Eq. (15) is shown in Table 4.3. Note here that for the V-notch specimen with notch angle $2\beta=160^{\circ}$, relative error is −50.5%. The author considers that the fracture test data of the V-notch is perhaps unbelievable because the minimum size, 306 mm, of the specimen length from the notch specimen geometry shown in Figure 4.5 is more than the specimen length, 192 mm. After deleting the unusual test data, the calculation error indexes are: M.E. = −3.0%, M.A.E. = 4.3%, E.R. = [−18.4,4.1]%, from which it is seen that the calculated results of the notch fracture toughness are in good agreement with those measured experimentally.
(3) From Eq.(16), we have K_{Ic} = 1.857 MPam$^{0.5}$, which is very close to K_{vc} = 1.866 MPam$^{0.4996}$ at notch angle $2\beta = 20^{\circ}$.

For V-notches made of duraluminum, similar results to those shown in Table 4.3 are obtained (see Table 4.4), with the calculation error indexes: M.E. = 0.1%, M.A.E. = 2.5%, E.R. = [−8.9,3.8]%, which shows that the theoretical results are in excellent agreement with the experimental ones.

For duraluminum, K_{Ic} obtained from Eq.(16) is 53.28 MPam$^{0.5}$, which is very close to K_{vc} = 53.51 MPam$^{0.4996}$ at notch angle $2\beta = 20^{\circ}$.

Seweryn [11] stated "The Plexiglas fracture was of *brittle* character and no plastic strains were observed. Cracking surfaces were found to be perpendicular to

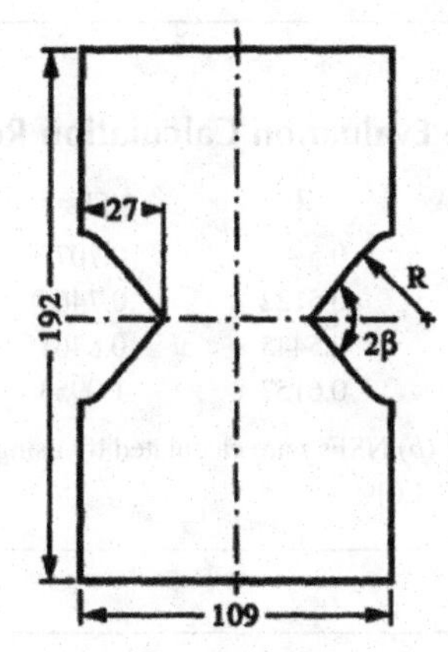

FIGURE 4.5 The geometric size of V-notch DENT specimens.

Source: (Data from Seweryn [11])

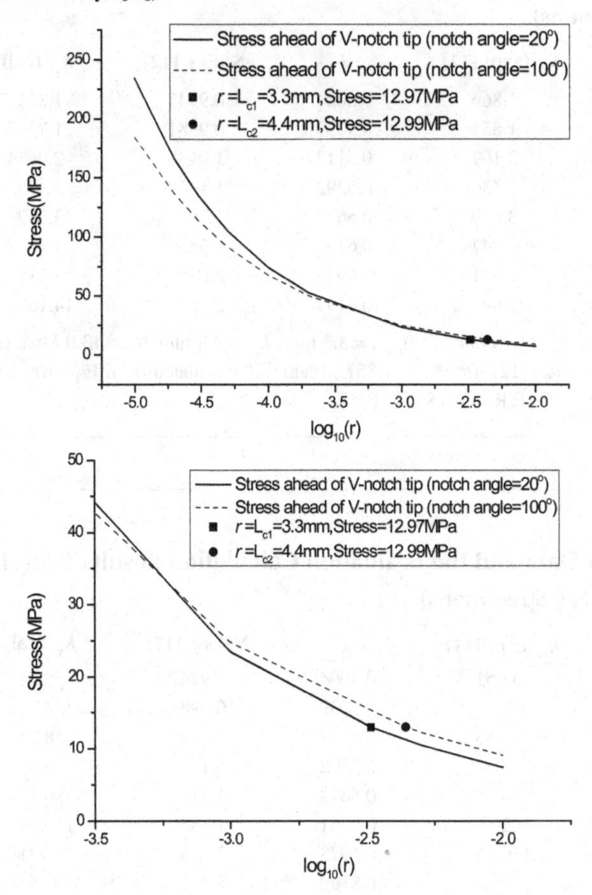

FIGURE 4.6 Stress ahead of V-notch under fracture load, and the critical stress and the critical distance determined by using the local stress field failure model: (a) values of r from 0.00001 m to 0.01 m; (b) values of r from 0.0002 m to 0.01 m.

TABLE 4.2
Fracture Test Data and the Evaluation Calculation Results

γ (degrees)	K_{vc} (exp) [1]	λ	NSFs f	K_{vc} (cal)	R.E.(%)
0	1.02	0.5	0.7078	1.02	0
60	1.44	0.5122	0.7465	1.20	−16.2
90	2.22	0.5445	0.8461	1.84	−17.1
120	4.78	0.6157	1.0953	4.53	−5.2

Note: (*a*) Unit of K_{vc} is $MPam^{1-\lambda}$; (*b*) NSFs f are calculated by using formula (23); (*c*) M.E. = −9.6%, M.A.E. = 9.6%, E.R. = [−17.1,0]%

TABLE 4.3
Fracture Test Data and the Evaluation Calculation Results (Plexiglas) (V-Notch DENT Specimens)

β (degrees)	K_{vc} (exp) [11]	λ	NSFs f [12]	K_{vc} (cal)	R.E.(%)
10	1.866	0.5004	0.9747	1.866	0
20	1.851	0.5039	0.9981	1.9263	4.1
30	2.167	0.5117	1.049	2.0634	−4.8
40	2.436	0.5292	1.146	2.375	−2.5
50	3.059	0.5642	1.311	3.059	0
60	4.347	0.6161	1.585	4.36	0.3
70	8.861	0.6978	2.055	7.235	−18.4
80	28.6	0.8199	2.936	14.166	−50.5

Note: (*a*) Unit of K_{vc} is $MPam^{1-\lambda}$; (*b*) L_{c1} = 3.3 mm, L_{c2} = 4.4 mm, σ_0 = 13.0 Mpa; (*c*) the NSFs f are taken from Ref.[12]; (*d*) K_{Ic} = 1.857 $MPam^{0.5}$ is calculated from Eq.(16); (*e*) M.E. = −3.0%, M.A.E. = 4.3%, E.R. = [−18.4,4.1]%

TABLE 4.4
Fracture Test Data and the Evaluation Calculation Results (Duraluminum) (V-Notch DENT Specimens)

β (degrees)	K_{vc}(exp) [11]	λ	NSFs f [12]	K_{vc} (cal)	R.E.(%)
10	53.51	0.5004	0.9747	53.51	0
20	57.1	0.5039	0.9981	54.97	3.7
30	60.53	0.5117	1.049	58.23	3.8
40	66.34	0.5292	1.146	65.38	1.5
50	80.15	0.5642	1.311	80.15	0
60	102	0.6161	1.585	106.16	−4.1
70	150.44	0.6978	2.055	156.91	−4.3
80	291.81	0.8199	2.936	258.5	11.4

Note: (*a*) Unit of K_{vc} is $MPam^{1-\lambda}$; (*b*) L_{c1} = 13.5 mm, L_{c2} = 18.1 mm, σ_0 = 183.7 Mpa; (*c*) K_{Ic} = 53.28 $MPam^{0.5}$ is calculated from Eq.(16); (*d*) M.E. = 0.1%, M.A.E. = 2.5%, E.R. = [−8.9.3.8]%

specimens' faces. On the contrary, in the duraluminum specimens visible *plastic* deformations were present (as the overall specimens' dimensions changed) and fracture surfaces were inclined at a 45° angle to specimens' faces".

From this study, thus, it is found that fracture analysis for the V-notch DENT specimens made of both *brittle* Plexiglas and *plastic* duraluminum can be performed by using the local stress field failure model proposed in this study based on the linear elastic stress field ahead of the V-notch.

Example 3 Fracture Analysis of V-Notched Samples of Annealed Tool Steel

In 2002, Strandberg tested V-notched samples of annealed tool steel, AISI 01, at -50°C. Single-edge notched tension and three-point bend specimens (see Figure 4.7), with notch angles ranging from 0° to 140°, were used.

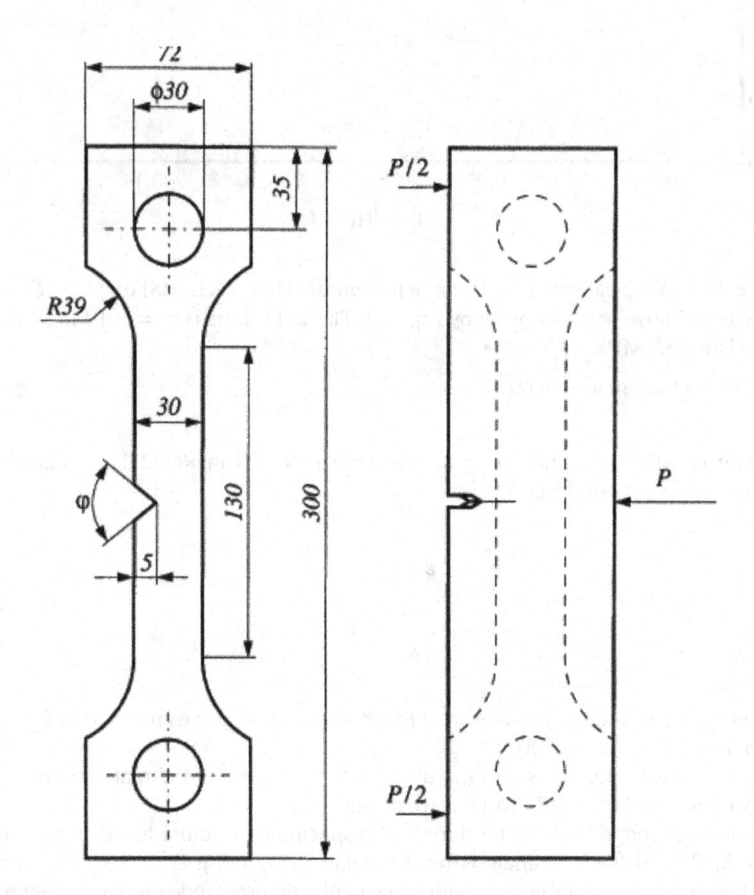

FIGURE 4.7 The dimensions of the SENT specimens. The right-hand figure shows how the SENT specimen with $\varphi = 0^\circ$ was manufactured. The thickness was 16.6 mm for all specimens.

Source: (Data from Strandberg [22]).

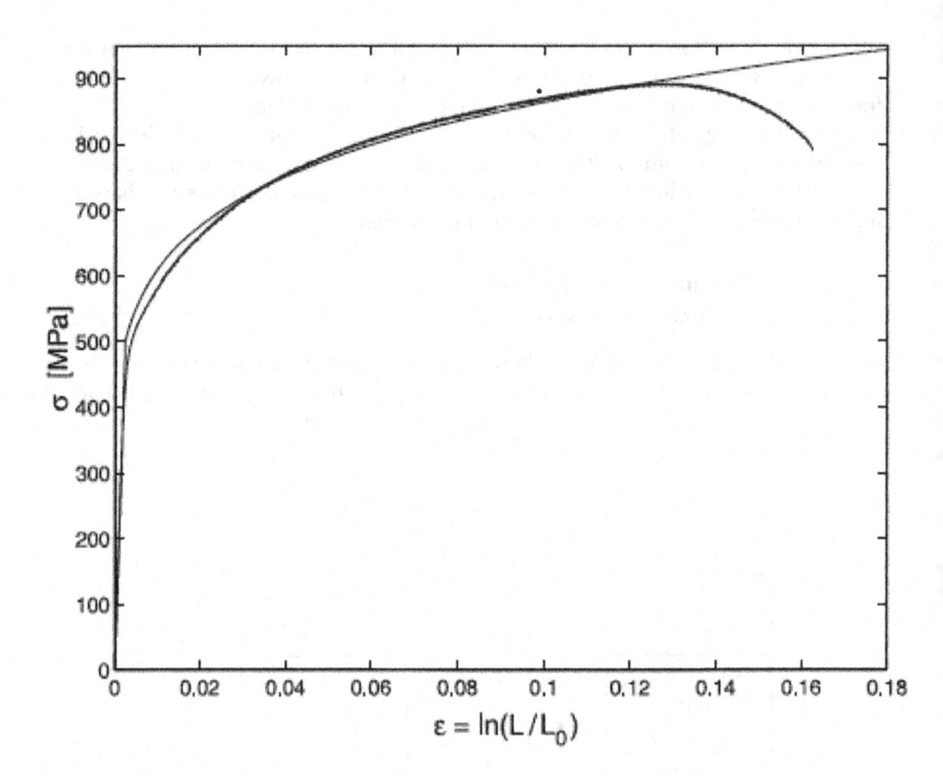

FIGURE 4.8 A typical stress-strain curve for annealed tool steel, AISI 01, at -50°*C* together with the constitutive behavior given by Eq. (24). The yield stress is $\sigma_y = 501$ MPa with standard deviation 5.7 MPa.

Source: (Data from Strandberg [22]).

The constitutive behavior of the material reported in Ref.[22], as shown in Figure 4.8, was well described by

$$\varepsilon = \begin{cases} \dfrac{\sigma}{E}, & \varepsilon \le \varepsilon_y \\ \dfrac{\sigma}{E} + \varepsilon_0\{(\dfrac{\sigma}{\varepsilon_y})^n - 1\}, & \varepsilon > \varepsilon_y \end{cases} \tag{24}$$

where ε_y is the strain at yielding. A fit to the stress-strain curves resulted in $E = 205$ GPa, $n = 6.6$ and $\varepsilon_0 = 0.0027$.

The fracture toughness and the ultimate tensile stress of the material are $K_{Ic} =$ 52 MPam$^{0.5}$ and $\sigma_t = 1173$ Mpa, respectively.

For Single-edge notched tension (SENT) specimens, fracture test data reported in Ref.[22] and fracture analysis results obtained by using Eq.(17) and by using Eq.(15) are shown in Table 4.5, with the calculation error indexes: M.E. = –9.6%, M.A.E. = 9.8%, E.R. = [–14.0,0.4]% and M.E. = 1.7%, M.A.E. = 4.2%, E.R. = [–5.0,16.0]%, respectively, which shows that the fracture analysis results obtained by using Eq.(15) are better than those by using Eq. (17).

TABLE 4.5
Fracture Test Data Reported in Ref.[22] and the Evaluation Calculation Results (Annealed Tool Steel, AISI 01, at -50°C) (SENT V-Notch Specimens)

Notch angle φ (degrees)	K_{vc} (exp) [22]	λ	NSFs f [7]	K_{vc} (cal 1)	R.E.1 (%)	K_{vc} (cal2)	R.E.2 (%)
0	51.8	0.5	1.085	52	0.4	60.07	16
30	61.1	0.5015	1.097	52.92	−13.4	61.1	0
60	67.3	0.5122	1.169	59.51	−11.6	68.38	1.6
90	96.2	0.5445	1.366	82.71	−14	93.7	−2.6
120	176.6	0.6157	1.804	160.83	−8.9	176.6	0
140	350.2	0.6972	2.150	314.11	−10.3	332.79	−5

Note: (*a*) Unit of K_{vc} is $MPam^{1-\lambda}$; (*b*) K_{vc} (cal1) is calculated by using Eq.(17); (*c*) K_{vc} (cal2) is calculated by using Eq.(15); (*d*) $R.E.1 = \dfrac{(K_{vc}(cal1) - K_{vc}(\text{exp}))}{K_{vc}(\text{exp})} \times 100$; (*e*) $R.E.2 = \dfrac{(K_{vc}(cal2) - K_{vc}(\text{exp}))}{K_{vc}(\text{exp})} \times 100$;
(*f*) The error indexes of R.E.1 are: M.E. = −9.6%, M.A.E. = 9.8%, E.R. = [−14.0,0.4]%; (*g*) The error indexes of R.E.2 are: M.E. = 1.7%, M.A.E. = 4.2%, E.R. = [−5.0,16.0]%.

TABLE 4.6
Fracture Test Data and the Evaluation Calculation Results (Annealed Tool Steel, AISI 01, at -50°C) (TPB V-Notch Specimens)

Notch angle φ (degrees)	K_{vc}(exp) [22]	λ	NSFs f [7]	K_{vc} (pre)	R.E. (%)
0	52	0.5	1.767	52	0.0
90	94.7	0.5445	2.118	80.87	−14.6

Note: Unit of K_{vc} is $MPam^{1-\lambda}$.

For TPB V-notch specimens, fracture test data reported in Ref.[22] and fracture analysis results obtained by using Eq. (17) are shown in Table 4.6, from which it is found that the theoretical results are in good agreement with the experimental ones.

By the way, it is pointed out that, in fracture analysis of Single-edge notched tension specimens performed by using the local stress field failure model proposed in this study, two V-notches with notch angles $\varphi = 30^\circ$ and 120° are taken as notch 1 and notch 2, respectively.

4.5 CONCLUSIONS AND FINAL COMMENTS

From this study, the following conclusion can be made:

By using the concepts of the critical stress and the critical distance by Taylor [17], and by using the assumption that the critical distances of different notches are different and are connected closely with the variation of notched geometry, a local stress field failure model was proposed based on the linear elastic stress field ahead of the V-notch. The local stress field failure model has been proven to be both simple in computation form and highly accurate for fracture analysis of V-notched components made of both *brittle* materials (Plexiglas and PMMA) and *plastic* materials (duraluminum and annealed tool steel, AISI 01, at -50°C).

Final comments given here are as follows:

(A) Why can the fracture analysis for V-notch specimens made of *plastic* materials (duraluminum and annealed tool steel, AISI 01, at -50°C) be well performed by using the local stress field failure model proposed in this study? On this problem, Susmel and Taylor [25] said "it is much more difficult to understand why the **linear-elastic TCD** is so accurate also in estimating static strength in those situations in which notched materials undergo large-scale plastic deformations before final breakage occurs". Here, the following comments are given:

(1) The explanation on "**A practicability of establishing the unified prediction equation for a low/medium/high cycle fatigue of metallic materials",** which occurred in Ref.[26] (also see the Appendix *A*), is given to perhaps help the reader understand that it is practical that the fracture analysis for V-notch specimens made of *plastic* materials can be performed by using the local stress field failure model. Further, low-cycle test data of many metallic materials such as 16MnR steel [23,24], EN AW-2024-T3 aluminum alloy [34], A533B [35], Inconel 718 [36], SAE1045 [37], AISI304 [38], Ti-6Al-4V [39], 6061-T6 [40], 1Cr-18Ni-9Ti [41], AISI304 [42], AISIH11 [43], 34CrNiMo6 [44], RAFM steels [45], P92 ferritic-martensitic steel [46], Cr–Mo–V low alloy steel [47], S35C carbon steel and SCM 435 alloy steel [48,49], high-strength spring steel with different heat treatments [50], 316 L(N) stainless steel at room temperature [51], CLAM steel at room temperature [52]), Eurofer97 [53], F82H (a ferritic-martensitic steel) [54], En3B (a commercial cold-rolled low-carbon steel) [55], nodular cast iron [56] and low carbon gray cast iron [57] collected in Ref.[27] have proven that it is sure enough that the low-cycle fatigue life analysis of metallic materials is well performed by the well-known Wöhler Curve Method.

(2) The fracture analysis of notched components made of En3B performed by Susmel and Taylor [29] showed that the linear-elastic TCD can be successful in predicting static failures in notched components when the final breakage is preceded by large-scale plastic deformations.

(3) The fact that "A similar degree of accuracy was obtained when elasto-plastic stress analysis was used" reported by Susmel and Taylor [29] showed also that the local stress field failure model proposed in this study is used to perform the fracture analysis of notched components made of *plastic* materials.

(4) The author considers the fact that "A similar degree of accuracy was obtained when elasto-plastic stress analysis was used" reported by Susmel and Taylor [29] relies on the **stress intensity parameter** used

in the TCD being a "stress" parameter, not a "strain" or "energy" parameter. In Appendix *A*, the author stated: at a given loading, the "*stress quantity*" calculated using the *elastic-plastic stress analysis* almost is the same as that by the *linear-elastic stress analysis*. From a practical point of view, thus, for the LCF of metallic materials, the "*stress quantity*" *can be calculated using the linear-elastic stress analysis of the mechanical component.*

(5) The fact that "A similar degree of accuracy was obtained when elasto-plastic stress analysis was used" reported by Susmel and Taylor [29], in fact, can be a powerful verification for why the low-cycle fatigue life analysis of metallic materials is well performed by the well-known Wöhler Curve Method reported in Ref.[27].

(B) In the local approach, the assumption that the critical distances of different notches are different and are connected closely with the variation of notched geometry originated from a finding by Yang et al.[30] that, at a given fatigue life N_f, the products of the critical distances of different notched components determined by two fatigue failure curves of a smooth specimen and notched specimen, and their corresponding linear elastic stress concentration factors, k_t, are almost equal. Yang's idea was successfully used in the local stress field failure model of sharp notches under Mode I loading reported in Ref.[28]. For the pointed V-notches, from this study, the assumption (11) has been verified to be applicable. This applicability has been verified further in Ref.[28] for both Mode I and Mode III V-notches.

(C) From this study, the V-notch fracture toughness, denoted by K_{IVc}, is expressed mathematically to be

$$K_{IVc} = H_I(K_{Ic}, \sigma_o, \alpha) \tag{25}$$

For the V-notches under Mode III loading, similarly, the fracture toughness, denoted by K_{IIIVc}, is expressed mathematically to be

$$K_{IIIVc} = H_{III}(K_{IIIc}, \tau_o, \alpha) \tag{26}$$

For the V-notches under Mode I/III loading, the failure condition can be expressed in terms of an empirical failure equation similar to that employed in Refs [26,31,33]

$$S(K_{IV}, K_{IIIV}, K_{IVc}, K_{IIIVc}) = 0 \tag{27}$$

From Eqs. (25) and (26), Eq.(27) can be expressed as:

$$J(K_{IV}, K_{IIIV}, K_{Ic}, K_{IIIc}, \sigma_o, \tau_o, \alpha) = 0 \tag{28}$$

According to the empirical failure equation (27) similar to that employed in Refs [26,31,33], the failure condition (27) for Mode I and Mode III loading is simplified respectively as:

$$K_{IV} = K_{IVc} \tag{29}$$

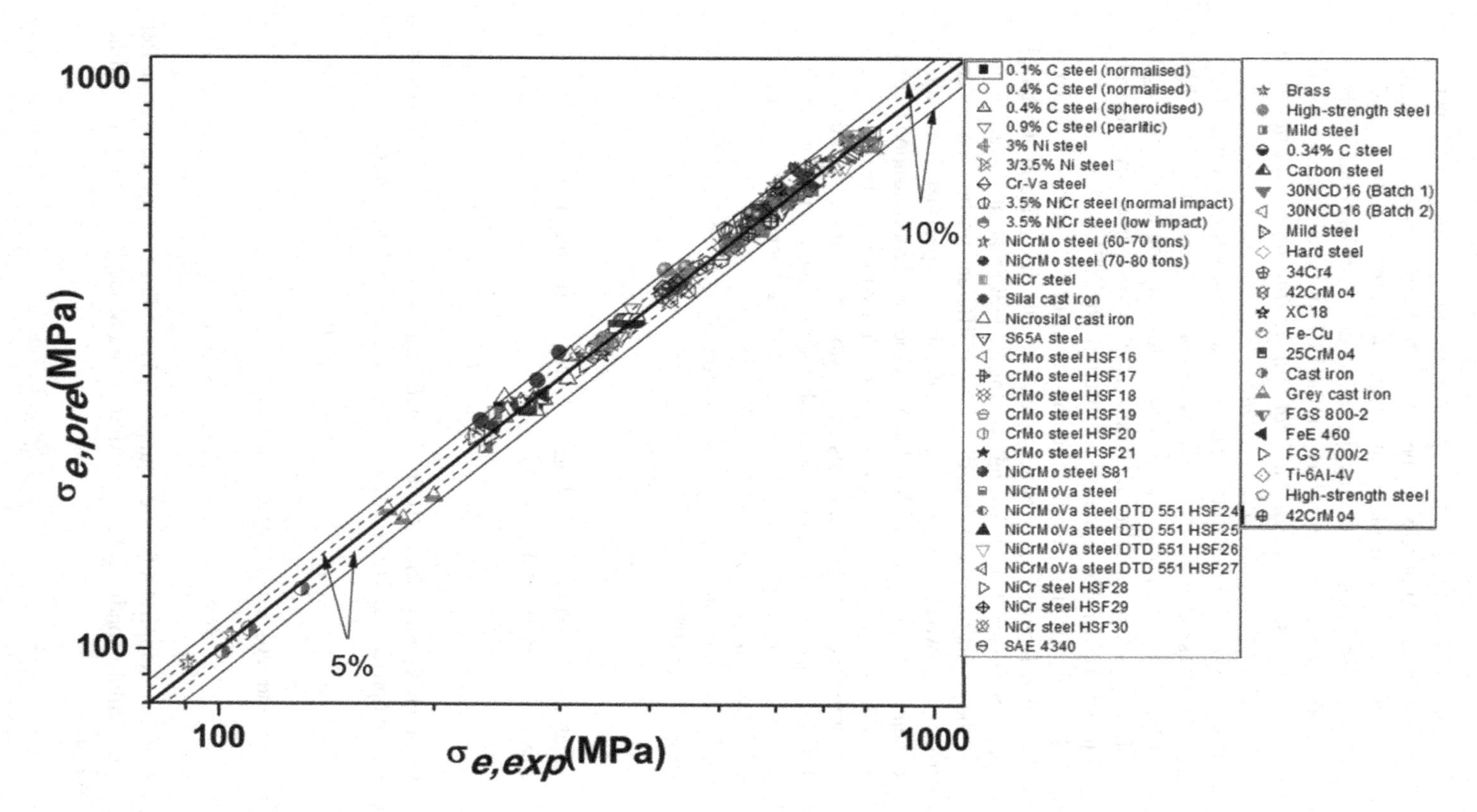

FIGURE 4.9 Comparison of experimental and predicted results of multiaxial fatigue limits.

and

$$K_{IIIV} = K_{IIIVc} \tag{30}$$

Obviously, Eq. (29) and Eq. (30) can be regarded as an extension of K criterion for cracks to notches.

(D) In our investigation subject on material failure equation at the multiaxial stresses state, including cracked or notched materials, the failure condition employed was an empirical failure equation that was originated from the multiaxial fatigue limit equation proposed by Liu and Yan [31] by using the well-known Wöhler equation and the multiaxial fatigue life equation by Liu and Yan [32]. By using 53 sets of experimental data from the literature (see Figure 4.9), experimental verifications given in Ref.[31] showed that the empirical failure equation is both simple in computation and high in accuracy falling within the ±10% interval. This accuracy was so exciting that the empirical failure equation was extended to the static failure model at multiaxial stresses state (see Refs [26,33]).

REFERENCES

1 Dunn, M.L., Suwtto, W., Cunningham, S. Fracture initiation at sharp notches: Correlation using critical stress intensities. *Int. J. Solids Structures* 34(29), 3873–3883 (1997).

2 Williams, M.L. Stress singularities resulting from various boundary conditions in angular corners of plates in extension. *Journal of Applied Mechanics* 74, 526–528 (1952).

3 England, A.H. On stress singularities in linear elasticity. *International Journal Engineering Science* 9, 571–585 (1971).

4 Seweryn, A., Molski, K. Elastic stress singularities and corresponding generalized stress intensity factors for angular corners under various boundary conditions. *Engineering Fracture Mechanics* 55, 529–556 (1996).

5 Lazzarin, P., Tovo, R. A unified approach to the evaluation of linear elastic stress fields in the neighbourhood of cracks and notches. *International Journal of Fracture* 78, 3–19 (1996).

6 Filippi, S., Lazzarin, P. and Tovo, R. Developments of some explicit formulas useful to describe elastic stress fields ahead of notches in plates. *International Journal of Solids and Structures* 39, 4543–4565 (2002).

7 Gross, B., Mendelson, A. Plane elastostatic analysis of V-notched plates. *International Journal of Fracture* 8, 267–276 (1972).

8 Knésl, Z. A criterion of V-notch stability. *International Journal of Fracture* 48, R79–R83 (1991).

9 Grenestedt, J.L., Hallström, S., Kuttenkeuler, J. On cracks emanating from wedges in expanded PVC foam. *Engineering Fracture Mechanics* 54, 445–456 (1996).

10 Strandberg, M. A numerical study of the elastic stress field arising from sharp and blunt V-notches in a SENT-specimen. *International Journal of Fracture* 100, 329–342 (1999).

11 Seweryn, A. Brittle fracture criterion for structures with sharp notches. *Engineering Fracture Mechanics* 47, 673–681 (1994).

12 Gómez, F.J., Elices, M. A fracture criterion for sharp V-notched samples. *International Journal of Fracture* 123, 163–175 (2003).

13 Carpinteri, A. Stress-singularity and generalized fracture toughness at the vertex of re-entrant corners. *Engineering Fracture Mechanics* 26, 143–155 (1987).

14 Neuber, H. Zur Theorie der technischen Formzahl. *Forschg Ing-Wes* 7, 271–281 (1936).
15 Neuber, H. *Theory of notch stresses: Principles for exact calculation of strength with reference to structural form and material.* 2nd ed. Berlin: Springer Verlag; 1958.
16 Peterson, R.E. Notch sensitivity. In: Sines, G., Waisman, J.L., editors. *Metal fatigue.* New York, USA: McGraw Hill; 1959. p. 293–306.
17 Taylor, D. *The theory of critical distances: A new perspective in fracture mechanics.* Oxford, UK: Elsevier; 2007.
18 Susmel, L. *Multiaxial notch fatigue: From nominal to local stress/strain quantities.* Cambridge: Woodhead & CRC; 2009.
19 Berto, F., Lazzarin, P. Recent developments in brittle and quasi-brittle failure assessment of engineering materials by means of local approaches. *Materials Science and Engineering R* 75, 1–48 (2014).
20 Lazzarin, P., Zambardi, R. A finite-volume-energy based approach to predict the static and fatigue behaviour of components with sharp V-shaped notches. *International Journal of Fracture* 112, 275–298 (2001).
21 Lazzarin, P., Berto, F. Some expressions for the strain energy in a finite volume surrounding the root of blunt V-notches. *International Journal of Fracture* 135, 161–185 (2005).
22 Strandberg, M. Fracture at V-notches with contained plasticity. *Engineering Fracture Mechanics* 69, 403–415 (2002).
23 Gao, Z., Zhao, T., Wang, X., Jiang, Y. Multiaxial fatigue of 16MnR steel. *ASME J Press Vess Technol* 131(2), 021403 (2009).
24 Gao Z., Qiu, B., Wang, X., Jiang, Y. An investigation of fatigue of a notched member. *International Journal of Fatigue* 32, 1960–1969 (2010).
25 Susmel, L., Taylor, D. The theory of critical distances to estimate the static strength of notched samples of Al6082 loaded in combined tension and torsion. Part II: Multiaxial static assessment. *Engineering Fracture Mechanics* 77, 470–478 (2010).
26 Yan, X.Q. An empirical fracture equation of mixed mode cracks. *Theoretical and Applied Fracture Mechanics* 116, 103146 (2021).
27 Yan, X.Q. Applicability of the Wöhler Curve Method for a low/medium/high cycle fatigue of metallic materials. Submitted to *Application in Science and Engineering* for publication.
28 Yan, X.Q. A local stress field failure model of sharp notches. To be Published in *Engineering Fracture Mechanics.*
29 Susmel, L., Taylor, D. On the use of the theory of critical distances to predict static failures in ductile metallic materials containing different geometrical features. *Engineering Fracture Mechanics* 75, 4410–4421 (2008).
30 Yang, X.G. Wang, J.K., Liu, J. L. High temperature LCF life prediction of notched DS Ni-based superalloy using critical distance concept. *International Journal of Fatigue* 33, 1470–1476 (2011).
31 Liu, B.W., Yan, X.Q. A multiaxial fatigue limit prediction equation for metallic materials. *ASME Journal of Pressure Vessel Technology* 142, 034501 (2020).
32 Liu, B.W., Yan, X.Q. A new model of multiaxial fatigue life prediction with influence of different mean stresses. *Int. J. Damage Mech.* 28(9), 1323–1343 (2019).
33 Szusta, J., Seweryn, A. Fatigue damage accumulation modelling in the range of complex low-cycle loadings: The strain approach and its experimental verification on the basis of EN AW-2007 aluminum alloy. *Int J Fatigue* 33, 255–264 (2011).
34 Szusta, J., Seweryn, A. Experimental study of the low-cycle fatigue life under multiaxial loading of aluminum alloy EN AW-2024-T3 at elevated temperatures. *Int J Fatigue* 33, 255–264 (2011).

35 Nelson, D.V., Rostami, A. Biaxial fatigue of A533B pressure vessel steel. *Transactions of the ASME, Journal of Pressure Vessel Technology* 119, 325–331 (1997). doi: 10.1115/1.2842312
36 Socie, D.F., Kurath, P., Koch, J. A multiaxial fatigue damage parameter. In: Brown, M.W., Miller, K.J., editors. *Biaxial and multiaxial fatigue*, EGF 3. London: Mechanical Engineering Publications; 1989. p. 535–550.
37 Kurath, P., Downing, S.D., Galliart, D.R. Summary of non-hardened notched shaft: Round robin program. In: Leese, G.E., Socie, D.F., editors. *Multiaxial fatigue: Analysis and experiments*, SAE AE-14. Warrendale, PA: Society of Automotive Engineers; 1989. p. 13–32.
38 Socie, D.F. Multiaxial fatigue damage models. *Transactions of the ASME, Journal of Engineering Materials and Technology* 109, 293–298 (1987).
39 Kallmeyer, A.R., Krgo, A., Kurath, P. Evaluation of multiaxial fatigue life prediction methodologies for Ti–6Al–4V. *Transactions of the ASME, Journal of Engineering Materials and Technology* 124, 229–237 (2002). doi: 10.1115/1.1446075
40 Lin, H., Nayeb-Hashemi, H., Pelloux, R.M. Constitutive relations and fatigue life prediction for anisotropic Al–6061–T6 rods under biaxial proportional loadings. *International Journal of Fatigue* 14, 249–259 (1992). doi: 10.1016/0142-1123(92)90009-2
41 Chen, X., An, K., Kim, K.S. Low-cycle fatigue of 1Cr–18Ni–9Ti stainless steel and related weld metal under axial, torsional and 90 out-of-phase-loading. *Fatigue and Fracture of Engineering Materials and Structures* 27, 439–448 (2004). doi: 10.1111/j.1460-2695.2004.00740.x
42 Itoh, T., Sakane, M., Ohnami, M., Socie, D.F. Nonproportional low cycle fatigue criterion for type 304 stainless steel. *Transactions of the ASME, Journal of Engineering Materials and Technology* 117, 285–292 (1995).
43 Du, W., Luo, Y., Wang, Y., Chen, S., Yu, D. A new energy-based method to evaluate low-cycle fatigue damage of AISI H11 at elevated temperature. *Fatigue Fract Engng Mater Struct* 40, 994–1004 (2017).
44 Branco, R., Costa, J.D., Antunes, F.V. Low-cycle fatigue behaviour of 34CrNiMo6 high strength steel. *Theoretical and Applied Fracture Mechanics* 58, 28–34 (2012).
45 Shankar, V., Mariappan, K., Nagesha, A., Prasad Reddy, G.V., Sandhya, R., Mathew, M.D., Jayakumar, T. Effect of tungsten and tantalum on the low cycle fatigue behavior of reduced activation ferritic/martensitic steels. *Fusion Engineering and Design* 87, 318–324 (2012).
46 Zhang, Z., Hu, Z.F., Schmauder, S., Zhang, B.S., Wang, Z.Z. Low cycle fatigue properties and microstructure of P92 ferritic-martensitic steel at room temperature and 873 K. *Materials Characterization* 157, 109923 (2019).
47 Li, Z.Q., Han, J.M., Li, W.J., Pan, L.K. Low cycle fatigue behavior of Cr–Mo–V low alloy steel used for railway brake discs. *Materials and Design* 56, 146–157 (2014).
48 Hatanaka, K. Cyclic stress-strain responds and low-cycle fatigue life in metallic materials. *JSME, International Journal*, Series 1 33(1), 13–25 (1990).
49 Hatanaka, K., Fujimisu, T. The cyclic stress-strain response and strain life behavior of metallic materials. *Proc. Of Fatigue '84* 1, 93 (1984).
50 Li, D.M., Kim, K.W., Lee, C.S. Low cycle fatigue data evaluation for a high strength spring steel. *Int. J. Fatigue* 19(8–9), 607–612 (1997).
51 Roy, S.C., Goyal, S., Sandhy, R., Ray, S.K. Low cycle fatigue life prediction of 316 L(N) stainless steel based on cyclic elasto-plastic response. *Nuclear Engineering and Design* 253, 219–225 (2012).
52 Hu, X., Huang, L.X., Wang, W.G., Yang, Z.G., Sha, W., Wang, W., Yan, W., Shan, Y.Y. Low cycle fatigue properties of CLAM steel at room temperature. *Fusion Engineering and Design* 88, 3050–3059 (2013).

53 Marmy, P., Kruml, T. Low cycle fatigue of Eurofer 97. *Journal of Nuclear Materials* 377, 52–58 (2008).
54 Stubbins, J.F., Gelles, D.S. Fatigue performance and cyclic softening of F82H, a ferritic-martensitic steel. *Journal of Nuclear Materials* 233–237, 331–335 (1996).
55 Atzori, B., Meneghetti, G., Susmel, L., Taylor, D. The modified Manson–Coffin method to estimate low-cycle fatigue damage in notched cylindrical bars. In: U. S. Fernando, editor. Proceedings of the 8th International Conference on Multiaxial Fatigue and Fracture. Sheffield, UK, 2007 July.
56 Šamec, B., Potrcˇ, I., Šraml, M. Low cycle fatigue of nodular cast iron used for railway brake discs. *Engineering Failure Analysis* 18, 1424–1434 (2011).
57 Pevec, M., Oder, G., Potrcˇ, I., Šraml, M. Elevated temperature low cycle fatigue of grey cast iron used for automotive brake discs. *Engineering Failure Analysis* 42, 221–230 (2014).

APPENDIX A: THE PRACTICABILITY OF ESTABLISHING THE UNIFIED PREDICTION EQUATION FOR A LOW/MEDIUM/ HIGH CYCLE FATIGUE OF METALLIC MATERIALS

(Please See Appendix *A* in Chapter 1)

5 A Local Stress Field Failure Model for Sharp Notches

5.1 INTRODUCTION

V-notches, which are widely used in mechanical components such as bolts, nuts and screws, decrease dramatically the load-bearing capacity of components due to the concentration of stress at the vicinity of their tips. A reliable prediction of the mechanical failure like crack formation and growth in the vicinity of V-notches has been a topic of great interest to researchers.

Since the beginning of the last century by Neuber [1, 2] and by Peterson [3], a lot of advance of local approaches to deal with brittle failure of notched components has been made, in particular, the CD approach [4,5] (often called TCD, Theory of Critical Distance) and the volume-based SED (strain energy density) approach [6–8]. In the two well-known local approaches, the size of Neuber's elementary volume is taken to be a material constant. For brittle materials (such as ceramics [9]), for example, the critical distance according to Taylor's TCD is determined by

$$L = \frac{1}{\pi}\left(\frac{K_{Ic}}{\sigma_0}\right)^2 \tag{1}$$

where K_{Ic} is the plane strain fracture toughness and σ_0 is taken to be the ultimate tensile stress, σ_{UTS}. Whereas when static failures are preceded by a certain amount of plasticity, σ_0 takes a value which is higher than the plain material strength and both parameters in TCD (i.e., the critical stress σ_0 and the critical distance L) can be determined only carrying out ad hoc experimental investigations [10,11]. For example, the critical stress and the critical distance of Perspex are the values of the point of intersection of stress-distance curves for the sharp notch and blunt notch (3 mm hole) at the failure load, as shown in Figure 5.1. According to Taylor's PM (point method), thus, failure is predicted to occur when the critical stress σ_0 is present at a critical distance $r = r_c = \frac{L}{2}$ from the notch root (see Figure 5.2). This failure condition is expressed as:

$$\sigma_\theta(r = r_c, \theta = 0) = \sigma_0 \tag{2}$$

where σ_θ is the circumference stress, and the coordinate r is the distance from notch tip.

DOI: 10.1201/9781003356721-5

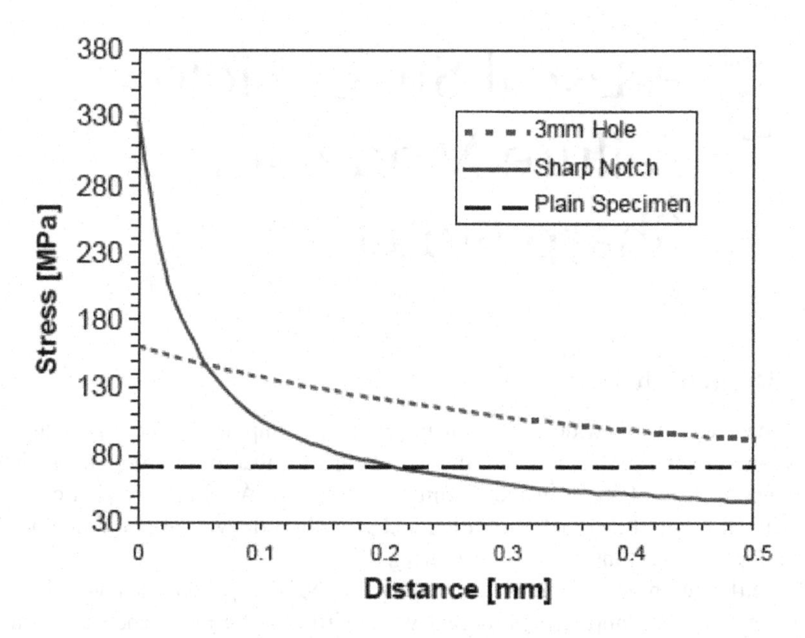

FIGURE 5.1 Stress-distance curves at the failure load for the sharp notch, 3 mm hole and plain specimen, in Perspex.

Source: (Data from Taylor, *et al.* [10])

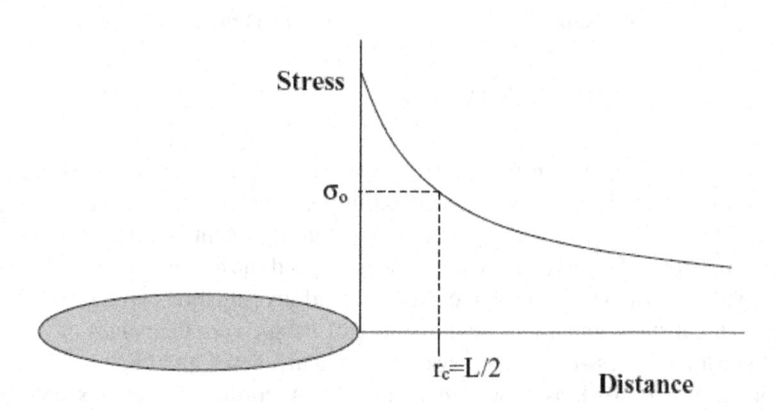

FIGURE 5.2 The theory of critical distances—the point method: failure is predicted to occur when the critical stress σ_0 is present at the critical distance $L/2$ from the notch root.

Source: (Data from Taylor, *et al.* [10]).

Here, it is worthy to mention the fatigue life prediction of notched specimens reported in Refs [12–15]. In 2010, using the concept of the TCD and on the basis of notch elastic-plastic stress analysis, the Manson-Coffin equation together with SW T parameter was employed by Susmel and Taylor [12] to perform LCF assessment of notched components. In 2011, using the approach by Susmel and Taylor [12], Yang et al. [15] performed high-temperature LCF life prediction of notched DS Ni-based superalloy. Yang et al. [15] found that, using two calibration fatigue failure curves of a smooth specimen and a notched specimen, the critical distance obtained is different rather than taking a determinate value at a given fatigue life N_f. Thus it could be speculated that the equation between the critical distance and fatigue life N_f obtained according to the approach by Susmel and Taylor [13,14],

$$L(N_f) = A \cdot N_f^B \tag{3}$$

together with the plain specimen fatigue life equation may not be applied to other notch types quite well. At the same time, Yang et al. [15] found also that, at a given fatigue life N_f, the products of the critical distances of different notched components determined by two fatigue failure curves of a smooth specimen and notched specimen, and their corresponding linear elastic stress concentration factors, k_t, are almost equal. Thus it could be speculated that the equation between the product of the critical distance and k_t and fatigue life N_f

$$k_t \cdot L(N_f) = A' \cdot N_f^{B'} \tag{4}$$

together with the plain specimen fatigue life equation may be applied to other notch types quite well. Fatigue fracture test data given by Yang et al. [15] proved that prediction results obtained by means of the equation (4) together with the plain specimen fatigue life equation are much better than those by the equation (3) together with the plain specimen fatigue life equation.

From the study of Taylor and his coworkers [9–14] and Yang et al. [15], the aim of this investigation is to propose a new local approach to prompt the accuracy of the failure assessment of notched components by means of the two parameters failure equation (2). In the present local approach, the size of the critical distance is not taken to be a material constant, whereas it is assumed that the critical distances of different notches are different and are connected closely with the variation of notched geometry.

The contents of this study are outlined as follows:

Part 1

2 A brief description of local stress field ahead of rounded V-notches
3 A local stress field failure model
4 A concept of the stress concentration factor eigenvalue k^*
5 Experimental verifications
6 Conclusions of Part 1

Part 2

7 Effect of notch angle on k^*
8 Effect of notch depth on k^*
9 Effect of different materials on k^*
10 Comments of Part 2

PART 1

This part is a basis of the local stress field failure model proposed in this study.

On the basis of the linear elastic notch tip stress fields, in fact, a local approach for fracture analysis of notched specimens is proposed in this study. In part 1, a brief description of the linear elastic notch tip stress fields are given first. Then basic equations involved in the local approach are proposed in Section 3. In the whole local stress field failure model proposed in this study, an important concept, the stress concentration factor eigenvalue (denoted by k^*), is introduced in Section 4 and at the same time, an approach of determining k^* is proposed in Section 4. Basic experimental verifications on the local stress field failure model proposed in this study are given in Section 5. At the end of Part 1, comments of Part 1 are given in Section 6.

5.2 A BRIEF DESCRIPTION OF LOCAL STRESS FIELD AHEAD OF ROUNDED V-NOTCHES

An analytical approach for describing the stress field ahead of rounded V-notches was proposed by Filippi et al. [16]. With reference to the coordinate system shown in Figure 5.3, Mode I stresses are [16]:

$$\sigma_{ij} = a_1 r^{\lambda_1 - 1}[f_{ij}(\theta,\alpha) + (\frac{r}{r_0})^{\mu_1 - \lambda_1} g_{ij}(\theta,\alpha)] \tag{5}$$

where $\lambda_1 > \mu_1$ and the parameter a_1 can be expressed by means of the elastic maximum notch stress σ_{tip} in the case of blunt V-notches. In Eq. (5), r_0 is the distance evaluated on the notch bisector line between the V-notch tip and the origin of the local coordinate system; r_0 depends both on the notch root radius ρ and the opening angle 2α (see Figure 5.3), according to the expression

$$r_0 = \frac{\pi - 2\alpha}{2\pi - 2\alpha}\rho$$

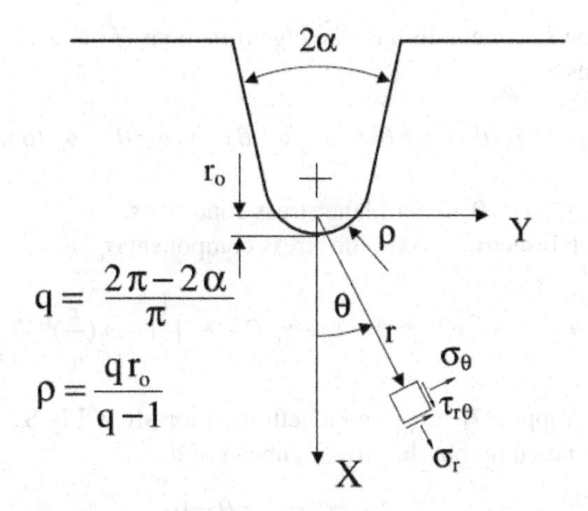

FIGURE 5.3 Coordinate system and symbols used for the stress field components.

The angular functions f_{ij} and g_{ij} are given in Ref. [16]:

$$\begin{Bmatrix} f_{\theta\theta} \\ f_{rr} \\ f_{r\theta} \end{Bmatrix} = \frac{1}{1+\lambda_1+\chi_{b1}(1-\lambda_1)} \left(\begin{Bmatrix} (1+\lambda_1)\cos(1-\lambda_1)\theta \\ (3-\lambda_1)\cos(1-\lambda_1)\theta \\ (1-\lambda_1)\sin(1-\lambda_1)\theta \end{Bmatrix} + \chi_{c1}(1-\lambda_1) \begin{Bmatrix} \cos(1+\lambda_1)\theta \\ -\cos(1+\lambda_1)\theta \\ \sin(1+\lambda_1)\theta \end{Bmatrix} \right) \tag{6}$$

and

$$\begin{Bmatrix} g_{\theta\theta} \\ g_{rr} \\ g_{r\theta} \end{Bmatrix} = \frac{q}{4(q-1)[1+\lambda_1+\chi_{b1}(1-\lambda_1)]} \left(\chi_{d1} \begin{Bmatrix} (1+\mu_1)\cos(1-\mu_1)\theta \\ (3-\mu_1)\cos(1-\mu_1)\theta \\ (1-\mu_1)\sin(1-\mu_1)\theta \end{Bmatrix} + \chi_{c1} \begin{Bmatrix} \cos(1+\mu_1)\theta \\ -\cos(1+\mu_1)\theta \\ \sin(1+\mu_1)\theta \end{Bmatrix} \right) \tag{7}$$

The eigenfunctions f_{ij} depend only on Williams' eigenvalue, λ_1, which controls the sharp solution for zero notch radius. The eigenfunctions g_{ij} mainly depend on eigenvalue μ_1, but are not independent from λ_1. Since $\lambda_1 > \mu_1$, the contribution of μ-based terms in Eq. (7) rapidly decreases with the increase of the distance from the notch tip. All parameters in Eqs. (5)–(7) have closed-form expressions (see Ref.[16]).

Under the plane strain conditions, the eigenfunctions f_{ij} and g_{ij} satisfy the following expressions:

$$f_{zz}(\theta)=\nu(f_{rr}(\theta)+f_{\theta\theta}(\theta)), \qquad g_{zz}(\theta)=\nu(g_{rr}(\theta)+g_{\theta\theta}(\theta)) \tag{8}$$

whereas $f_{zz}(\theta)=g_{zz}(\theta)=0$ under plane stress conditions.

Along the notch bisector ($\theta = 0$), the stress component $\sigma_{\theta\theta}$ is

$$\sigma_{\theta\theta}(r,\theta=0)=\lambda_1 r^{\lambda_1-1}a_1[1+\lambda_1+\chi_{b1}(1-\lambda_1)]\left\{1+\omega_1(\frac{r}{r_0})^{\mu_1-\lambda_1}\right\} \tag{9}$$

Lazzarin and Filippi [17] suggested a definition for Mode I N-SIFs involving not the stresses at the notch tip but the stresses ahead of it

$$K_{\nu\rho}=\sqrt{2\pi}r^{1-\lambda_1}\frac{\sigma_{\theta\theta}(r,\theta=0)}{1+\omega_1(\frac{r}{r_0})^{\mu_1-\lambda_1}} \tag{10}$$

which is usually called a generalized notch stress intensity factor (G-N-SIFs). Using this parameter, one has [18] (Note: in the following, for brevity, $\sigma_{\theta\theta}$, λ_1 and μ_1 are denoted by σ_θ, λ and μ, respectively.)

$$\sigma_\theta(r,0)=\frac{K_{\nu\rho}}{\sqrt{2\pi}r^{1-\lambda}}[1+(\frac{r_0}{r})^{\lambda-\mu}\frac{A(\alpha)}{1+B(\alpha)}] \tag{11}$$

From (11), one has [18],

$$\sigma_\theta(r_0,0)=\frac{K_{\nu\rho}}{\sqrt{2\pi}r_0^{1-\lambda}}[1+\frac{A(\alpha)}{1+B(\alpha)}] \tag{12}$$

and

$$\begin{aligned} K_{\nu\rho}&=\sqrt{2\pi}r_0^{1-\lambda}\sigma_\theta(r_0,0)[\frac{1+B(\alpha)}{1+B(\alpha)+A(\alpha)}] \\ &=\sqrt{2\pi}r_0^{1-\lambda}\sigma_\theta(r_0,0)f(\alpha) \end{aligned} \tag{13a}$$

or [16,17]

$$K_{\nu\rho}=\sqrt{2\pi}\sigma_{\max}\frac{r_0^{1-\lambda}}{1+\omega_1}=\sigma_{\max}\frac{\sqrt{2\pi}}{1+\omega_1}(\frac{q-1}{q}\rho)^{1-\lambda} \tag{13b}$$

Table 5.1 shows values of $f(\alpha)$ and related parameters (λ, μ, A and B) for different notch angles. A linear elastic calculation will provide us with the value of $\sigma_\theta(r_0,0)$ at the rounded notch tip, as a function of the external load. In some cases, knowledge of the stress concentration factor [19–28] may expedite the computations.

TABLE 5.1
Values of the Parameters λ, μ, A, B and f, as a Function of the Opening Angle 2α (Data from Filippi *et al.* [16])

2α	λ	μ	A	B	$f(2\alpha)$
0	0.5000	−0.5000	1.000	2.000	0.500
30	0.5015	−0.4561	1.075	2.104	0.492
45	0.5050	−0.4319	1.178	2.110	0.496
60	0.5122	−0.4057	1.339	2.086	0.508
90	0.5445	−0.3449	1.914	1.928	0.552
120	0.6157	−0.2678	3.133	1.578	0.637
135	0.6736	−0.2194	4.305	1.310	0.698
150	0.7520	−0.1624	6.513	0.968	0.776

5.3 A LOCAL STRESS FIELD FAILURE MODEL

It is assumed that, at a pointed V-notch (notch angle α, Williams' eigenvalue, λ), there are two branch notches with the notch radii, ρ_1 and ρ_2, the stress concentration factors, k_1 and k_2, the fracture toughness, $K_{\rho c1}$ and $K_{\rho c2}$, and the critical distances, L_{c1} and L_{c2}. Here, the two branch notches are denoted by notch 1 and notch 2, and the critical stress of the notched material is denoted by σ_0.

For notch 1, it is assumed from Eq. (11) that the failure occurs when the following condition is satisfied:

$$\sigma_\theta = \frac{K_{\rho c1}}{\sqrt{2\pi}\,(r_0 + L_{c1})^{1-\lambda}}[1 + (\frac{r_0}{(r_0 + L_{c1})})^{\lambda-\mu}\frac{A(\alpha)}{1+B(\alpha)}] = \sigma_0 \quad (14a)$$

Obviously, the failure equation (14a) is wholly the same as that described by PM in TCD.

For the brevity, Eq.(14a) is written to be:

$$F(K_{\rho c1}, L_{c1}, \rho_1, \alpha, \sigma_0) = 0 \quad (14b)$$

For notch 2, a similar equation to Eq. (14b) can be written to be

$$F(K_{\rho c2}, L_{c2}, \rho_2, \alpha, \sigma_0) = 0 \quad (15)$$

According to Yang's finding [15], it is assumed here that the following condition is valid:

$$k_2 \cdot L_{c2} = k_1 \cdot L_{c1} \quad (16)$$

After $K_{\rho c1}$ and $K_{\rho c2}$ are determined by experimental measurement, σ_0, L_{c1} and L_{c2} can be obtained by solving Eqs. (14b), (15) and (16) provided that k_1 and k_2

are known (e.g., the stress concentration factors are calculated by means of FEM [25–27] or BEM [19–24]).

For any notch 3 (the notch radius, ρ_3, the stress concentration factors, k_3, the notch angle α), its failure condition can be assumed to be:

$$F(K_{\rho c3}, L_{c3}, \rho_3, \alpha, \sigma_0) = 0 \tag{17}$$

and

$$k_3 \cdot L_{c3} = k_1 \cdot L_{c1} \tag{18}$$

After the critical distance of notch 3, L_{c3}, is calculated from Eq.(18), then the fracture toughness of notch 3, $K_{\rho c3}$, is predicted easily by substituting L_{c3} obtained from Eq.(18) into Eq.(17).

By the way, it is pointed out that, on one hand, the accuracy and validation of the model proposed here can be proven by comparing the predicted $K_{\rho c3}$ with the measured $K_{\rho c3}$, and on the other hand, using the predicted $K_{\rho c3}$, the maximum stress of notch 3, σ_{max}, is predicted from Eq.(13). Using the predicted σ_{max}, thus, the failure assessment of notch 3 can be easily carried out.

5.4 A CONCEPT OF THE STRESS CONCENTRATION FACTOR EIGENVALUE k^*

Based on the previous investigations on fatigue and static failure of notch specimens, it is found in the present study that there is a notch radius ρ^* for a given pointed V-notch. A stress concentration factor of the notch with the notch radius ρ^* is called a stress concentration factor eigenvalue of the pointed V-notch and denoted by k^*. In the following sections, it will be seen that k^* is a material notch geometric characteristic parameter and is very important for the failure assessment of different notches.

In this section, first the existing of k^* is illustrated based on the previous investigations on fatigue and static failure of notch specimens. Then an approach of determining k^* is given.

5.4.1 On the Existence of k^* (or ρ^*)

In 2002, a fracture toughness was evaluated by Lee et al. [29] using small notched specimens. The material used was SA 508 class 3 steel. Standard size ($10 \times 10 \times 55$mm) Charpy specimens were used for both dynamic and static fracture toughness tests. Notches with root radii ranging from 60 to 280 um were machined mechanically. Their root radii were measured with an optical microscope. The ratio of notch depth to the specimen width (*a/W*) was set at 0.5 for all specimens. A pre-cracked specimen was also prepared to obtain the plane strain fracture toughness. Fatigue pre-crack was made according to the ASTM E399 method [30]. Dynamic fracture toughness testing was performed using a pendulum instrumented impact tester. Static fracture toughness testing was performed using 10 ton capacity universal tester (Instron Model 5582). The specimens were loaded in three-point bending at a crosshead speed of 0.5 mm min−1. The plane strain fracture toughness values were obtained from pre-cracked specimens, satisfying plane strain conditions, under each loading state. The values for dynamic K_{Id} and static K_{Ic} were 32.78 MPa $m^{0.5}$ and 46.70 MPa $m^{0.5}$, respectively. Apparent fracture toughness values obtained from notched specimens (see Figure 5.4 and Figure 5.5), are also shown as a function of the square

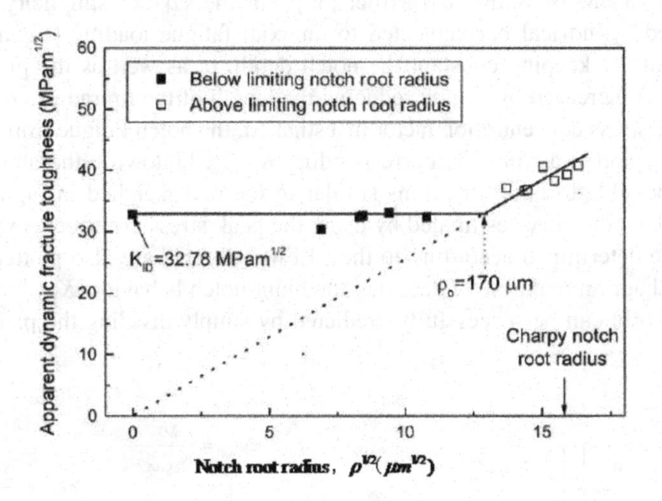

FIGURE 5.4 Apparent fracture toughness as a function of the square root of notch root radius ($\rho^{1/2}$) under dynamic loading.

Source: (Data from Lee *et al.* [29])

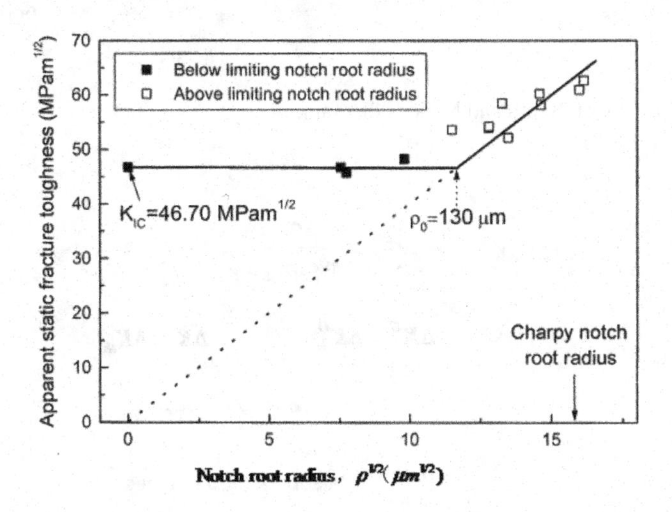

FIGURE 5.5 Apparent fracture toughness as a function of the square root of notch root radius ($\rho^{1/2}$) under static loading.

Source: (Data from Lee *et al.* [29])

root of ρ. There appears to be a limiting notch root radius (ρ_0) below which the fracture toughness is independent of ρ. Above ρ_0, apparent fracture toughness values are directly proportional to the square root of ρ. This observation is also shown in other previous research [31–33].

The last schematization linking the sharp-to-blunt notch behavior, as is well known to us, is the one proposed by Frost [34,35] and subsequently reinterpreted in terms of LEFM by Smith and Miller [36]. For the sake of simplicity, consider a U-notched cylindrical bar subjected to uniaxial fatigue loading (Figure 5.6(a)). Assume that, by keeping constant the notch depth, a, as well as the gross diameter, $K_{t,gross}$ is increased by simply reducing the length of the tip radius, r_n. For any value of the stress concentration factor investigated, the notch fatigue limit obtained can then be plotted against the corresponding $K_{t,gross}$. Following the above procedure it is possible to build diagrams similar to the one sketched in Figure 5.6(b). In such a chart the curve estimated by using the peak stress criterion as well as the straight line determined according to the LEFM concepts are also plotted. Such a schematic diagram makes it evident that the blunt-notch behavior ($K_{t,gross} < K^*_{t,gross}$) in Figure 5.6(b) can be successfully predicted by simply dividing the plain fatigue

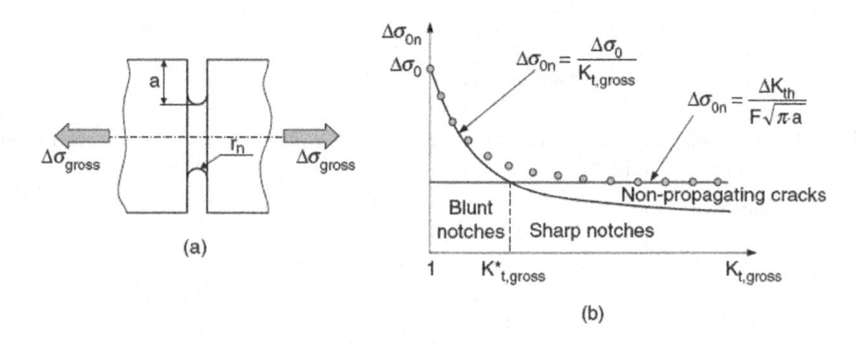

FIGURE 5.6 Frost, Smith and Miller's diagram.

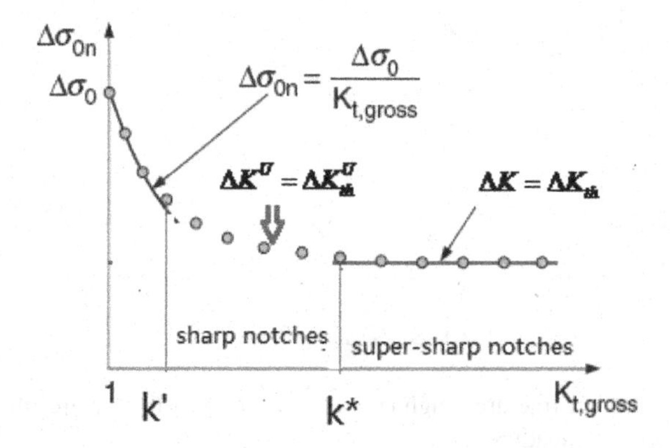

FIGURE 5.7 Schematic illustration of three regions in Frost, Smith and Miller's diagram.

limit, $\Delta\sigma_0$, by the stress concentration factor. On the contrary, for $K_{t,gross}$ values larger than $K^*_{t,gross}$, predictions made by using the peak stress criterion become too conservative, because in that region notches behave like long cracks. In other words, sharp-notch fatigue limits can be accurately estimated by simply using the following relationship [36]: $\Delta\sigma_{0n} = \Delta K_{th} / (F\sqrt{\pi a})$.

On the basis of the comparison of the fracture test data with the model predictions (peak criterion and LEFM criterion), here, model predictions are divided into three regions (see Figure 5.7). Two stress concentration factors, k' and k^*, are introduced (see Figure 5.7): notches with $k_{t,gross} > k^*$ are called super-sharp notches; notches with $k_{t,gross} < k'$ are called blunt notches and notches with $k' \geq k_{t,gross} \geq k^*$ are called sharp notches. For blunt notches, the peak stress criterion can usually be applied. For super-sharp notches, the crack stress field criterion or the pointed V-notch stress field criterion is applied, whereas for sharp notches, the local notch stress field criterion is applied.

The author considers that k^* shown in Figure 5.7 corresponds to ρ_0 (i.e., ρ^* here) in Figure 5.5.

On the basis of the investigation on the Mode I fatigue limit of components containing notches or defects by Atzori *et al.* [37], $\Delta K = \Delta K_{th}$ and $\Delta K^U = \Delta K^U_{th}$ in Figure 5.7 can be replaced by $\Delta K^V = \Delta K^V_{th}$ and $\Delta K^{\rho,V} = \Delta K^{\rho,V}_{th}$ for V-notches.

In 2006, Lazzarin and Filippi [38] stated "In the presence of sharp (zero radius) V-shaped notches the notch stress intensity factors (N-SIFs) quantify the intensities of the asymptotic linear elastic stress distributions. They are proportional to the limit of the Mode I stress components multiplied by the distance powered $1-\lambda$ from the notch tip, λ being Williams's eigenvalues. When the notch tip radius is different from zero, the definition is no longer valid from a theoretical point of view and the characteristic, singular, sharp-notch field diverges from the rounded-notch solution very next to the notch. Nevertheless, NSIFs continue to be used as parameters governing fracture if the notch root radius is sufficiently small with respect to the notch depth".

The author considers here that "the notch root radius is sufficiently small with respect to the notch depth" stated by Lazzarin and Filippi [38] means that there is a notch radius ρ^* for a given pointed V-notch, and that for a notch with notch radius $\rho < \rho^*$, a failure assessment of the notched component can be carried out by means of the theory of NSIFs.

Variation of nondimensional fracture toughness of U-notches with nondimensional notch radius pictured by Berto and Lazzarin [6] (see Figure 5.8), appears to reveal the existence of ρ^*.

The static failure behavior of Perspex described by Taylor *et al.* [10] is shown in Figure 5.9: "Three different regions of behavior occur. At very high k_t values, σ_f becomes independent of k_t, but will depend on the length of the notch: in this region all notches behave as if they were cracks. TCD methods such as PM will predict this effect. At intermediate values of k_t, the TCD prediction lines curve upwards so that, as k_t decreases, we approach the condition $\sigma_f = \sigma_0 / k_t$, which will be a straight line of slope −1 on a log/log plot. A third region occurs at low k_t, when this line crosses the horizontal line corresponding to $\sigma_f = \sigma_u$".

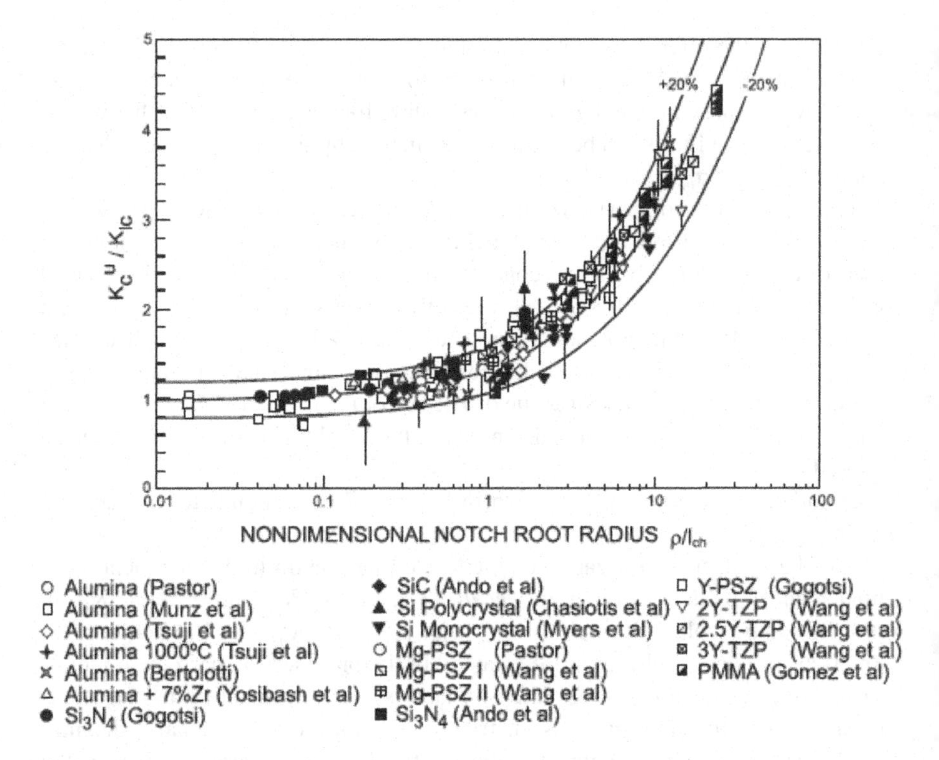

FIGURE 5.8 Normalized values of the generalized stress intensity factor to failure.

Source: (Data from Berto and Lazzarin [6])

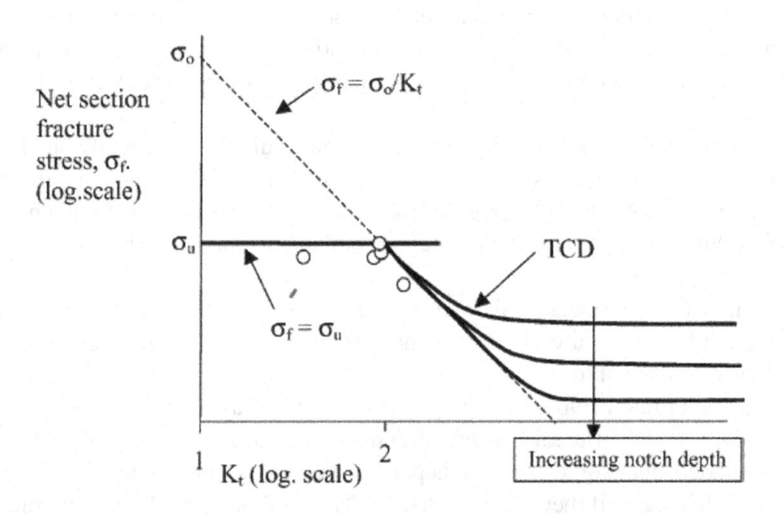

FIGURE 5.9 Schematic drawing showing the effect of k_t on net-section fracture stress. The TCD (modified PM) method gives accurate predictions until it intersects with the line $\sigma_f = \sigma_u$ (note: σ_u is the ultimate tensile stress, σ_{UTS}).

Source: (Data from Taylor *et al.* [10]).

Obviously the static failure behavior of Perspex described by Taylor et al. [10] appears to be same as that by model predictions proposed here (see Figure 5.7).

By the way, a condition $\sigma_f = \sigma_u$ and $\sigma_f = \sigma_0 / k_t$, i.e., $k_t = \sigma_0 / \sigma_u$ given by Taylor *et al.*[10] is perhaps taken as k' in Figure 5.7.

5.4.2 An Approach to Determining k^*

Here, the fracture toughness of the pointed V-notch (with notch angle α) is denoted by K_{vc}. For a notch with notch radius ρ^* and notch angle α, obviously, its failure condition can be expressed to be:

$$F(K_{vc}, L_c^*, \rho^*, \alpha, \sigma_0) = 0 \tag{19}$$

where L_c^* and ρ^* can be determined by trial and error by means of the following conditions:

$$k^* \cdot L_c^* = k_i \cdot L_{ci}, \quad (i = 1, 2, \ldots, M) \tag{20}$$

and

$$F(K_{pci}, L_{ci}, \rho_i, \alpha, \sigma_0) = 0, \quad (i = 1, 2, \cdots, M) \tag{21}$$

By comparing $\{K_{pci}\}(i = 1, 2, \ldots, M)$ obtained from Eqs. (19) to (21) and $\{K_{\rho ci}\}(i = 1, 2, \ldots, M)$ measured experimentally, L_c^*, ρ^* and k^* can be determined. M in (20) and (21) is the number of notches with different notch radii. For brittle materials such as ceramics, the critical stress σ_0 can be taken to be the ultimate tensile stress, σ_{UTS}, $M \geq 1$. Whereas when static failures are preceded by a certain amount of plasticity, $M \geq 2$.

The parameters L_c^*, ρ^* and k^*, obtained by using Eqs. (19) to (21) are called the geometric material characteristic parameters of the pointed V-notch (called GMPs for short). It will be seen from this study that the GMPs are very important in failure assessments of notched components performed by means of the local stress field failure model proposed here.

5.5 EXPERIMENTAL VERIFICATIONS

Based on the test data of fracture failure of notched components from the literature, in this section, the local approach of failure assessment of notched components proposed in this study will be verified. In order to quantitatively evaluate the accuracy of the failure analysis, the following error indexes are defined:

$$R.E. = \frac{(\mathrm{K}_{v\rho c}(cal) - \mathrm{K}_{v\rho c}(\exp)) \times 100}{\mathrm{K}_{v\rho c}(\exp)} \tag{22}$$

$$M.E. = \frac{1}{n}\sum_{i=1}^{n}(R.E.)_i \tag{23}$$

$$M.A.E. = \frac{1}{n}\sum_{i=1}^{n} abs(R.E.)_i \quad (24)$$

$$E.R. = [\min(R.E.)_i, \max(R.E.)_i] \quad (25)$$

where n is the number of test cases, $K_{v\rho c}(\exp)$ is the experimentally measured fracture toughness; and $K_{v\rho c}(cal)$ are the calculated fracture toughness by means of the local approach proposed in this study.

For the sake of clear discussions, the experimental verifications given here are described in the forms of examples, respectively.

Example 1: Fracture Analysis of Notched Components Made of PMMA at −60°

In 2005, Gómez *et al.* [39] studied a static failure of round-notched samples. The chosen material was a glassy polymer (PMMA) that at −60° C behaves as linear elastic. More than 150 notched samples were tested at −60° C, until fracture. Notch acuity was changed by changing the notch tip radii. Loading of the samples ranged from pure tensile up to bending.

Fracture toughness measured from cracked samples is K_{Ic} = 1.7 MPam$^{0.5}$. The ultimate tensile strength measured from unnotched samples is σ_{UTS} = 128.4 MPa.

In this example, the critical stress σ_0 of the studied material is taken to be the ultimate tensile strength σ_{UTS}.

Geometric size of U-notch TPB specimens is shown in Figure 5.10. Two notch depths were tested: a = 14 mm and a = 5 mm. In the two cases, seven different notch root radii were considered: 0.2 mm, 0.3 mm, 0.5 mm, 1.0 mm, 1.5 mm, 2.0 mm and 4.0 mm. The average relevant values of the notch geometries and rupture loads are summarized in Tables 5.2 and 5.3. In Tables 5.2 and 5.3, some calculation results are also given including the stress concentration factor, k, the net nominal stress, σ_n, and the notched fracture toughness, $K_{\rho c}$. Here, the stress concentration factor, k, is calculated by means of the equation reported in Ref. [27]

$$k = m(a/\rho)^n \quad (26)$$

where m and n are constants [27]. The fracture toughness of notched components, $K_{\rho c}$, is calculated by using Eq.(13).

Using the approach of determining k^* proposed in this study, the stress concentration factor eigenvalue, k^*, of U-notched TPB specimens made of PMMA at

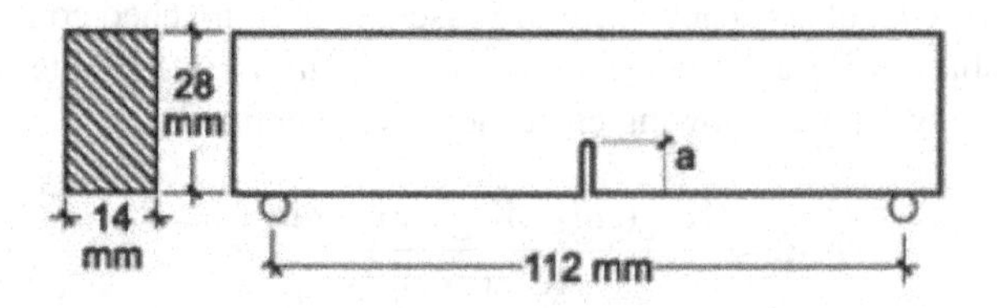

FIGURE 5.10 Geometric size of U-notch TPB specimens.

Source: (Data from Gómez *et al.*[39])

TABLE 5.2
Fracture Test Data of U-Notch TPB Specimens and Some Calculation Results (Notch Depth a = 5 mm)

Notch Depth a (mm)	Notch Radius ρ (mm)	Failure Load F_c (N)	SCFs k	σ_n (MPa)	σ_{max} (MPa)	$K_{\rho c}$ (MPam$^{0.5}$)
5.21	0.19	1080	7.27	28.07	204.15	2.49
4.95	0.34	1690	5.38	42.94	230.86	3.77
5.01	0.52	1840	4.41	47	207.48	4.19
5.01	0.94	2220	3.33	56.7	188.72	5.13
4.95	1.47	2510	2.67	63.78	170.49	5.79
4.98	1.97	2670	2.33	68.02	158.58	6.24
5.08	3.98	3200	1.68	82.23	138.37	7.74

Note: (a) In this table, notch depth a, notch radius ρ and failure load F_c are taken from Ref.[39]; (b) σ_n is the net nominal stress; (c) The stress concentration factor, $k = m(a/\rho)^n$, where m = 1.49768, n = 0.4772; (d) Values of $K_{\rho c}$ are calculated using Eq.(13).

TABLE 5.3
Fracture Test Data of U-Notch TPB Specimens and Some Calculation Results (Notch Depth a = 14 mm)

Notch Depth a (mm)	Notch Radius ρ (mm)	Failure Load F_c (N)	SCFs k	σ_n (MPa)	σ_{max} (MPa)	$K_{\rho c}$ (MPam$^{0.5}$)
14.1	0.18	430	6.47	30.05	194.49	2.31
13.94	0.34	690	4.8	47.12	226.24	3.7
14.03	0.53	730	3.92	50.5	198.13	4.04
14	0.93	920	3.02	63.37	191.61	5.18
13.93	1.46	1060	2.45	72.29	177.09	6
14	1.97	1150	2.14	79.21	169.39	6.66
14.04	3.98	1300	1.55	90.05	139.39	7.79

Note: (a) In this table, notch depth a, notch radius ρ and failure load F_c are taken from Ref.[39]; (b) σ_n is the net nominal stress; (c) The stress concentration factor, $k = m(a/\rho)^n$, where $m = 0.8651$, $n = 0.4615$; (d) Values of $K_{\rho c}$ are calculated using Eq.(13).

−60° C are obtained by trial and error by taking notch radii (denoted by ρ') to be 0.012 mm, 0.015 mm, 0.0175 mm, 0.020 mm, 0.025 mm and 0.030 mm for notch depth a = 5 mm, and 0.010 mm, 0.012 mm, 0.016 mm, 0.020 mm,0.025 mm and 0.035 mm for notch depth a = 14 mm. The calculated stress concentration factors (denoted by k'), the critical distances (denoted by L_c') and the relative errors are given in Tables 5.4 and 5.5. From Tables 5.4 and 5.5, it can be seen that values of k^* of U-notch TPB specimens made of PMMA at −60° C are 20.88 and 19.71 for notch depths a = 5 mm and 14 mm, respectively.

TABLE 5.4
Some Computation Data in Determining k^* for U-Notch TPB Specimens Made of PMMA at −60° C (Notch Depth a = 5 mm)

No.	1	2	3	4	5	6
ρ' (mm)	0.012	0.015	0.0175	**0.020**	0.025	0.030
k'	26.64	23.95	22.25	**20.88**	18.77	17.21
L'_c (mm)	0.0315	0.032	0.0323	**0.0326**	0.0329	0.033
R.E (%)	−29	−23.9	−20.6	−17.9	−13.3	−9.6
	−1.3	2.3	4.7	6.6	9.8	12.4
	−3.2	0.1	2.3	4.0	7.0	9.3
	−2.2	0.6	2.5	3.9	6.4	8.4
	−5.8	−3.3	−1.6	−0.3	1.9	3.7
	−9.2	−6.9	−5.3	−4.1	−2.0	−0.4
	−15.3	−13.4	−12.2	−11.2	−9.5	−8.2
M.E.(%)	−4.3	−3.4	−1.6	−0.2	2.26	4.2
E.R.(%)	[−15.3,−1.3]	[−13.4,2.3]	[−12.2,4.7]	[−11.2,6.6]	[−9.5,9.8]	[−8.2,12.4]
M.A.E.(%)	6.2	4.4	4.8	5	6.1	7.1

Note: the error indexes are calculated by deleting the unusual data.

TABLE 5.5
Some Computation Data in Determining k^* for U-Notch TPB Specimens Made of PMMA at −60° C (Notch Depth a = 14 mm)

No.	1	2	3	4	5	6
ρ' (mm)	0.010	0.012	**0.016**	0.020	0.025	0.035
k'	24.49	22.51	**19.71**	17.79	16.04	13.74
L_c (mm)	0.0311	0.0315	**0.0321**	0.0325	0.0329	0.033
R.E (%)	−40.3	−35.5	−28.9	−24.1	−19.4	−12.7
	−4	−1.1	3.3	6.4	9.5	13.9
	−7.9	−5.4	−1.4	1.6	4.4	8.5
	−1.2	1.3	4.5	6.8	9.1	12.3
	−1.9	0.1	3	5	6.9	9.7
	−2	−0.3	2.3	4.1	5.8	8.3
	−14.2	−12.7	−10.6	−9	−7.5	−5.4
M.E.(%)	−5.2	−3	0.17	2.5	4.7	7.9
E.R.(%)	[−14.2,−1.2]	[−12.7,1.3]	[−11,3.9]	[−9,6.8]	[−7.5,9.5]	[−5.4,13.9]
M.A.E.(%)	5.2	3.5	4.2	5.5	7.2	9.7

Note: the error indexes are calculated by deleting the unusual data.

Double-edge tensile of U-notch specimens (see Figure 5.11(b)) made of PMMA at −60°C were carried out by Gómez et al. [39]. Here, two notch depths were tested: a = 8 mm and a = 11 mm. In the two cases, four notch root radii were considered: 0.50 mm, 1.0 mm, 1.5 mm and 2.0 mm. Fracture test data reported in Ref. [39] and some calculation results given here, including the stress concentration factor, k, the net nominal stress, σ_n , and the notched fracture toughness, K_{pc}, are listed in Tables 5.6 and 5.7. Here, the stress concentration factor, k, is calculated by means of the equation proposed by Shin *et al.* [28]

$$k = 1 + 2F\sqrt{\frac{a}{\rho}} \tag{27}$$

where F is the shape function of the corresponding crack geometry, i.e.

$$K_I = F \cdot \sigma\sqrt{\pi a} \tag{28}$$

By using the approach of determining k^* proposed in this study, the MGPs obtained for the U-notch DENT specimens made of PMMA at −60° C with notch

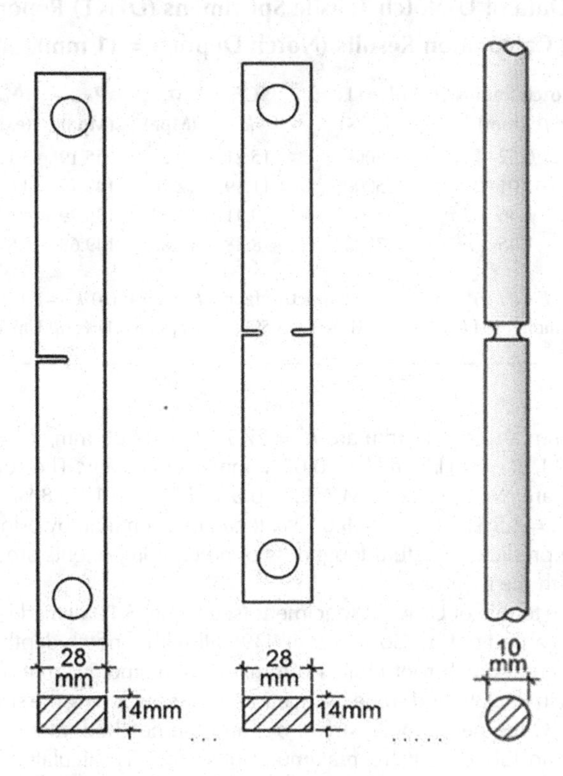

FIGURE 5.11 Geometric size of tensile tests with notched samples: (a) Single-edge notched tensile (SENT) specimens. (b) Double-edge notched tensile (DENT) specimens. (c) Round notched cylindrical (RNC) specimens.

Source: (Data from Gómez *et al.*[39]).

TABLE 5.6
Fracture Test Data of U-Notch Tensile Specimens (DENT) Reported in Ref. [39] and Some Calculation Results (Notch Depth a = 8 mm)

Notch Depth a (mm)	Notch Radius ρ (mm)	Failure Load F_c (N)	SCFs k	σ_n (MPa)	σ_{max} (MPa)	$K_{\rho c}$ (exp)	$K_{\rho c}$ (cal)	R.E (%)
7.95	0.53	6400	10.3	16.33	168.21	3.43	3.41	0.8
8.07	0.97	7600	7.96	19.39	154.27	4.26	4.31	−1.1
7.95	1.46	8940	6.63	22.81	151.17	5.12	5.1	0.4
7.93	1.91	9890	5.89	25.23	148.61	5.76	5.71	0.8

Note: (*a*) $k=1+2\cdot F\cdot(a/\rho)^{0.5}$, where the geometric factor F = 1.203; (*b*) k^* = 27.5, ρ^* = 0.066 mm, L_c^* = 0.031 mm; (*c*) M.E. = 0.2%, M.A.E. = 0.8%, E.R. = [−1.1,0.8]%; (*d*) Unit of $K_{\rho c}$ is MPam$^{0.5}$

TABLE 5.7
Fracture Test Data of U-Notch Tensile Specimens (DENT) Reported in Ref. [39] and Some Calculation Results (Notch Depth a = 11 mm)

Notch Depth a (mm)	Notch Radius ρ (mm)	Failure Load F_c (N)	SCFs k	σ_n (Mpa)	σ_{max} (Mpa)	$K_{\rho c}$ (exp)	$K_{\rho c}$ (cal)	R.E (%)
11.01	0.52	4000	15.21	10.2	155.19	3.14	3.1	1.2
11.01	1.01	5000	11.19	12.76	142.79	4.02	4.11	−2.2
10.96	1.46	5700	9.41	14.54	136.79	4.63	4.84	−4.5
11.06	1.96	7000	8.38	17.86	149.68	5.87	5.51	6.1

Note: (*a*) $k=1+2\cdot F\cdot(a/\rho)^{0.5}$, where the geometric factor F = 1.54; (*b*) k^* = 31.84, ρ^* = 0.11 mm, L_c^* = 0.0247 mm; (*c*) M.E. = 0.15%, M.A.E. = 3.5%, E.R. = [−4.5,6.1]%; (*d*) Unit of $K_{\rho c}$ is MPam$^{0.5}$

depth a = 8 mm and a = 11 mm are k^* = 27.5, ρ^* = 0.066 mm, L_c^* = 0.031 mm and k^* = 31.84, ρ^* = 0.11 mm, L_c^* = 0.0247 mm, respectively. The corresponding error indexes are: M.E. = 0.2%, M.A.E. = 0.8%, E.R. = [−1.1,0.8]%, and M.E. = 0.15%, M.A.E. = 3.5%, E.R. = [−4.5,6.1]%. It can be seen that, by using the MGPs obtained, the predicted fracture toughness of notch components are in excellent agreement with the measured ones.

Single-edge tensile of U-notch specimens (see Figure 5.11(a)) made of PMMA at −60° C were carried out by Gómez et al. [39]: all with a notch depth a = 14 mm and three different notch root radii: 1.0 mm, 1.5 mm, and 2.0 mm. Fracture test data reported in Ref.[39] and some calculation results, including the stress concentration factor, k, the net nominal stress, σ_n, and the notched fracture toughness, $K_{\rho c}$, are listed in Table 5.8. Here, maximum stress, σ_{max}, is calculated by using the following equation [40]:

$$\sigma_{max}=\frac{K_V}{\sqrt{2\pi}}\cdot R_I\cdot\rho^{\lambda-1} \tag{29}$$

where R_l is the stress rounding factor (for U-notch, $R_l = 2.992$), K_V is the notch stress intensity factor (pointed V-notch), ρ is the notch root radius. The values of K_V for a single edge notched specimen in pure tension were given by Gross and Mendlson [41]. The values of $K_{\rho c}$ in Table 5.8 are calculated by using (13). The stress concentration factor, k, shown in Table 5.8, is defined as a ratio of the maximum stress, σ_{max}, to the net nominal stress, σ_n.

By using the approach of determining k^* proposed in this study, the MGPs obtained for U-notch SEN specimens made of PMMA at −60° C are: $k^* = 43.7$, $\rho^* = 0.06$ mm, $L_c^* = 0.0316$ mm, with the calculated error indexes: M.E. = 0.1%, M.A.E. = 7.4%, E.R. = [−11,6.2]%.

Round notched cylindrical (RNC) specimens reported by Gómez *et al.* [39] are shown in Figure 5.11(c). All samples have an external diameter of 10 mm. One set has an inner diameter of 4 mm and notch radii of 2 mm, 4 mm and 7 mm. The other set of inner diameter 8 mm were machined with notch radii of 0.2 mm, 2 mm, 4 mm and 7 mm. The fracture test data and some calculation results are given in Tables 5.9 and 5.10. Here, the stress concentration factors, k, are calculated by means of equation proposed in Ref.[20].

TABLE 5.8
Fracture Test Data of U-Notch SEN Specimens Made of PMMA at −60° C and Some Calculation Results

Notch Depth a (mm)	Notch Radius ρ (mm)	Failure Load F_c (N)	SCFs k	σ_n (Mpa)	σ_{max} (Mpa)	$K_{\rho c}$ (exp)	$K_{\rho c}$ (cal)	R.E (%)
13.92	0.97	2860	11.41	14.51	165.59	4.57	4.34	5.1
13.94	1.47	2900	9.26	14.73	136.39	4.63	5.14	−11
14.02	1.99	3900	7.91	19.93	157.65	6.23	5.85	6.2

Note: (*a*) $k^* = 43.7$, $\rho^* = 0.06$ mm, $L_c^* = 0.0316$ mm; (*b*) M.E. = 0.1%, M.A.E. = 7.4%, E.R. = [−11,6.2]%; (*c*) Unit of $K_{\rho c}$ is $MPam^{0.5}$

TABLE 5.9
Fracture Test Data of U-Notch RNC Specimens Made of PMMA at −60° C and Some Calculation Results (Notch Depth *a* = 1 mm, Inner Diameter = 8 mm)

Notch Depth a (mm)	Notch Radius ρ (mm)	SCFs k	Failure Load F_c (N)	σ_{max} (MPa)	$K_{\rho c}$ (cal)	$K_{\rho c}$ (exp)	R.E. (%)
0.97	0.19	4.01	2500	196.52	2.21	2.4	−8.1
0.96	2.13	1.62	4680	147.51	5.91	6.03	−2
0.93	4.06	1.36	5200	135.82	7.91	7.67	3.2
0.98	7.07	1.24	5200	126.65	10.23	9.44	8.4

Note: (*a*) $k^* = 10.96$, $\rho^* = 0.09$ mm, $L_c^* = 0.028$ mm; (*b*) M.E. = 0.4%, M.A.E. = 5.4%, E.R. = [−8.1,8.4]%; (*c*) Unit of $K_{\rho c}$ is $MPam^{0.5}$

TABLE 5.10

Fracture Test Data of U-Notch RNC Specimens Made of PMMA at −60°C and Some Calculation Results (Notch Depth *a* = 3 mm, Inner Diameter = 4 mm)

Notch Depth a (mm)	Notch Radius ρ (mm)	SCFs k	Failure Load F_c (N)	σ_{max} (MPa)	$K_{\rho c}$ (cal)	$K_{\rho c}$ (exp)	R.E. (%)
2.98	2.1	1.413	1460	160.93	6.37	6.54	−2.6
3.01	4	1.209	1450	140.91	8.36	7.9	5.8
2.99	7.05	1.183	1580	147.27	10.72	10.96	−2.1

Note: (*a*) k^* = 15.71, ρ^* = 0.04 mm, L_c^* = 0.033 mm; (*b*) M.E. = 0.4%, M.A.E. = 3.5%, E.R. = [−2,5.8]%; (*c*) Unit of $K_{\rho c}$ is MPam$^{0.5}$

By using the approach of determining k^* proposed in this study, the MGPs obtained for U-notch RNC specimens made of PMMA at −60° C are: k^* = 10.96, ρ^* = 0.09 mm, L_c^* = 0.028 mm, with the calculated error indexes: M.E. = 0.4%, M.A.E. = 5.4%, E.R. = [−8.1,8.4]% for the set of inner diameter of 4 mm, and k^* = 15.71, ρ^* = 0.04 mm, L_c^* = 0.033 mm, with the calculated error indexes: M.E. = 0.4%, M.A.E. = 3.5%, E.R. = [−2,5.8]% for the set of inner diameter of 8 mm.

Example 2: Fracture Analysis of U-Notch Specimens Made of PMMA at Room Temperature

As is well known to us, PMMA exhibits a nonlinear behavior at room temperature (see Figure 5.12). Here, a fracture test of U-notch specimens performed by Gómez *et al.* [42] is described briefly. Two types of specimens were tested: three-point bend (TPB) beams and single-edge-notched (SEN) tensile specimens (see Figure 5.13). In all the specimens, the thickness was 14 mm.

To explore the influence of the U-notch radius, *R*, seven sets of TPB beams (size *D* = 28 mm, notch depth *a* = 14 mm), with *R* = 0.11, 0.13, 0.20, 0.50, 1.0, 1.5 and 2.0 mm, and three other sets of TPB beams (*D* = 28 mm, *a* = 2 mm), with *R* = 0.50, 1.0 and 2.0 mm, were tested. Four sets of SEN tensile specimens (*D* = 28 mm, *a* = 14 mm) with *R* = 0.50, 1.0, 1.5 and 2.0 mm, were also tested.

The fracture toughness K_{Ic} and the ultimate tensile strength σ_{UTS} of PMMA at room temperature given in Ref.[42] is 1.0 MPam$^{0.5}$ and 75 MPa.

Fracture test data of U-notch TPB specimens performed by Gómez et al [42] and some calculation results given here are listed in Table 5.11. The calculation retails are as follows:

First, the critical stress σ_0 is determined by means of the local approach proposed in this study (see Figure 5.14): two test points with notch radii 0.5 mm and 2.0 mm are chosen, the critical stress σ_0 obtained by solving equations (14) to (16) is 86 MPa. By the way, it is pointed out here that stress concentration factor, *k*, is calculated by means of equation (26), where *m* = 0.8651, *n* = 0.4615 and the fracture toughness, $K_{\rho c}$, of notched specimens is calculated by using equation (13).

Second, the MGPs of U-notch TPB specimens are determined by using the approach of determining k^* proposed in this study. The values of the GMPs obtained are: k^* = 27.15, ρ^* = 0.008 mm, L_c^* = 0.0239 mm.

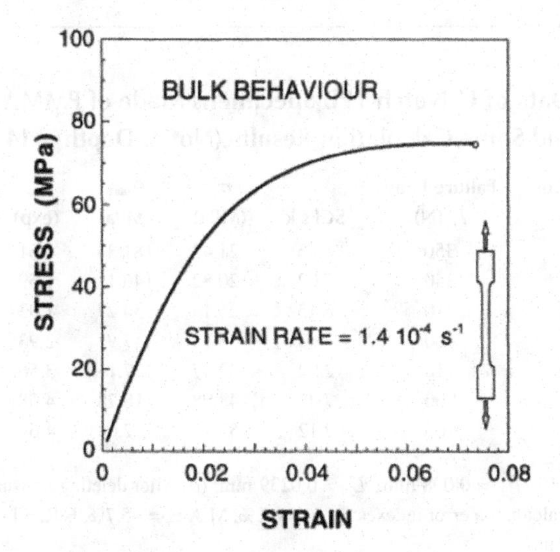

FIGURE 5.12 Stress–strain curve of PMMA at room temperature.

Source: (Data from Gómez *et al.* [42])

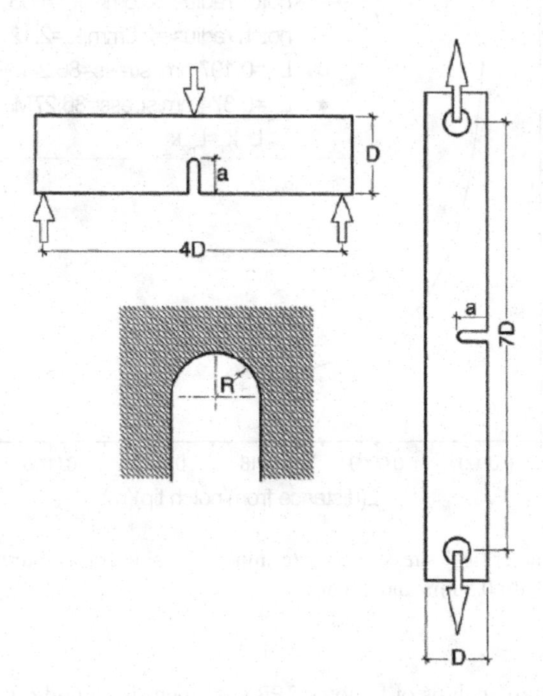

FIGURE 5.13 Two types of test specimens: three-point bend (TPB) beams and single-edge-notched (SEN) tensile specimens.

Source: (Data from Gómez *et al.* [42])

TABLE 5.11
Fracture Test Data of U-Notch TPB Specimens Made of PMMA at Room Temperature and Some Calculation Results (Notch Depth = 14 mm)

No.	Notch Radius ρ (mm)	Failure Load F_c (N)	SCFs k	σ_n (MPa)	σ_{max} (MPa)	K_{pc} (exp)	K_{pc} (cal)	R.E (%)
1	0.10	350	8.46	21.43	181.34	1.61	1.75	−8.75
2	0.15	340	7.02	20.82	146.1	1.59	1.94	**−22.06**
3	0.20	410	6.15	25.1	154.27	1.93	2.09	−8.05
4	0.5	600	4.03	36.73	147.91	2.93	2.73	6.83
5	1.0	710	2.92	43.47	127.11	3.56	3.44	3.49
6	1.5	800	2.43	48.98	118.78	4.08	3.98	2.45
7	2.0	900	2.12	55.1	117.02	4.64	4.43	4.45

Note: (*a*) $k^* = 27.15$, $\rho^* = 0.008$ mm, $L_c^* = 0.0239$ mm; (*b*) After deleting unusual data with R.E. = −22.06, the calculation error indexes: M.E. = 0.1%, M.A.E. = −5.7%, E.R. = [−8.8,6.8]%; (*c*) Unit of K_{pc} is MPam$^{0.5}$

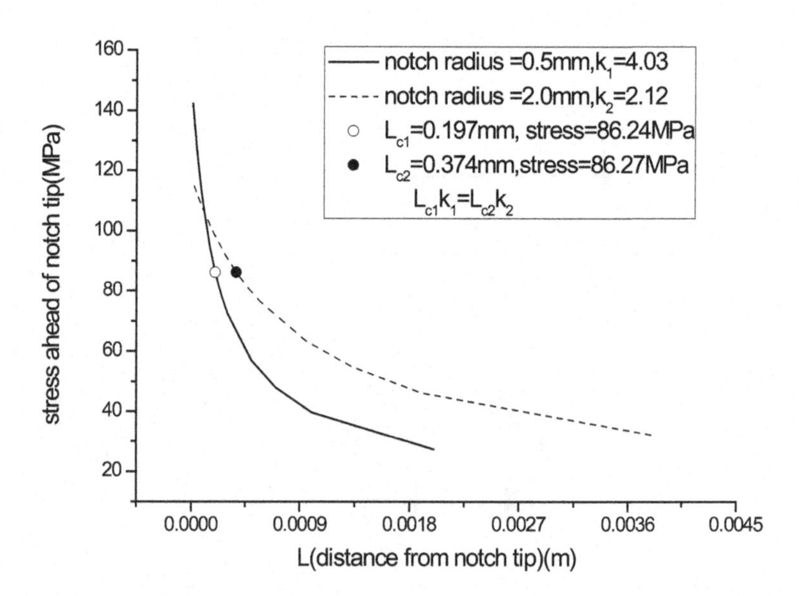

FIGURE 5.14 The critical stress σ_0 is determined by using fracture test data of two test points with notch radii 0.5 mm and 2.0 mm.

Third, fracture analysis of U-notch TPB specimens is carried out by using the obtained GMPs and by means of the local stress field failure model proposed in this study. The failure assessment results are given in Table 5.11, with error indexes: M.E. = 0.1%, M.A.E. = −5.7%, E.R. = [−8.8,6.8]%.

TABLE 5.12
Fracture Test Data of U-Notch SENT Specimens at Room Temperature and Some Calculation Results (Notch Depth = 14 mm)

No.	Notch Radius ρ (mm)	Failure Load F_c (N)	SCFs k	σ_n (MPa)	σ_{max} (MPa)	$K_{\rho c}$ (exp)	$K_{\rho c}$ (pre)	R.E (%)
1	0.5	1800	15.81	9.18	145.16	2.88	2.95	2.6
2	1	2400	11.18	12.24	136.85	3.84	3.69	−3.7
3	1.5	2600	9.13	13.27	121.05	4.15	4.26	2.4
4	2	3000	7.9	15.31	120.96	4.79	4.72	−1.5

Note (*a*) $k^* = 133.58$, $\rho^* = 0.007$ mm, $L_c^* = 0.0236$ mm; (*b*) the calculation error indexes: M.E. = 0.04%, M.A.E. = 2.5%, E.R. = [−3.7,2.6]%; (*c*) Unit of $K_{\rho c}$ is MPam$^{0.5}$

For U-notch SEN specimens made of PMMA at room temperature, fracture test data and some calculation results given here are listed in Table 5.12. Some calculation retails are as follows:

(1) The maximum stress, σ_{max}, and the stress concentration factor, k, shown in Table 5.12, are calculated by using the methods described in Example 1 in this section.
(2) By using the critical stress σ_0 = 86 MPa obtained above and the fracture toughness K_{Ic} = 1.0 MPam$^{0.5}$ given in Ref.[42], the GMPs of U-notch SEN specimens are determined by means of the approach of determining k^* proposed in this study. The values of the GMPs obtained are: k^* = 133.58, ρ^* = 0.007, L_c^* = 0.0236 mm.
(3) Fracture analysis of U-notch SEN specimens is carried out by using the obtained GMPs and by means of the local stress field failure model proposed in this study. The failure assessment results are given in Table 5.12, with error indexes: M.E. = 0.04%, M.A.E. = 2.5%, E.R. = [−3.7, 2.6]%.

Example 3: Fracture Analysis of U-Notch Specimens Made of Two Ceramics, SI3N4 and Y-TZP

In Ref.[43], fracture toughness of two ceramics (Silicon nitride ceramics Si3N$_4$ and Sintered (Y-TZP) Yttria Tetragonal Zirconia Polycrystal ceramics) was determined by a single edge V-notched beam (SEVNB) method. The specimens were tested under three-point or four-point bending, according to Standards ASTM C1421-99 and DIN 51-109, respectively.

Geometric sizes of U-notch TPB specimens made of SI3N4 are length L = 16 mm, width W = 4 mm, thickness B = 3 mm and notch depth a = 1.88 mm. And σ_{UTS} and K_{Ic} reported in Ref.[43] are 700 MPa and 5.35 MPam$^{0.5}$, respectively.

Fracture test data of U-notch TPB specimens made of SI3N4 are listed in Table 5.13. Because all fracture toughness values of U-notch specimens with very small notch radii listed in Table 5.13 are less than the one of K_{Ic} given in Ref.[43], the mean value of fracture toughness values of U-notch specimens with very small notch radii listed in Tables 5.13 and 5.17, MPam$^{0.5}$, is taken to be K_{Ic} in this study.

Here, some calculation details are described briefly as follows:

(1) Fracture test data of order numbers, 5 and 13, shown in Table 5.13, are used to calculate the critical stress σ_0 by using Eqs. (14) to (16). Its value is 560 MPa.
(2) For U-notch TPB specimens, the stress concentration factor, k, is calculated using Eq.(26), where $m = 0.8978$ and $n = 0.4631$ are taken from Ref.[27].
(3) By using the approach of determining k^* proposed in this study, the GMPs of U-notch TPB specimens made of SI3N4 are calculated. The obtained GMPs are: $k^* = 9.33$, $\rho^* = 0.012$ mm, $L_c^* = 0.01598$ mm.
(4) By using the values of the GMPs obtained, fracture analysis of U-notch TPB specimens made of SI3N4 is performed by means of the local stress field failure model proposed in this study. The variations of the predicted fracture toughness $K_{\rho c}$ and the critical distance L_c with notch radii are shown in Table 5.14. By comparing the predicted $K_{\rho c}$ with the measured $K_{\rho c}$, errors obtained are listed in Table 5.14, with error indexes: M.E. = −0.6%, M.A.E. = 5.1%, E.R. = [−7, 9.5] %.

Geometric sizes of U-notch FPB specimens made of Y-TZP reported in Ref. [43] are $S_1 = 40$ mm, $S_2 = 20$ mm, $W = 4.62$ mm, $B = 3$ mm, $a = 2.36$ mm. σ_{UTS} and K_{Ic} reported in Ref.[43] are 425 MPa and 5.9 MPam$^{0.5}$, respectively.

TABLE 5.13
Fracture Test Results of U-Notch TPB Specimens Made of SI3N4 (Taken from [6])

No.	Notch Radius ρ (um)	Failure Load F_c (N)	$K_{\rho c}$ (MPam$^{0.5}$)
1	2.5	110	5.11
2	3.5	110	5.2
3	3.5	112	5.14
4	4	107	5.24
5	10	97.4	5.34
6	14	102	5.69
7	14.5	94.7	5.23
8	20	102	5.47
9	33.5	122	6.3
10	33.5	110	5.99
11	36.5	129	6.86
12	36.5	117	6.37
13	98.5	162	9.34
14	98.5	157	9.3
15	98.5	172	9.23
16	98.5	155	9.11
17	98.5	152	8.62

TABLE 5.14
Some Calculation Results of U-Notch TPB Specimens Made of SI3N4

No.	SCFs k	Notch Radius ρ (mm)	L_c (mm)	$K_{\rho c}$ (exp) (MPam$^{0.5}$)	$K_{\rho c}$ (cal) (MPam$^{0.5}$)	R.E. (%)
5	10.1	0.01	0.014686	5.34	4.97	−7
6	8.7	0.014	0.017162	5.69	5.35	−6
7	8.5	0.0145	0.017444	5.23	5.39	3.1
8	7.4	0.02	0.020245	5.47	5.8	6.1
9	5.8	0.0335	0.025707	6.3	6.56	4.1
10	5.8	0.0335	0.025707	5.99	6.56	9.5
11	5.6	0.0365	0.026749	6.86	6.7	−2.3
12	5.6	0.0365	0.026749	6.37	6.7	5.2
13	3.5	0.0985	0.042361	9.34	8.74	−6.4
14	3.5	0.0985	0.042361	9.3	8.74	−6
15	3.5	0.0985	0.042361	9.23	8.74	−5.3
16	3.5	0.0985	0.042361	9.11	8.74	−4.1
17	3.5	0.0985	0.042361	8.62	8.74	1.4

Note: (*a*) $k = m(a/\rho)^n$, where $m = 0.8978, n = 0.4631$; (b) $k^* = 9.33$, $\rho^* = 0.012$ mm, $L_c^* = 0.01598$ mm; (*c*) $K_{Ic} = 5.17$ MPam$^{0.5}$, $\sigma_0 = 560$ MPa; (*d*) M.E. = −0.6%, M.A.E. = 5.1%, E.R. = [−7,9.5] %.

TABLE 5.15
Fracture Test Data of U-Notch FPB Specimens Made of Y-TZP (Taken from [6])

No.	Notch Radius ρ (um)	F_c (N)	$K_{\rho c}$
1	2	80.8	<u>**6.1**</u>
2	2.5	81.8	<u>**5.3**</u>
3	3	85.5	<u>**7.06**</u>
4	4	127	<u>**6.96**</u>
5	7.5	124	7.41
6	7.5	125	7.64
7	18	147	9.47
8	41.5	169	10.8
9	53	190	12.2
10	72	141	11.3
11	80	161	11.4
12	179	297	17.4
13	403.5	345	22.2
14	403.5	363	23.9

TABLE 5.16
Some Calculation Results of U-Notch FPB Specimens Made of Y-TZP

No.	SCFs k	Notch Radius ρ (mm)	L_c (mm)	$K_{\rho c}$ (exp) ($MPam^{0.5}$)	$K_{\rho c}$ (pre) ($MPam^{0.5}$)	R.E. (%)
5	13.81	0.0075	0.013	7.41	7.31	−1.3
6	13.81	0.0075	0.013	7.64	7.31	−4.3
7	9.04	0.018	0.0198	9.47	8.95	−5.5
8	6.04	0.0415	0.0297	10.8	11.01	1.9
9	5.37	0.053	0.0334	12.2	11.74	−3.8
10	4.63	0.072	0.0387	11.3	12.76	13
11	4.4	0.08	0.0407	11.4	13.15	15.3
12	2.98	0.179	0.0601	17.4	16.75	−3.7
13	2.01	0.4035	0.089	22.2	22.03	−0.8
14	2.01	0.4035	0.089	23.9	22.03	−7.8

Note: (*a*) $k = m(a/\rho)^n$, where m = 0.8573, and n = 0.4832; (*b*) k^* = 18.71, ρ^* = 0.004 mm, L_c^* = 0.00858 mm; (*c*): M.E = 0.3%, M.A.E = 5.7%, E.R. = [−7.8,15.3]%

Fracture test data of U-notch FPB specimens made of Y-TZP are listed in Table 5.15. As done for SI3N4, here, the mean value of fracture toughness values of U-notch specimens with very small notch radii listed in Table 5.15, 6.35 $MPam^{0.5}$, is taken to be K_{Ic} in this study.

Here, some calculation details are described briefly.

(1) Fracture test data of order numbers, 7 and 12, shown in Table 5.15, are used to calculate the critical stress σ_0 by using Eqs. (14) to (16). Its value is 873 MPa.
(2) For U-Notch FPB specimens, the stress concentration factor, k, is calculated using Eq.(26), where m = 0.8573, and n = 0.4832 are taken from Ref.[27].
(3) By means of the approach of determining k^* proposed in this study, the GMPs of U-notch FPB specimens made of Y-TZP are calculated. The obtained GMPs are: k^* = 18.71, ρ^* = 0.004 mm, L_c^* = 0.00858 mm.
(4) By using the values of the GMPs obtained, fracture analysis of U-notch FPB specimens made of Y-TZP is performed by means of the local stress field failure model proposed in this study. The variations of the predicted fracture toughness $K_{\rho c}$ and the critical distance L_c with notch radii are shown in Table 5.16. By comparing the predicted $K_{\rho c}$ with the measured $K_{\rho c}$, errors obtained are listed in Table 5.16, with error indexes: M.E = 0.3%, M.A.E = 5.7%, E.R. = [−7.8,15.3]%.

Example 4: Fracture Analysis of U-Notch TPB Specimens Made of a Steel

In 2011, fracture assessment of U-notch TPB specimens were performed by Berto and Barati [44]. Material chosen is a steel with a ferritic–pearlitic structure tested at −40°C.

The Fracture toughness, K_{Ic}, and the ultimate tensile strength, σ_{UTS}, of the steel at −40°C reported in Ref.[44] are 12.3 $MPam^{0.5}$ and 502 MPa.

The geometric sizes of U-notch TPB specimens, the span length, the width and the thickness were S = 220 mm, W = 40 mm, and B = 20 mm, respectively. Two different values of the notch depth were considered a = 10 and 20 mm, while the notch radius ranged from 0.5 to 2 mm.

For notch depth a = 20 mm, fracture test data and some calculation results are given in Table 5.17. Some calculation retails are as follows:

(1) For U-notch TPB specimens, the stress concentration factor, k, is calculated using Eq.(26), where m = 0.8651 and n = 0.4615 are taken from Ref.[27].
(2) The values of the measured notch fracture toughness, $K_{\rho c}$, are calculated by using Eq.(13).
(3) Fracture test data of U-notch TPB specimens with notch radii 0.5 mm and 1.5 mm shown in Table 5.17 are used to calculate the critical stress σ_0 by using Eqs. (14) to (16). Its value is 475 MPa.
(4) By means of the approach of determining k^* proposed in this study, the GMPs of U-notch TPB specimens made of the steel at −40°C are calculated. The obtained GMPs are: $k^* = 6.42$, $\rho^* = 0.26$ mm, $L_c^* = 0.118$ mm.
(5) By using the obtained GMPs, fracture analysis of U-notch TPB specimens is performed by means of the local stress field failure model proposed in this study. The obtained results are listed in Table 5.17, with the calculation error indexes: M.E. = 0.3%, M.A.E. = 3.8%, E.R. = [−3.8,6.3]%.

For notch depth a = 10 mm, fracture test data and some calculation results are given in Table 5.18. Some calculation retails are as follows:

(1) The stress concentration factor, k, is calculated using Eq.(26), where m = 1.3055, and n = 0.4772 are taken from Ref.[27].
(2) By using the critical stress σ_0 = 475 MPa obtained above and the fracture toughness K_{Ic} = 12.3 MPam$^{0.5}$ given in Ref. [44], the GMPs of U-notch TBP specimens are determined by means of the approach of determining k^* proposed in this study. The obtained GMPs are: $k^* = 7.32$, $\rho^* = 0.27$ mm, $L_c^* = 0.117$ mm.
(3) By using the obtained GMPs, fracture analysis of U-notch TPB specimens is performed by means of the local stress field failure model proposed in this study. The obtained results are listed in Table 5.18, with the calculation error indexes: M.E. = 0.2%, M.A.E. = 1.9%, E.R. = [−2.8,2.3]%.

TABLE 5.17
Fracture Test Data and Some Calculation Results (Notch Depth = 20 mm)

No.	Notch Radius (mm)	Failure Load F_c (N)	SCFs k	σ_{max} (MPa)	$K_{\rho c}$ (exp) (MPam$^{0.5}$)	$K_{\rho c}$ (pre) (MPam$^{0.5}$)	E.R. (%)
1	0.5	4010	4.75	785.21	15.56	14.96	−3.8
2	1.0	4450	3.45	632.81	17.73	18.85	6.3
3	1.5	5570	2.86	656.91	22.55	21.81	−3.3
4	2.0	5830	2.5	602.09	23.86	24.31	1.9

Note: (a) $k = m(a/\rho)^n$, where 0.8651 and n = 0.4615; (b) $k^* = 6.42$, $\rho^* = 0.26$ mm, $L_c^* = 0.118$ mm; (c) M.E. = 0.3%, M.A.E. = 3.8%, E.R. = [−3.8,6.3]%

TABLE 5.18
Fracture Test Data and Some Calculation Results (Notch Depth = 10 mm)

No.	Notch Radius (mm)	Failure Load F_c (N)	SCFs k	σ_{max} (MPa)	$K_{\rho c}$ (exp) (MPam$^{0.5}$)	$K_{\rho c}$ (pre) (MPam$^{0.5}$)	E.R. (%)
1	0.5	7730	5.45	772.77	15.31	14.88	−2.8
2	1	9140	3.92	656.39	18.4	18.82	2.3
3	1.5	10,800	3.23	639.16	21.94	21.82	−0.5
4	2	11,690	2.81	603.09	23.9	24.34	1.8

Note: (a) $k = m(a/\rho)^n$, where $m = 1.3055$, and $n = 0.4772$; (b) $k^* = 7.32$, $\rho^* = 0.27$ mm, $L_c^* = 0.117$ mm; (c) M.E. = 0.2%, M.A.E. = 1.9%, E.R. = [−2.8,2.3]%

Example 5: Fracture Analysis of Center U-Notched Graphite Plates

In 2012, brittle fracture of center U-notched graphite (isostatic polycrystalline graphite) plate specimens was studied by Berto et al.[45]. Main dimensions of the tested center U-notched graphite plates are width W = 50 mm, thickness B = 10 mm, center notch depth $2a$ = 10 mm. The ultimate tensile strength, σ_{UTS}, and the fracture toughness, K_{Ic}, of the material studied are 46 MPa and 1.06 MPam$^{0.5}$, respectively.

Fracture test data reported in Ref.[45] and some calculation results given in this study are shown in Table 5.19. Here, some calculation details are as follows:

(1) The stress concentration factor, k, is calculated by using the equation (27) proposed by Shin et al [28], where $F = 1.025$.

(2) The fracture toughness of notch specimens, $K_{\rho c}$ (exp), shown in Table 5.19, is calculated by using Eq.(13).

(3) Fracture test data of order numbers, 1 and 4, shown in Table 5.19, are used to calculate the critical stress σ_0 by using Eqs. (14) to (16). Its value is 13.86 MPa.

(4) By using the critical stress σ_0 = 13.86 MPa obtained above and the fracture toughness K_{Ic} = 1.06 MPam$^{0.5}$ given in Ref. [45], the GMPs of center U-notch plate specimens studied are determined by means of the approach of determining k^* proposed in this study. The obtained GMPs are: $k^* = 14.23$, $\rho^* = 0.12$ mm, $L_c^* = 0.98$ mm.

(5) By using the obtained GMPs, fracture analysis of center U-notch plate specimens can be performed by means of the local stress field failure model proposed in this study. The obtained results are shown in Table 5.19, with the calculation error indexes: M.E. = 0.2%, M.A.E. = 1.9%, E.R. = [−17.1,8.2]%.

TABLE 5.19
Fracture Test Data Reported in [45] and Some Calculation Results (Center U-Notched Graphite Plates)

No.	Notch Radius (mm)	Failure Load F_c (N)	SCFs k	σ_{max} (MPa)	$K_{\rho c}$ (exp) ($MPam^{0.5}$)	$K_{\rho c}$ (pre) ($MPam^{0.5}$)	E.R. (%)
1	0.25	4426	10.17	90.01	1.26	1.24	−1.7
2	0.5	4505	7.48	67.42	1.34	1.43	6.9
3	1.0	4814	5.58	53.76	1.51	1.63	8.2
4	2.0	5516	4.24	46.79	1.85	1.84	−0.5
5	4.0	6789	3.29	44.7	2.51	2.08	−17.1

Note: (a) $k^* = 14.23$, $\rho^* = 0.12$ mm, $L_c^* = 0.98$ mm; (b) M.E. = 0.2%, M.A.E. = 1.9%, E.R. = [−17.1,8.2]%

5.6 CONCLUSIONS OF PART 1

From the study of Part 1, the following conclusions can be made:

(1) The approach of determining k^* proposed in this study is feasible.
(2) By using the GMPs, the failure assessment analysis of notched specimens performed by means of the local stress field failure model proposed in this study has high accuracy.
(3) The stress concentration factor eigenvalue, k^*, introduced in this study, is a parameter that depends both on the geometry of notched specimens and on the material of notched specimens (see Table 5.20).

PART 2

From the study of Part 1, it can be seen that the stress concentration factor eigenvalue, k^*, is a very important parameter in the failure assessment analysis of notched specimens. From Table 5.20, moreover, it can be observed that the eigenvalue, k^*, is such a parameter that depends both on the geometry of notched specimens and on the material of notched specimens. In order to perform fracture analysis for any V-notch with high accuracy, in Part 2, therefore, an attempt is made to propose some models to predict k^*, including the effect of different notch angles on k^*, the effect of different notch depths on k^* and the effect of different materials on k^*.

5.7 EFFECT OF NOTCH ANGLES ON k^*

In failure assessment analysis of notched specimens, notch angle effect is to be considered. In this section, such a model of the effect of notch angles on k^* will be proposed.

TABLE 5.20
Summary of the GMPs of U-Notch Specimens for Different Materials

Materials	Notch Specimen Type	Geometric Size (mm)	k^*	ρ^* (mm)	L_c^* (mm)	Error Indexes (%)
PMMA at −60°C, K_{Ic} = 1.7 MPam$^{0.5}$ σ_0 = 128.4 MPa	TPB	S = 126 mm, B = 14 mm W = 28 mm, a = 14 mm	19.71	0.016	0.0321	M.E = 0.12 M.A.E = 4.2 [−11,3.9]
	TPB	S = 126 mm, B = 14 mm W = 28 mm, a = 5 mm	20.88	0.02	0.0326	M.E = −0.2 M.A.E = 5 [−11.2,6.6]
	DEN	W = 28 mm, B = 14 mm a = 8 mm	27.5	0.066	0.031	M.E = 0.2 M.A.E = 0.8 [−1.1,0.8]
	DEN	W = 28 mm, B = 14 mm a = 11 mm	31.84	0.11	0.0247	M.E = 0.8 M.A.E = 1 [−0.2,2.3]
	SEN	W = 28 mm, B = 14 mm a = 14 mm	43.7	0.06	0.0316	M.E = 0.1 M.A.E = 7.4 [−11,6.2]
	RNC	D = 10 mm a = 1 mm	10.96	0.09	0.028	M.E. = 0.4 M.A.E. = 5.4 E.R. = [−8.1,8.4]
	RNC	D = 10 mm a = 3 mm	15.71	0.04	0.033	M.E. = 0.4 M.A.E. = 3.5 E.R. = [−2.6,5.8]
PMMA at room temperature, K_{Ic} = 1 MPam$^{0.5}$ σ_0 = 86 Mpa	TPB	S = 126 mm, B = 14 mm W = 28 mm, a = 14 mm	27.15	0.008	0.0239	M.E. = 0.1 M.A.E. = −5.7 E.R. = [−8.8,6.8]
	SEN	W = 28 mm, B = 14 mm a = 14 mm	133.58	0.007	0.0236	M.E = −0.04 M.A.E = 2.5 [−3.7,2.6]
		W = 28 mm, B = 14 mm a = 5 mm	73	0.005	0.0239	
A steel with ferritic pearlitic structure tested at −40°C K_{Ic} = 12.37 MPam$^{0.5}$ σ_0 = 475 MPA	TPB	S = 220 mm, B = 20 mm W = 40 mm, a = 20 mm	6.42	0.26	0.118	M.E. = −0.26, M.A.E. = 3.8, E.R. = [−3.8,6.3]
	TPB	S = 220 mm, B = 20 mm W = 40 mm, a = 10 mm	7.32,	0.27	0.117	M.E. = 0.2, M.A.E. = 1.9, E.R. = [−2.8,2.3]

SI3N4 K_{Ic} = 5.17 MPam$^{0.5}$ σ_0 = 560 MPam$^{0.5}$	TPB	S = 16 mm, W = 4 mm, B = 3 mm, a = 1.88 mm	9.33	0.012	0.01598	M.E = 0.5 M.A.E = 5.1 [–9.5,7]
Y-TZP K_{Ic} = 6.35 MPam$^{0.5}$ σ_0 = 873.1 MPa	FPB	S_1 = 40 mm, S_2 = 20 mm W = 4.62 mm, B = 3 mm, a = 2.36 mm	18.71	0.004	0.00958	M.E = –0.3 M.A.E = 5.7 [–15,5.5]
Graphite K_{Ic} = 1.06 MPam$^{0.5}$ σ_0 = 13.86 Mpa	CN	W = 50 mm, 2a = 10 mm B = 10 mm	14.23	0.12	0.98	M.E = 0.8 M.A.E = 6.9 [–6.9,17.1]

5.7.1 A Model of the Effect of Notch Angles on k^*

It is assumed that there are two pointed V-notches with notch angles α_1 and α_2 (Williams' eigenvalues λ_1 and λ_2), which are denoted by pointed V-notch 1 and pointed V-notch 2, respectively. The two pointed V-notches have the GMPs of (k_1^*, ρ_1^*, L_{c1}^*) and (k_2^*, ρ_2^*, L_{c2}^*) and the fracture toughness, K_{vc1} and K_{vc2}.

By the way, it is pointed out here that, if the fracture toughness K_{Ic} is known, the fracture toughness, K_{vc1} and K_{vc2}, can be calculated by means of the model in Ref.[46].

For pointed V-notch 1, the following equation is valid from the local stress field failure model proposed in this study (see equation (14b)):

$$F(K_{vc1}, L_{c1}^*, \rho_1^*, \alpha_1, \sigma_0) = 0 \tag{30}$$

For pointed V-notch 2, a similar equation to Eq. (30) can be written as:

$$F(K_{vc2}, L_{c2}^*, \rho_2^*, \alpha_2, \sigma_0) = 0 \tag{31}$$

Here, the following equation is assumed to be valid:

$$G(\alpha_2) \cdot L_{c2}^* = G(\alpha_1) \cdot L_{c1}^* \tag{32}$$

where $G(\alpha_2)$ and $G(\alpha_1)$ are the **notch geometric shape factors** (NSFs) of point V-notch 2 and point V-notch 1, respectively. According to Ref.[41], the notch stress intensity factor of point V-notch with notch angle α (Williams' eigenvalue λ) is expressed as:

$$K_v = \sigma_{nom,n} \cdot G(\alpha) \cdot W^{1-\lambda} \tag{33}$$

where $\sigma_{nom,n}$ is the net nominal stress, W is the width of the notched specimen, and λ is Williams' eigenvalue.

If the GMPs of pointed V-notch 1 (k_1^*, ρ_1^*, L_{c1}^*) have been obtained by means of the approach of determining k^* proposed in this study, the GMPs of pointed V-notch 2 (k_2^*, ρ_2^*, L_{c2}^*) can be calculated as follows:

(1) L_{c2}^* is calculated from Eq.(32);
(2) By substituting L_{c2}^* obtained into Eq.(31), ρ_2^* can be calculated;
(3) From the obtained ρ_2^* and the geometry of pointed V-notch 2, it is easy to obtain the solution of k_2^* by using the Finite Element Method or using related references (e.g. Refs [19–22,24–27]).

5.7.2 Experimental Verifications

Here, the model of the effect of notch angles on k^* is verified by using the fracture test data of TPB V-notch specimens made of PMMA at -60°C reported in Ref.[39].

In the Part 1 of this study, the GMPs of U-notch TPB specimens made of PMMA at -60°C have been obtained by means of the approach of determining k^* proposed in this study (see Table 5.20). By using the fracture toughness K_{Ic} and the critical stress σ_0 of the studied material, the fracture toughness, K_{vc}, of point V-notches with notch angles 60°, 90°, 120° and 150° can be calculated by using the model reported in Ref.[46]. Further, the GMPs of these point V-notches can be calculated by using the model of the effect of notch angles on k^* proposed in this section. The obtained results are given in Table 5.21.

For any point V-notch TPB specimen (with notch angle α) made of PMMA at -60°C, the failure assessment analysis of V-notch TPB specimens with different notch radii can be performed by means of the local stress field failure model proposed in this study after the GMPs corresponding to the point V-notch have been obtained. For the V-notch TPB specimens with different notch radii (with notch angle 90°), for example, fracture test data and some calculated results are given in Table 5.22, with the calculated error indexes: M.E. = −4.5%, MA.E. = 4.5%, E.R. = [−8.91,−0.4]%. The predicted fracture toughness K_{vpc} is in excellent agreement with that measured experimentally. For notch angles α = 120°, 150° and 60°, the comparisons of fracture test data and the predicted results are given in Tables 5.23 to 5.25. As done for notch angle $\alpha = 90^\circ$, the predicted fracture toughness K_{vpc} is in excellent agreement with that measured experimentally.

TABLE 5.21
The GMPs of Pointed V-Notch TPB Specimens Made of PMMA at -60°C (a/W = 0.5)

Notch Angle (°)	λ [6]	NSFs G [41]	K_{vc} (MPam$^{1-\lambda}$)	L_c^* (mm)	ρ_c^* (mm)	k^*
0	0.5000	1.767	1.71	0.6324	0.016	19.71
60	0.5122	1.857	1.94	0.0305	0.108	8.57
90	0.5448	2.118	2.71	0.0268	0.087	7.50
120	0.6157	2.77	5.50	0.0205	0.080	7.85
150	0.7520	4.483	18.94	0.0127	0.0005	14.42

TABLE 5.22
Fracture Test Data and Some Calculated Results of V-Notch TPB Specimens Made of PMMA at −60°C (Notch Angle 90°, a/W = 0.5)

No.	Failure Load F_c (N)	SCFs k	Notch Radius ρ (mm)	L_c (mm)	$K_{v\rho c}$ (exp)	$K_{v\rho c}$ (cal)	R.E (%)
1	660	4.1	0.47	0.0575	4.20	4.57	−8.9
2	880	3.1	0.84	0.0748	5.60	5.62	−0.4
3	510	6.8	0.15	0.0342	3.24	3.16	2.4
		6.0	0.20	0.039	3.24	3.45	−6.5

Note: (a) Unit of $K_{v\rho c}$ is $MPam^{1-\lambda}$ ($\lambda = 0.5448$); (b) The third case, $\rho = 0.04$ mm [39]< $\rho_c^* = 0.087$ mm, while the notch facture toughness $K_{v\rho c} = 3.24 > K_{vc} = 2.71$, which is not reasonable; (c) For the third case, here, notch radius is taken to be 0.15 mm and 0.20 mm, the predicted fracture toughness $K_{v\rho c}$ is in excellent agreement with that measured experimentally; (d) M.E. = −4.5%, MA.E. = 4.5%, E.R. = [−8.9,−0.4]%

TABLE 5.23
Fracture Test Data and Some Calculated Results of V-Notch TPB Specimens Made of PMMA at −60°C (Notch Angle 120°, a/W = 0.5)

No.	Failure Load F_c (N)	SCFs k	Notch Radius ρ (mm)	L_c (mm)	$K_{v\rho c}$ (exp)	$K_{v\rho c}$ (cal)	R.E (%)
1	830	3.7	0.43	0.0391	8.9	8.39	5.8
2	960	2.9	0.83	0.0503	10.3	10.22	0.7
3	570	5.7	0.14	0.0254	6.11	6.23	−1.8
		5.8	0.13	0.0247	6.11	6.12	−0.1
		6	0.12	0.0239	6.11	6	1.8
		6.2	0.11	0.0231	6.11	5.88	3.8
		6.5	0.10	0.0223	6.11	5.76	5.9

Note: (a) Unit of $K_{v\rho c}$ is $MPam^{1-\lambda}$ ($\lambda = 0.6157$); (b) The third case, $\rho = 0.06$ mm [39]< $\rho_c^* = 0.08$ mm, while notch facture toughness $K_{v\rho c} = 6.11 > K_{vc} = 5.5$, which is not reasonable. (c) For the third case, here, notch radius is taken to be 0.14 mm, 0.13 mm, 0.12 mm, 0.11 mm and 0.10 mm, the predicted fracture toughness $K_{v\rho c}$ is in excellent agreement with that measured experimentally; (d) M.E. = 3.1%, MA.E. = 3.1%, E.R. = [0.7,5.8]%.

TABLE 5.24
Fracture Test Data and Some Calculated Results of V-Notch TPB Specimens Made of PMMA at −60°C (Notch Angle 150°, a/W = 0.5)

No.	Failure Load F_c (N)	SCFs k	Notch Radius ρ (mm)	L_c (mm)	$K_{v\rho c}$ (exp)	$K_{v\rho c}$ (cal)	R.E (%)
1	990	4.28	0.06	0.0424	29.97	27.05	9.7
2	1400	1.80	2.0	0.1012	40.92	41.77	−2.1
3	1500	1.51	2.4	0.1201	43.83	43.66	0.4

Note: (a) Unit of $K_{v\rho c}$ is $MPam^{1-\lambda}$ ($\lambda = 0.7520$); (b) M.E. = 2.7%, MA.E. = 4.1%, E.R. = [−2.1,9.7]%.

TABLE 5.25
Fracture Test Data and Some Calculated Results of V-Notch TPB Specimens Made of PMMA at −60°C (Notch Angle 60°, $a/W = 0.5$)

Failure Load F_c (N)	SCFs k	Notch Radius ρ (mm)	L_c (mm)	$K_{v\rho c}$ (exp)	$K_{v\rho c}$ (cal)	R.E (%)
450	7.3	0.15	0.0358	2.29	2.13	6.6
	6.8	0.175	0.0386	2.29	2.24	2.1
	6.6	0.185	0.0397	2.29	2.28	0.3
	6.3	0.20	0.0412	2.29	2.33	−2.2

Note: (a) Unit of $K_{v\rho c}$ is $\mathrm{MPam}^{1-\lambda}$ ($\lambda = 0.5122$); (b) $\rho = 0.04$ mm [39] $< \rho_c^* = 0.108$ mm, while notch facture toughness $K_{v\rho c} = 2.29 > K_{vc} = 1.94$, which is not reasonable; (c) Here, notch radius is taken to be 0.15 mm, 0.175 mm, 0.185 mm and 0.20 mm, the predicted fracture toughness $K_{v\rho c}$ is in excellent agreement with that measured experimentally.

5.8 EFFECT OF NOTCH DEPTH ON K^*

It is assumed that there are two point V-notches with the same notch angle α (Williams' eigenvalue λ), and with different notch depths a_1 and a_2, which are denoted by pointed V-notch 1 and pointed V-notch 2, respectively. It is assumed also that the GMPs of pointed V-notch 1 (k_1^*, ρ_1^*, L_{c1}^*) have been obtained by using the approach to determine k^* proposed in this study. Naturally the notch fracture toughness K_{vc} and the critical stress σ_0 are known. Now an attempt is made to propose a model to calculate the GMPs of pointed V-notch 2 (k_2^*, ρ_2^*, L_{c2}^*) from the GMPs of pointed V-notch 1 (k_1^*, ρ_1^*, L_{c1}^*).

According to the local stress field failure model proposed in this study, the following equations are valid:

$$F(K_{vc}, L_{c1}^*, \rho_1^*, \alpha, \sigma_0) = 0 \tag{34}$$

and

$$F(K_{vc}, L_{c2}^*, \rho_2^*, \alpha, \sigma_0) = 0 \tag{35}$$

It is found from the GMPs of different notch geometries listed in Table 5.20 that a different notch depth model is proposed for a different notch geometry.

5.8.1 Notch Depth Model for TPB Notch Specimens and Experimental Verifications

By observing the GMPs of U-notch TPB specimens made of PMMA at −60°C shown in Table 5.20, it is assumed here that k^* is not varied with notch depths, i.e.,

$$k^* = c \tag{36}$$

where c is a constant.

Here, U-notch TPB specimens made of PMMA at room temperature reported in Ref.[42] are taken to verify the assumption of (36) to be valid.

For U-notch TPB specimens made of PMMA at room temperature, the GMPs have been obtained at the geometry of S = 126 mm, B = 14 mm, W = 28 mm, a = 14 mm (see Table 5.20), in which $k^* = 27.16$. The notch is taken to be pointed V-notch 1, i.e., $k_1^* = 27.16$. According to the assumption (36), $k_2^* = 27.16$ for any pointed V-notch 2. From the geometry of pointed V-notch 2 and $k_2^* = 27.16$, ρ_2^* can be obtained. Note here that the stress concentration factors of U-notch TPB specimens are calculated using Eq.(26). Further, L_{c2}^* is calculated by substituting ρ_2^* obtained into Eq.(35). The calculation process data are given in Table 5.26.

After the GMPs of pointed V-notch 2 have been calculated, the failure assessment of U-notch TPB specimens with different notch radii can be performed by means of the local stress field failure model proposed in this study. The obtained results and fracture test data reported in Ref.[42] are listed in Table 5.27, from which it can be seen that the predicted notch fracture toughness is in excellent agreement with that measured experimentally.

TABLE 5.26
Calculation Process Data of Notch Depth Effect for U-Notch TPB Specimens Made of PMMA at Room Temperature

Notch Depth (mm)	m	n	ρ_2^* (mm)	L_{c2}^* (mm)
2	2.0076	0.4765	0.00845	0.0239
5	1.5201	0.4772	0.012	0.0245
10	1.0701	0.4752	0.011	0.0244
14	0.8651	0.4615	0.008	0.0239

Note: (*a*) $k_2^* = 27.16$; (*b*) m, n in this table are constants in Eq. (26).

TABLE 5.27
Fracture Test Data Reported in Ref.[42] and Evaluation Calculation Results (Notch Radius ρ = 1.5 mm)

Notch Depth (mm)	Failure Load F_c (N)	SCFs k	$\sigma_{non,n}$ (MPa)	$K_{\rho c}$ (cal)	$K_{\rho c}$ (exp)	R.E. (%)
2	2700	2.30	47.93	3.79	4.03	6.4
5	2000	2.70	45.37	3.90	4.20	−7.3
10	1100	2.64	40.74	3.92	3.68	6.4
14	800	2.43	48.98	3.98	4.08	−2.4

Note: (a) Main sizes of U-notch TPB specimen: S = 126 mm, B = 14 mm, W = 28 mm; (b) Unit of $K_{\rho c}$ is MPam$^{0.5}$

5.8.2 Notch Depth Model for SEN Notch Specimens and Experimental Verifications

By observing the GMPs of U-notch SEN specimens made of PMMA at room temperature shown in Table 5.20, it is assumed here that the critical distance L_c^* is not varied with notch depths,

$$L_c^* = c \tag{37}$$

where c = 0.0236 mm at the geometry of W = 28 mm, B = 14 mm, a = 14 mm. Now, the notch is taken to be pointed V-notch 1, i.e., $L_{c1}^* = 0.0236$ mm. According to the assumption (37), for any pointed V-notch 2, $L_{c2}^* = 0.0236$ mm. Then ρ_2^* can be calculated from Eq.(35). From the geometry of pointed V-notch 2 and the value of ρ_2^*, further, k_2^* can be calculated by using the method described in Section 5 for U-notch SEN specimens. After the GMPs of pointed V-notch 2 have been calculated, the failure assessment of V-notch SEN specimens with different notch radii can be performed by means of the local stress field failure model proposed in this study. The obtained results and fracture test data reported in Ref. [42] are listed in Table 5.28, from which it can be seen that the predicted notch fracture toughness $K_{\rho c}$ is in good agreement with that measured experimentally.

5.8.3 Notch Depth Model for DEN Notch Specimens and Experimental Verifications

From the study on the GMPs of U-notch DEN specimens made of PMMA at −60°C (see Table 5.20), here, the following relation is assumed to be valid for U-notch DEN specimens,

$$F(a_1) \cdot L_{c1}^* = F(a_2) \cdot L_{c2}^* \tag{38}$$

TABLE 5.28
Fracture Test Data Reported in Ref.[42] and Evaluation Calculation Results (W = 28 mm, B = 14 mm)

Notch Depth (mm)	Notch Radius ρ (mm)	SCFs k	Failure Load F_c (N)	σ_n (Mpa)	σ_{max} (Mpa)	$K_{\rho c}$ (cal) (MPam$^{0.5}$)	$K_{\rho c}$ (exp) (MPam$^{0.5}$)	R.E. (%)
5	1	5.2	8700	27.01	140.18	3.73	3.93	−5.1
10	1	8.3	4600	18.25	151.50	3.73	4.25	−12.1

Note: (a) When notch depth a = 5 mm, $k^* = 63.9$, $\rho^* = 0.0066$ mm, $L_c^* = 0.0236$ mm; (b) When notch depth a = 10 mm, $k^* = 102.2$, $\rho^* = 0.0066$ mm, $L_c^* = 0.0236$ mm.

where F is the geometric shape function (GSF) of crack in LEFM. For DEN notch specimens, thus, the GMPs (k_2^* , ρ_2^* , L_{c2}^*) of pointed V-notch 2 are calculated as follows:

(1) The L_{c2}^* is calculated from Eq.(38);
(2) By substituting the obtained L_{c2}^* into Eq.(35), ρ_2^* can be calculated;
(3) From the geometry of pointed V-notch 2 and the value of ρ_2^*, k_2^* can be determined by using the Finite Element Method or related references (e.g., Refs [19–22,24–27]).

Here, U-notch DEN specimens made of PMMA at −60°C reported in Ref.[39] are taken to verify the notch depth model in this subsection. For two notch depths a = 8 mm and a = 11 mm, the GMPs of U-notch DEN specimens have been obtained in section 5 (see also Table 5.29). In Table 5.29, the values of $F \cdot L_c^*$ for the two notch depths are given. From $(F(a_2)\cdot L_{c2}^*)/(F(a_1)\cdot L_{c1}^*) = 0.98$, the assumption (38) can be considered to be valid.

The following is the details in which the notch depth model is completed:

(1) From Eq.(38), the calculated L_{c2}^* = 0.02415 mm;
(2) By substituting L_{c2}^* = 0.02415 mm into Eq.(35), the calculated ρ_2^* = 0.112 mm;
(3) From the geometry of pointed V-notch 2 and ρ_2^* = 0.112 mm, the obtained k_2^* = 31.56.

By using the GMPs of pointed V-notch 2, the failure assessment analysis of notch 2 with different notch radii is performed by means of the local stress field failure model proposed in this study. The obtained results and fracture test data reported in Ref.[39] are given in Table 5.30, from which it can be seen that the calculated fracture toughness $K_{\rho c}$ is in excellent agreement with that measured experimentally, with the calculation error indexes: M.E. = −0.6%, M.A.E. = 3.5%, E.R. = [−6.3,4.1]%

As another example, U-notch DEN specimens with notch depths a = 11 and a = 8 are taken here as notch 1 and notch 2, respectively. Similar results to Table 5.30 can be obtained (see Table 5.31), with the calculation error indexes: M.E. = −1%, M.A.E. = 1%, E.R. = [−2.3,−0.1]%.

TABLE 5.29

The GMPs of U-Notch DEN Specimens Made of PMMA at −60°C (W = 28 mm, B = 14 mm)

Notch Depth a (mm)	F	k^*	ρ^* (mm)	L_c^* (mm)	$F \cdot L_c^*$ (mm)	Notch
8	1.201	27.5	0.066	0.031	0.0380	Notch 1
11	1.542	31.84	0.11	0.0247	0.0372	Notch 2

Note: F in this table is the geometric shape function (GSF) of crack in LEFM.

TABLE 5.30
Fracture Test Data of U-Notch DEN Specimens Made of PMMA at −60°C and Evaluation Calculation Results (Notch Depth *a* = *11* mm)

Notch Radius (mm)	Failure Load (N)	SCFs k	Critical Distance (mm)	$K_{\rho c}$ (cal) ($MPam^{0.5}$)	$K_{\rho c}$ (exp) ($MPam^{0.5}$)	E.R. (%)
0.52	4000	15.2	0.0501	3.08	3.14	−1.7
1.01	5000	11.2	0.0681	4.10	4.02	1.8
1.46	5700	9.4	0.0810	4.82	4.63	4.1
1.96	7000	8.4	0.0910	5.50	5.87	−6.3

Note: (a) $k^* = 31.56$, $\rho^* = 0.112$ mm, $L_c^* = 0.02415$ mm; (b) M.E. = −0.6%, M.A.E. = 3.5%, E.R. = [−6.3,4.1]%

TABLE 5.31
Fracture Test Data of U-Notch DEN Specimens Made of PMMA at −60°C and Evaluation Calculation Results (Notch Depth *a* = *8* mm)

Notch Radius (mm)	Failure Load (N)	SCFs k	Critical Distance (mm)	$K_{\rho c}$ (cal) ($MPam^{0.5}$)	$K_{\rho c}$ (exp) ($MPam^{0.5}$)	E.R. (%)
0.53	6400	10.3	0.0317	3.46	3.43	−0.8
0.97	7600	8.0	0.0886	4.36	4.26	−2.3
1.46	8940	6.6	0.1147	5.15	5.12	−0.6
1.91	9890	5.9	0.1377	5.76	5.76	−0.1

Note: (a) $k^* = 28.8$, $\rho^* = 0.06$ mm, $L_c^* = 0.0317$ mm; (b) M.E. = −1%, M.A.E. = 1%, E.R. = [−2.3,−0.1]%

5.8.4 Notch Depth Model for RNT Notch Specimens and Experimental Verifications

From the study on the GMPs of U-notch RNT specimens made of PMMA at −60°C (see Table 5.20), here, the following relation is assumed to be valid for RNT notch specimens

$$\frac{F_{DEN}}{L_c^*} = c \tag{39}$$

The GMPs of U-notch RNT specimens made of PMMA at −60°C obtained in Section 5 with notch depths $a = 1$ mm and $a = 3$ mm are listed in Table 5.32. In Table 5.32, the geometric shape functions, F_{RNC} (Round notched cylindrical (RNC) specimens) and F_{DEN} (Double Edge Crack) and the values of $\frac{F_{DEN}}{L_c^*}$ are also shown. From values of $\frac{F_{DEN}}{L_c^*}$, the assumption (39) can be considered to be valid.

As an example, here, the U-notches with notch depths $a = 3$ mm and $a = 1$ mm are taken as pointed V-notch 1 and pointed V-notch 2, respectively. Then we have

TABLE 5.32
GMPs of U-Notch RNT Specimens Made of PMMA at –60°C

Notch Depth a (mm)	Diameter ϕ (mm)	$2a/\phi$	F_{RNC}	F_{DEN}	k^*	ρ^* (mm)	L_c^* (mm)	$\frac{F_{DEN}}{L_c^*}$
3	10	0.6	2.516	1.226	15.71	0.04	0.033	37.15
1	10	0.2	1.261	1.118	10.96	0.09	0.028	39.93

Note: the geometric shape functions: F_{RNC} (Round notched cylindrical (RNC) specimens) and F_{DEN} (Double Edge Crack)

TABLE 5.33
Fracture Test Data of U-Notch RNT Specimens Made of PMMA at –60°C and Evaluation Calculation Results (Notch Depth a = 1 mm)

Notch Depth a (mm)	Notch Radius (mm)	SCFs k	Failure Load (N)	σ_{max} (MPa)	$K_{\rho c}$ (cal) (MPam$^{0.5}$)	$K_{\rho c}$ (exp) (MPam$^{0.5}$)	R.E. (%)
0.97	0.19	4.01	2500	196.52	2.31	2.4	–3.8
0.96	2.13	1.62	4680	147.51	6.03	6.03	–0.1
0.93	4.06	1.36	5200	135.82	8.03	7.67	4.7
0.98	7.07	1.24	5200	126.65	10.35	9.44	9.6

Note: (a) k^* = 12.0, ρ^* = 0.075 mm, L_c^* = 0.030 mm; (b) M.E. = 2.6%, M.A.E. = 4.6%, E.R. = [–3.8,9.6]%

k_1^* = 15.71, ρ_1^* = 0.04 mm, L_{c1}^* = 0.033 and $\frac{F_{DEN}}{L_c^*}$ = 37.15. Thus from Eq.(39), L_{c2}^* = 0.030 mm can be obtained. Further from Eq.(35), ρ_2^* = 0.075 mm can be obtained. From the geometry of pointed V-notch 2 and ρ_2^* = 0.075 mm, finally, k_2^* can be determined by using the Finite Element Method or related references (e.g., Refs [19–22,24–27]). Its value is 12.0.

By using the obtained GMPs of pointed V-notch 2, the failure assessment analysis of notch 2 with different notch radii is performed by means of the local stress field failure model proposed in this study. The obtained results and fracture test data of U-notch RNT specimens made of PMMA at –60°C reported in Ref.[39] are given in Table 5.33, from which it can be seen that the calculated notch fracture toughness $K_{\rho c}$ is in excellent agreement with that measured experimentally, with the calculated error indexes: M.E. = 2.6%, M.A.E. = 4.6%, E.R. = [–3.8,9.6]%.

As another example, U-notch RNT specimens with notch depths a = 1 mm and a = 3 mm are taken here as pointed V-notch 1 and pointed V-notch 2, respectively. Similar results to Table 5.33 can be obtained (see Table 5.34), with the calculated error indexes: M.E. = –3.8%, M.A.E. = 4.8%, E.R. = [–7.7,1.5]%.

TABLE 5.34
Test Data of U-Notch RNT Specimens Made of PMMA at −60°C and Evaluation Calculation Results (Notch Depth a = 3 mm)

Notch Depth a (mm)	Notch Radius (mm)	SCFs k	Failure Load Fc (N)	σ_{max} (MPa)	$K_{\rho c}$ (cal) (MPam$^{0.5}$)	$K_{\rho c}$ (exp) (MPam$^{0.5}$)	R.E. (%)
2.98	2.1	1.413	1460	160.93	6.03	6.54	−7.7
3.01	4	1.209	1450	140.91	8.02	7.9	1.5
2.99	7.05	1.183	1580	147.27	10.38	10.96	−5.3

Note: (a) $k^* = 11.87$, $\rho^* = 0.07$ mm, $L_c^* = 0.0307$ mm; (b) M.E. = −3.8%, M.A.E. = 4.8%, E.R. = [−7.7,1.5]%

5.9 EFFECT OF DIFFERENT MATERIALS ON k^*

From Table 5.20, it is found that k^* depends both on notch geometries and on notched materials. In Sections 7 and 8, a model of the effect of notch geometry on k^* was proposed. In this section, an effect of different materials on k^* will be the focus.

In Section 5, the GMPs of U-notch specimens (TPB, SEN, DEN and RNT) made of PMMA at −60°C have been obtained by using the approach of determining k^* proposed in this study.

Here, PMMA at −60°C is denoted by material 1 and the other material studied is denoted by material 2.

For material 1, obviously, the GMPs of U-notch specimens (TPB, SEN, DEN and RNT), the fracture toughness and the critical stress can be known from this study.

Here, the GMPs of U-notch TPB specimens made of material 1 (denoted by k^*_{TPB-1}, ρ^*_{TPB-1}, L^*_{TPB-1}), the fracture toughness and the critical stress of material 1 (denoted by K_{IC-1} and σ_{0-1}), have been obtained (see Table 5.20).

Here, it is assumed that the GMPs of U-notch TPB specimens made of material 2 (denoted by k^*_{TPB-2}, ρ^*_{TPB-2}, L^*_{TPB-2}) and the fracture toughness (denoted by K_{IC-2}) and the critical stress (denoted by σ_{0-2}) have been obtained.

Here, the GMPs of U-notch SEN specimens made of material 1 (denoted by k^*_{SEN-1}, ρ^*_{SEN-1}, L^*_{SEN-1}) have been obtained (see Table 5.20).

Here, the GMPs of U-notch SEN specimens made of material 2 are denoted by k^*_{SEN-2}, ρ^*_{SEN-2}, L^*_{SEN-2}, which are unknown.

In this section, a model of determining (k^*_{SEN-2}, ρ^*_{SEN-2}, L^*_{SEN-2}) is proposed from (k^*_{TPB-1}, ρ^*_{TPB-1}, L^*_{TPB-1}), (k^*_{TPB-2}, ρ^*_{TPB-2}, L^*_{TPB-2}) and (k^*_{SEN-1}, ρ^*_{SEN-1}, L^*_{SEN-1}).

According to Eq.(14b), the following equations are valid

$$F(K_{IC-1}, L^*_{TPB-1}, \rho^*_{TPB-1}, \alpha, \sigma_{0-1}) = 0 \tag{40}$$

$$F(K_{IC-2}, L^*_{TPB-2}, \rho^*_{TPB-2}, \alpha, \sigma_{0-2}) = 0 \tag{41}$$

and

$$F(K_{IC-1}, L^*_{SEN-1}, \rho^*_{SEN-1}, \alpha, \sigma_{0-1}) = 0 \tag{42}$$

$$F(K_{IC-2}, L^*_{SEN-2}, \rho^*_{SEN-2}, \alpha, \sigma_{0-2}) = 0 \tag{43}$$

where ρ^*_{SEN-2} and L^*_{SEN-2} in Eq.(43) are unknown and α in Eqs (40) to (43) is notch angle (α = 0, for U-notch). In order to determine ρ^*_{SEN-2} and L^*_{SEN-2} in Eq.(43), here, an equation must be added.

Here, the following parameter is introduced:

$$L_{ch} = (\frac{K_{Ic}}{\sigma_0})^2 \tag{44}$$

which has the unit of length. At the same time, the following parameters are defined for U-notch TPB specimens

$$\gamma_{1TPB} = L_{ch1} / L^*_{TPB-1} \tag{45}$$

$$\gamma_{2TPB} = L_{ch2} / L^*_{TPB-2} \tag{46}$$

and

$$\eta_{TPB} = \gamma_{1TPB} / \gamma_{2TPB} \tag{47}$$

For U-notch SEN specimens, similar parameters to those of (45) to (47) are defined:

$$\gamma_{1SEN} = L_{ch1} / L^*_{SEN-1} \tag{48}$$

$$\gamma_{2SEN} = L_{ch2} / L^*_{SEN-2} \tag{49}$$

and

$$\eta_{SEN} = \gamma_{1SEN} / \gamma_{2SEN} \tag{50}$$

Here, it is assumed that the following relation is valid:

$$\eta_{SEN} = \eta_{TPB} \tag{51}$$

From Eqs. (51) and (43), ρ^*_{SEN-2} and L^*_{SEN-2} can be calculated. From the geometry of U-notch SEN specimens made of material 2 and the obtained ρ^*_{SEN-2}, further, k^*_{SEN-2} can be determined by using the Finite Element Method or related references (e.g., Refs [19–22,24–27]).

As an example, here, material 2 is taken PMMA at room temperature (R.T). For U-notch specimens (TPB and SEN) made of material 1 and material 2, the GMPs and parameters γ and η are listed in Table 5.35, from which it can be seen that η_{TPB} = 0.974 is almost equal to η_{SEN} = 0.977, which shows that the assumption (51) is valid for material 1 and material 2 studied here.

From (48) and (51), the following relation can be obtained

$$L^*_{SEN-2} = L_{ch2}\eta_{TPB} / \gamma_{1SEN} \tag{52}$$

TABLE 5.35
GMPs of U-Notch PAMM Specimens at Room Temperature (R.T.) and at −60°C (a/W = 0.5)

Materials	Type of Specimens	k^*	ρ^* (mm)	L_c^* (mm)	γ	η
PMMA(R.T)	TPB	27.15	0.008	0.0239	5.605	0.974
PMMA(−60°C)	TPB	19.72	0.016	0.0279	5.461	
PMMA(R.T)	SEN	133.58	0.007	0.0236	5.676	0.977
PMMA(−60°C)	SEN	43.7	0.060	0.0316	5.547	

Note: (a) PMMA (−60°C), K_{Ic} = 1.7 Mpam$^{1/2}$, σ_0 = 128.4 MPa, L_{ch} = 0.175 mm; (b) PMMA(R.T), K_{Ic} = 1.0 Mpam$^{1/2}$, σ_0 = 86.4 MPa, L_{ch} = 0.134 mm

TABLE 5.36
Fracture Test Data of U-Notch SEN Specimens at Room Temperature and Evaluation Calculation Results (W = 28 mm, B = 14 mm, a/W = 0.5)

Failure Load (N)	SCFs k	Notch Radius ρ (mm)	Critical Distance L_c (mm)	$K_{\rho c}$ (exp) (Mpam$^{1/2}$)	$K_{\rho c}$ (Pre) (Mpam$^{1/2}$)	R.E (%)
1800	15.81	0.5	0.172	2.88	2.99	−3.9
2400	11.18	1.0	0.243	3.84	3.73	2.6
2600	9.13	1.5	0.297	4.15	4.30	−3.4
3000	7.90	2.0	0.343	4.79	4.76	0.6

Note: (a) L^*_{SEN-2} = 0.02353 mm, ρ^*_{SEN-2} = 0.065 mm, k^*_{SENB-2} = 138.6; (b) M.E. = −1%, M.A.E. = 2.6%, E.R. = [−3.9.2.6]%

From L_{ch2} = 0.134 mm, η_{TPB} = 0.974 and γ_{1SEN} = 5.547, here, we have L^*_{SEN-2} = 0.02353 mm. From Eq.(43), further, we have ρ^*_{SEN-2} = 0.065 mm. From the geometry of U-notch SEN specimen made of material 2 and ρ^*_{SEN-2} = 0.065 mm, finally, k^*_{SEN-2} = 138.6 can be obtained by using the method described in Section 5 for U-notch SEN specimens.

After the GMPs of U-notch SEN specimens made of material 2 have been determined, the failure assessment analysis of U-notch SEN specimens made of material 2 with different notch radii can be performed by means of the local stress field failure model proposed in this study. The obtained results and fracture test data reported in Ref.[42] are shown in Table 5.36, with the calculation error indexes: M.E. = −1%, M.A.E. = 2.6 % and E.R. = [−3.9.2.6]%, from which it can be seen that the predicted notch fracture toughness is in excellent agreement with that measured experimentally.

By the way, at the end of this section, it is pointed out that from the model formulas (see (40) to (51)) proposed in this section, similarly, it is easy to propose the model formulas of predicting the GMPs of the U-notch DEN and RNT specimens made of material 2.

5.10 COMMENTS OF PART 2

On the basis of study on the notch angle effect on k^* proposed in Section 7 and the obtained values of the GMPs of the V-notch TPB specimens made of PMMA at -60°C (material 1), it is easy to propose the model formulas of predicting the GMPs of the V-notch SEN, DEN and RNT specimens made of material 2.

On the basis of study on the notch depth effect on k^* in Section 8, further, for a V-notch (SEN, DEN, and RNT) specimen (with any notch angle and any notch depth) made of material 2, its GMPs can be calculated by means of the models of notch angle effect and notch depth effect and different materials on k^* proposed in this study provided that the GMPs of the U-notch TPB specimen made of material 2 have been obtained by means of the approach of determining k^* proposed in this study.

So it can be seen here that although many examples are given in Section 5, the obtained GMPs (see Table 5.20) still are too few to verify the guess of this section. Further experimental work is very necessary to be done.

From the comments of this section, it can be seen that the GMPs listed in Table 5.20 are very important for the fracture analysis of the notched materials studied. Here, the geometry size effect on the fracture behavior of the notched materials should be taken into account. For the U-notch TPB specimen (with the geometry sizes: S = 126 mm, B = 14 mm, W = 28 mm, a = 14 mm) made of PAMM specimens at -60°C (denoted by specimen 1), for example, the GMPs (denoted by k_1^*, ρ_1^*, L_{c1}^*) are shown in Table 5.20 (also see Table 5.37). They satisfy the following failure equation:

$$F(K_{Ic}, L_{c1}^*, \rho_1^*, \alpha, \sigma_0) = 0 \tag{53}$$

Now it is assumed that there is a U-notch TPB specimen (denoted by Specimen 2 with such a geometry in proportion to the geometry of Specimen 1) made of PAMM specimens at -60°C, in the case that the specimen 2 satisfies the demands of some geometry condition(s), for example, such a condition that the plane strain fracture toughness is measured experimentally, and the GMPs of Specimen 2 (denoted by k_2^*, ρ_2^*, L_{c2}^*) can be calculated by using the following failure equation similar to Eq.(53):

$$F(K_{Ic}, L_{c2}^*, \rho_2^*, \alpha, \sigma_0) = 0 \tag{54}$$

Because of the geometry similarity between Specimens 1 and 2, it is assumed that the following relations are valid:

$$\rho_2^* = \gamma \rho_1^*, \; k_1^* = k_2^* \tag{55}$$

TABLE 5.37
Variation of the GMPs of U-Notch TPB Specimens Made of PAMM Specimens at –60°C with Notch Geometry Sizes

Material	Notch Specimen Type	Geometric Size (mm)	k^*	ρ^* (mm)	L_c^* (mm)	
PMMA at –60°C K_{Ic} = 1.7 MPam$^{0.5}$ σ_0 = 128.4 MPa	TPB	S = 126 mm, B = 14 mm W = 28 mm, a = 14 mm	19.71	0.016	0.0321	Specimen1
	TPB	S = 252 mm, B = 28 mm W = 56 mm, a = 28 mm	19.71	0.032	0.0331	Specimen2
	TPB	S = 63 mm, B = 7 mm W = 14 mm, a = 7 mm	19.71	0.008	0.0307	Specimen3

where γ is a proportion coefficient. By substituting $\rho_2^* = \gamma\rho_1^*$ into Eq.(54), thus, the value of L_{c2}^* can be calculated. As the proportion coefficient γ is taken to be 2 and 0.5, the obtained value of L_{c2}^* is 0.0331 mm and 0.0307 mm, respectively (see Table 5.37).

PART 3

See the comments of Part 2: for a V-notch (SEN, DEN, and RNT) specimen (with any notch angle and any notch depth) made of material 2, its GMPs can be calculated by means of the models of notch angle effect and notch depth effect and different materials on k^* proposed in this study, provided that the GMPs of the U-notch TPB specimen made of material 2 have been obtained by means of the approach of determining k^* proposed in this study. From the study of Part 1 and Part 2, the failure assessment analysis for the V-notch (SEN, DEN, and RNT) specimens with different notch radii performed by using the obtained GMPs and by means of the local stress field failure model proposed in this study has high accuracy. In this part, experimental verifications are given to further verify a wide application of the local stress field failure model proposed in this study.

5.11 AN EMPIRICAL EQUATION FOR PREDICTING THE FRACTURE TOUGHNESS K_{Ic}

Taking into account that the fracture toughness K_{Ic} is unknown at some circumstances, here, an empirical equation of determining K_{Ic} is proposed by using the tendency of wide test data of the fracture toughness, $K_{\rho c}$, of U-notch specimens shown in Figure 5.8.

$$\frac{K_{\rho c}}{K_{Ic}} = a_0 x^{0.5} + a_1 x + a_2 x^2 + a_3 x^3 + a_4 x^4 \quad (56a)$$

where $x = \rho / L_{ch}$, $a_0 = 0.130479$, $a_1 = 0.213032$, $a_2 = -0.00576$, $a_3 = -7.844E\text{-}07$, $a_4 = 2.59275E\text{-}06$ and L_{ch} is defined as:.

$$L_{ch} = \left(\frac{K_{Ic}}{\sigma_0}\right)^2 \quad (57a)$$

For V-notch specimens, Eqs. (56a) and (57a) are replaced by

$$\frac{K_{v\rho c}}{K_{vc}} = a_0 x^{0.5} + a_1 x + a_2 x^2 + a_3 x^3 + a_4 x^4 \quad (56b)$$

and

$$L_{ch} = \left(\frac{K_{vc}}{\sigma_0}\right)^{\frac{1}{1-\lambda}} \quad (57b)$$

For V-notch specimens with notch angle = 0, obviously, (56b) and (57b) become (56a) and (57a), respectively.

In the following, three examples are given to illustrate the empirical equation of predicting the fracture toughness proposed here to be valid.

Example 1: Tension Failure Analysis of Notched Cylindrical Bars Made of PMMA

In 2008, the experimental study of notched cylindrical bars was performed by Susmel and Taylor [47]. The material employed was a commercial PMMA. Four different cylindrical geometries were machined: the tested V-notched specimens had gross diameter equal to 12.8 mm, net diameter to 8.2 mm, notch angle equal to 60° and notch root radii equal to 0.2 mm, 0.4 mm, 1.2 mm and 4.0 mm, respectively. The ultimate tension strength was equal to about 67 MPa.

Tensile fracture test data of notched cylindrical bars reported in Ref.[47] were listed in Table 5.38. Here, the test data are used to determine the fracture

TABLE 5.38
Tensile Fracture Test Data of Notched Cylindrical Bars Reported in Ref.[47] and Evaluation Calculation Results (Notch Angle 60°)

No.	Notch Radius (mm)	Failure Load (KN)	SCFs k	σ_{max} (MPa)	$K_{v\rho c}$ (exp) ($MPam^{1-\lambda}$)	$K_{v\rho c}$ (cal) ($MPam^{1-\lambda}$)	R.E. (%)
1	0.2	1.78	4.7	158.42	2.02	2.24	−1.8
2	0.4	2.23	3.4	143.57	2.57	2.69	8.3
3	1.2	2.78	2.2	115.81	3.54	3.72	−5.7

Note: (a) Notch angle $\alpha = 60°$, gross diameter $d_g = 12.8$ mm, net diameter $d_n = 8.2$ mm; (b) $\sigma_0 = 74.5$ MPa, $K_{vc(60^\circ)} = 1.535$ $MPam^{1-\lambda}$ (where $\lambda = 0.5122$); (c) $k^* = 11.57$, $\rho^* = 0.04$ mm, $L_c^* = 0.016$ mm; (d) M.E. = 0.3%, M.A.E. = 5.3%, E.R. = [−5.7,8.3]%.

toughness K_{Ic} of the material studied by using the empirical equation of predicting the fracture toughness K_{Ic} proposed in this section. The details are as follows:

(1) According to Eq.(13), the values of K_{vpc} under various failure loads are calculated. The obtained results are given in Table 5.38;
(2) Fracture test data of order numbers, 1 and 3, shown in Table 5.38, on one hand, are used to calculate the critical stress σ_0 by using Eqs. (14) to (16), and on the other hand, are used to determine the value of K_{vc} (notch angle = 60°) by using the empirical Eq.(56b) proposed in this section. The obtained results are $\sigma_0 = 74.5$ MPa and $K_{vc} = 1.535$ MPam$^{1-\lambda}$ (where $\lambda = 0.5122$);
(3) By using the obtained values of the critical stress σ_0 and the pointed V-notch fracture toughness K_{vc} (notch angle = 60°), the GMPs of the pointed V-notch specimen (notch angle = 60°) are calculated by means of the approach of determining k^* proposed in this study. Their values are $k^* = 11.57$, $P = 0.04$ mm, $L_c^* = 0.016$ mm;
(4) By using the values of σ_0 and K_{vc} obtained above, the fracture toughness K_{Ic} can be calculated by means of the model of predicting the pointed V-notch fracture toughness proposed in Ref.[46]. Its value is $K_{Ic} = 1.36$ MPam$^{0.5}$;
(5) By using the GMPs of the pointed V-notch specimen (notch angle = 60°), the GMPs of U-notch specimen can be calculated by means of the model of predicting the effect of different notch angles on k^* proposed in this study. Their values are: $k^* = 10.98$, $\rho^* = 0.065$ mm, $L_c^* = 0.06515$ mm;
(6) By using the GMPs of the pointed V-notch specimen (notch angle = 60°) and the values of σ_0 and K_{vc} obtained above, the failure assessment of notched specimens with notch angle 60° is performed by means of the local stress field failure model proposed in this study. The predicted notch fracture toughness K_{vpc} (cal) are shown in Table 5.38. By comparing K_{vpc} (cal) with K_{vpc}(exp), the calculation error indexes are: M.E. = 0.3%, M.A.E. = 5.3% and E.R. = [−5.7,8.3]%;
(7) By using the GMPs of U-notch specimen and the values of σ_0 and K_{Ic} obtained above, the failure assessment of U-notched specimens can be performed by means of the local stress field failure model proposed in this study. The predicted result is given in Table 5.39, with relative error −7.6%.

TABLE 5.39
Tensile Fracture Test Data of Notched Cylindrical Bars Reported in Ref.[47] and Evaluation Calculation Results (U-Notch Specimen)

No.	Notch Radius (mm)	Failure Load (KN)	SCFs K	σ_{max} (MPa)	K_{pc} (exp) (MPam$^{0.5}$)	K_{pc} (cal) (MPam$^{0.5}$)	R.E. (%)
4	4	2.83	1.4	75.02	4.2	3.88	−7.6

Note: (a) notch angle $\alpha = 0°$, gross diameter $d_g = 12.8$ mm, net diameter $d_n = 8.2$ mm; (b) $\sigma_0 = 74.5$ MPa, $K_{Ic} = 1.36$ MPam$^{0.5}$; (c) $k^* = 10.98$, $\rho^* = 0.065$ mm, $L_c^* = 0.06515$ mm.

Example 2: Fracture Analysis of the RV-BD Specimens Made of PMMA

In 2010, the mixed mode fracture of notched components was studied by Ayatollahi and Torabi [48] by means of the RV-BD specimen made of PMMA (see Figure 5.15).

For all the RV-BD specimens, the disc diameter (D), the notch length (d/2) and the thickness (T) were 80 mm, 20 mm and 10 mm, respectively.

To study the effects of the notch angle and the notch tip radius on the fracture behavior of the RV-BD specimens, three values of notch angles $2\alpha = 30^{\circ}, 60^{\circ}, 90^{\circ}$ and three values of notch radius $\rho = 1, 2, 4$ mm were considered for preparing the specimens.

Here, fracture test data of notched components under Mode I failure loads (the loading angle $\beta = 0^{\circ}$), as done in Example 1 of this section, are used to determine the fracture toughness K_{Ic} of the material studied by using the empirical equation of predicting the fracture toughness K_{Ic} proposed in this section.

For the notch angle $2\alpha = 30^{\circ}$, fracture test data reported in Ref.[48] and the fracture analysis results given here are listed in Table 5.40. In Table 5.40, the values of the notch fracture toughness K_{vpc} at various failure loads were given in Ref.[48], the values of $\sigma_{\max}$ are calculated by using formula (13) from K_{vpc}, and the stress concentration factor, k, is defined as a ratio, $k = \sigma_{\max} / \sigma_{nom}$, $\sigma_{nom} = F_c / (T \cdot D / 2)$. The method of calculation the stress concentration factors by using the $\sigma_{\max}$ calculated

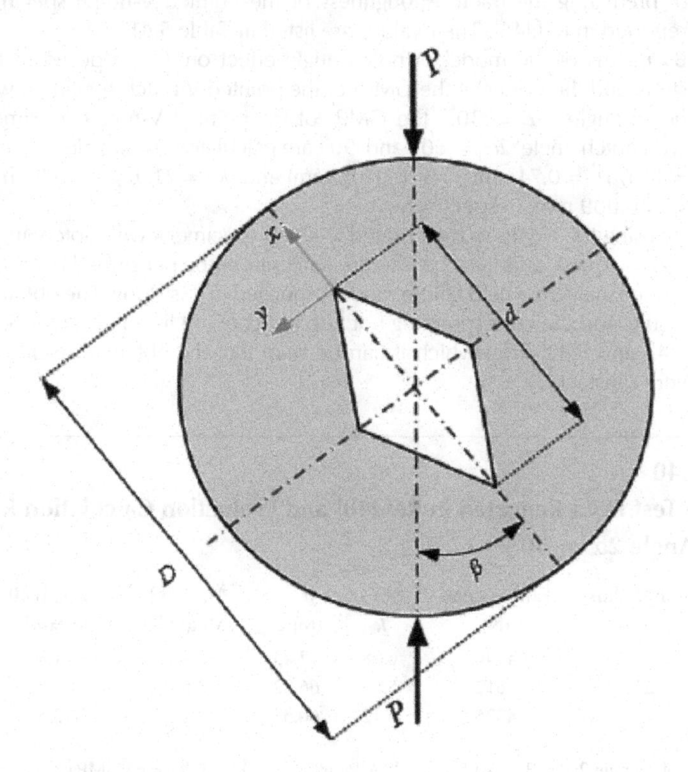

FIGURE 5.15 The RV-BD specimen.

Source: (Data from Ayatollahi and Torabi [48])

by using the generalized notch fracture toughness is called a generalized notch fracture toughness method (called a $K_{v\rho c}$ method for short).

The details are as follows:

(1) Fracture test data of order numbers 1 and 3 listed in Table 5.40, on one hand, are used to calculate the critical stress σ_0 by using Eqs. (14) to (16), and on the other hand, are used to determine the value of K_{vc} (the fracture toughness of the pointed V-notch specimen with notch angle $2\alpha = 30^{\circ}$) by using the empirical Eq.(56b) proposed in this section. The obtained results are $\sigma_0 = 64.6$ MPa and $K_{vc} = 1.48$ MPam$^{1-\lambda}$ (where $\lambda = 0.501$);

(2) By using the obtained values of σ_0 and K_{vc}, the GMPs of the pointed V-notch specimen with notch angle $2\alpha = 30^{\circ}$ can be calculated by means of the approach of determining k^* proposed in this study. Their values are $k^* = 12.3$, $\rho^* = 0.69$ mm, $L_c^* = 0.0108$ mm;

(3) By using $K_{vc} = 1.48$ MPam$^{1-\lambda}$ ($\lambda = 0.501$) and $\sigma_0 = 64.6$ MPa, the value of K_{Ic} can be calculated by means of the model of predicting the fracture toughness of the pointed V-notch specimen reported in Ref.[46]. Its value is $K_{Ic} = 1.46$ MPam$^{0.5}$.

(4) By using $K_{Ic} = 1.46$ MPam$^{0.5}$ and $\sigma_0 = 64.6$ MPa, the values of K_{vc} at the notch angles $2\alpha = 60^{\circ}$, 90° can be calculated by means of the model of predicting the fracture toughness of the pointed V-notch specimen reported in Ref.[46]. Their values are listed in Table 5.41.

(5) By means of the model of notch angle effect on k^* proposed in this study and the values of the GMPs of the pointed V-notch specimen with notch angle $2\alpha = 30^{\circ}$, the GMPs of the pointed V-notch specimens with notch angle $2\alpha = 60^{\circ}$ and 90° are calculated, Their values are ($k^* = 14.1$, $\rho^* = 0.74$ mm, $L_c^* = 0.0103$ mm) and ($k^* = 21.9$, $\rho^* = 0.81$ mm, $L_c^* = 0.009$ mm), respectively.

(6) By using the GMPs of the pointed V-notch specimens with notch angles $2\alpha = 30^{\circ}$,60° and 90°, the fracture analysis can be performed by means of the local stress field failure model proposed in this study. The obtained results and the corresponding fracture test data are listed in Tables 5.40, 5.42 and 5.43, from which it can be seen that the obtained results are very satisfactory.

TABLE 5.40

Fracture Test Data Reported in Ref.[48] and Evaluation Calculation Results (Notch Angle $2\alpha = 30^{\circ}$)

No.	Notch Radius (mm)	Failure Load (N)	SCFs k	σ_{max} (Mpa)	$K_{v\rho c}$ (exp) (MPam$^{1-\lambda}$)	$K_{v\rho c}$ (cal) (MPam$^{1-\lambda}$)	R.E. (%)
1	1	4440	6.98	77.43	2.05	1.8	−12.3
2	2	4612	5.8	66.82	2.5	2.5	0
3	4	4725	5.12	60.52	3.2	3.5	9.3

Note: (a) Notch angle $2\alpha = 30^{\circ}$; (b) $K_{vc} = 1.48$ MPam$^{1-\lambda}$ ($\lambda = 0.501$), $\sigma_0 = 64.6$ MPa; (c) $k^* = 12.3$, $\rho^* = 0.69$ mm, $L_c^* = 0.0108$ mm; (d) M.E. = −1%, M.A.E. = 7.2%, E.R. = [−12.3,9.3]%.

TABLE 5.41
The Predicted Fracture Toughness of Pointed V-Notch Specimens

Notch Angle (∘)	NSFs G [41]	λ	K_{vc} (MPam$^{1-\lambda}$)
0	1.767	0.5000	1.46
30	1.775	0.5010	1.48
60	1.857	0.5122	1.67
90	2.118	0.5455	2.37

Note: (*a*) For notch angle $2\alpha = 30^\circ$, $K_{vc} = 1.48$ MPam$^{1-\lambda}$ (where $\lambda = 0.501$) is obtained by using the empirical equation (56b); (*b*) The pointed V-notch fracture toughness at notch angle $2\alpha = 0^\circ, 60^\circ, 90^\circ$ are calculated by using the model of predicting the fracture toughness of the pointed V-notch specimen reported in Ref.[46].

TABLE 5.42
Fracture Test Data Reported in Ref.[48] and Evaluation Calculation Results (Notch Angle $2\alpha = 60^\circ$)

No.	Notch Radius (mm)	Failure Load (N)	SCFs k	σ_{max} (MPa)	$K_{v\rho c}$ (exp) (MPam$^{1-\lambda}$)	$K_{v\rho c}$ (cal) (MPam$^{1-\lambda}$)	R.E. (%)
1	1	3343	12.2	78.6	2.20	1.93	−12.1
2	2	4022	8.7	74.6	2.93	2.69	−8.1
3	4	4136	6.2	63.9	3.52	3.75	6.6

Note: (a) Notch angle $2\alpha = 60^\circ$; (b) $K_{vc} = 1.67$ MPam$^{1-\lambda}$ ($\lambda = 0.5122$), $\sigma_0 = 64.6$ MPa; (c) $k^* = 14.1$, $\rho^* = 0.74$ mm, $L_c^* = 0.0103$ mm; (d) M.E. = −4.5%, M.A.E. = 8.9%, E.R. = [−12.1,6.6]%.

TABLE 5.43
Fracture Test Data Reported in Ref.[48] and Evaluation Calculation Results (Notch Angle $2\alpha = 90^\circ$)

No.	Notch Radius (mm)	Failure Load F_c(N)	SCFs k	σ_{max} (MPa)	$K_{v\rho c}$ (exp)	$K_{v\rho c}$ (cal)	R.E. (%)
1	1	1617	19.9	73.39	2.65	2.6	−1.9
2	2	2206	14.5	72.71	3.6	3.54	−1.7
3	4	2700	10.6	71.44	4.85	4.83	−0.4

Note: (a) Notch angle $2\alpha = 90^\circ$; (b) $K_{vc} = 2.37$ MPam$^{1-\lambda}$ ($\lambda = 0.5455$), $\sigma_0 = 64.6$ MPa; (c) $k^* = 21.9$, $\rho^* = 0.81$ mm, $L_c^* = 0.009$ mm; (d) M.E. = −1.3%, M.A.E. = 1.3%, E.R. = [−1.9,−0.4]%.

TABLE 5.44
Fracture Test Data Reported in Ref.[50] and Evaluation Calculation Results (Aluminum Alloy, Al6082)

No.	Notch Radius ρ (mm)	d_n (mm)	α(°)	SCFs k	σ_{max} (MPa)	$K_{V\rho c}$ (exp)	$K_{V\rho c}$ (cal)	R.E. (%)
1	0.44	6.2	60	2.94	537.3	29.58	28.20	−4.7
2	0.50	6.1	60	2.76	551.6	20.35	30.14	−0.7
3	1.25	6.2	60	1.92	561.4	33.60	35.64	6.1
4	4.00	6.1	0	1.33	449.3	37.07	33.50	10.7

Note: (a) Unit of $K_{V\rho c}$ is $MPam^{1-\lambda}$ (Williams' eigenvalues λ).

Example 3: Fracture Analysis of Circumferentially Notched Samples Made of AS602

In 2010, the fracture of circumferentially notched samples of a commercial aluminum alloy, i.e., Al6082, was investigated by Susmel and Taylor [49,50]. The material was supplied in bars having diameter, d_g, equal to 10 mm. The plain samples used to determine the static properties of such an aluminum alloy were machined to obtain a gauge length of 5 mm with a diameter equal to 6 mm. The material was found to have an ultimate tensile stress, σ_{UTS}, equal to 367 MPa, a yield stress, σ_y, equal to 347 MPa and a Young's modulus, E, equal to 69,090 MPa. Figure 5.16 shows the stress–strain curve generated from plain specimens, plotted in terms of engineering quantities.

The tested V-notched cylindrical samples (Figure 5.17) had gross diameter, d_g, equal to 10 mm and net diameter, d_n, ranging between 6.1 mm and 6.2 mm. The notch angle, α, was equal to 60° and four different values of the notch root radius, ρ, were investigated, i.e., 0.44 mm (tensile stress concentration factor k = 2.94;); 0.50 mm (k = 2.76); 1.25 mm (k = 1.92) and 4.00 mm (k = 1.33).

Fracture test data reported in Ref.[50] and evaluation calculation results given here are listed in Table 5.44. In Table 5.44, the fracture toughness $K_{V\rho c}$ (exp) of V-notch specimens with notch angle 60° and U-notch specimen are calculated from formula (13). Fracture test data, as done in Example 1 of this section, are used to determine the fracture toughness K_{Ic} of the material studied by using the empirical equation of predicting the fracture toughness K_{Ic} proposed in this section. The details are as follows:

(1) Fracture test data of order numbers, 1 and 3, shown in Table 5.44, on one hand, are used to calculate the critical stress σ_0 by using Eqs. (14) to (16), and on the other hand, are used to determine the value of K_{Vc} (notch angle = 60°) by using the empirical Eq.(56b) proposed in this section. The obtained results are σ_0 = 358.7 MPa and K_{Vc} = 26.3 $MPam^{1-\lambda}$ (where λ = 0.5122);

(2) By using the obtained values of σ_0 and K_{Vc}, the GMPs of the pointed V-notch specimen with notch angle 60° are calculated by means of the approach of determining k^* proposed in this study. Their values are k^* = 3.73, ρ^* = 0.32 mm, L_c^* = 0.814 mm;

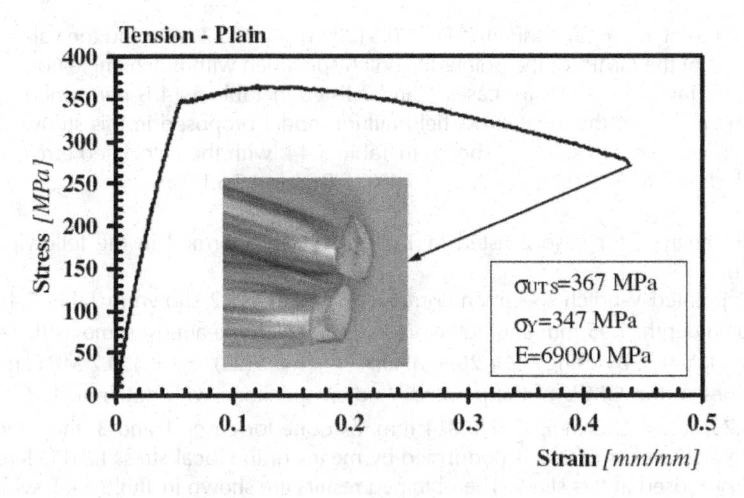

FIGURE 5.16 Plain stress–strain curve under tensile loading.

Source: (Data from Susmel and Taylor [49]).

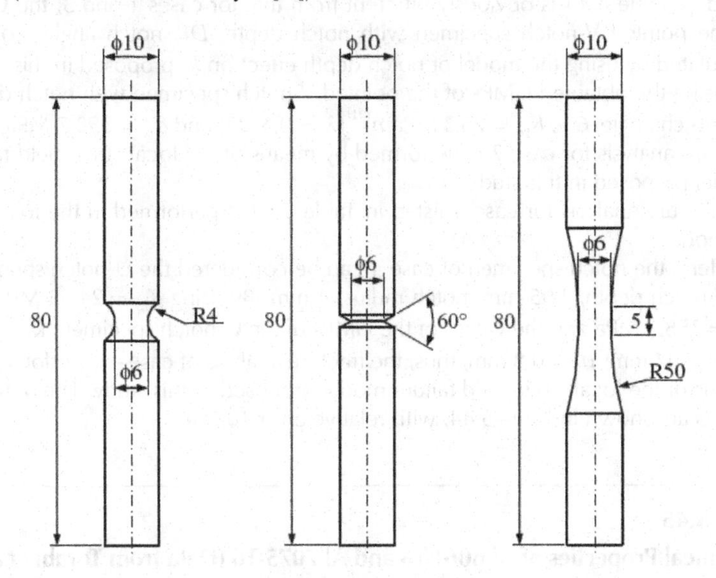

FIGURE 5.17 Geometries of the tested samples (dimensions in millimeters).

Source: (Data from Susmel and Taylor [49]).

(3) By using the obtained values of σ_0 and K_{vc}, the value of the fracture toughness, K_{Ic}, can be calculated by means of the model of predicting the fracture toughness of the pointed V-notch specimen reported in Ref. [46]. Its value is $K_{Ic} = 24.03\ MPam^{0.5}$.

(4) By using the values of the GMPs of the pointed V-notch specimen with notch angle 60°, the GMPs of the U-notch specimen are calculated by using the model of notch angle on k^* proposed in this study. Their values are $k^* = 2.97$, $L_c^* = 0.847$ mm, $\rho^* = 0.8$ mm;

(5) By using K_{vc} = 26.3 MPa$m^{1-\lambda}$ (λ = 0.5122), σ_0 = 358.7 MPa and the values of the GMPs of the pointed V-notch specimen with notch angle 60°, the fracture analysis for cases 1 and 3 listed in Table 5.44 is performed by means of the local stress field failure model proposed in this study. The obtained results are shown in Table 5.44, with the calculated error indexes: M.E. = 0.7%, M.A.E. = 5.4%, E.R. = [−4.7,6.1]%;

Fracture analysis for case 2 listed in Table 5.44 is performed in the following method:

The pointed V-notch specimen corresponding to case 2 shown in Table 5.44 has notch depth, 1.95 mm and notch angle, 60°, which are almost same as those of cases 1 and 3. By using K_{vc} = 26.3 MPa$m^{1-\lambda}$ (λ = 0.5122), σ_0 = 358.7 MPa and the values of the GMPs of the pointed V-notch specimen with notch angle 60°, k^* = 3.73, ρ^* = 0.32 mm, L_c^* = 0.814 mm, as done for cases 1 and 3, thus, the fracture analysis for case 2 is performed by means of the local stress field failure model proposed in this study. The obtained results are shown in Table 5.44, with relative error −0.7%.

By the way, it is pointed out that, if notch depth for case 2 (denoted by *DP*) listed in Table 5.44 is obviously different from that for cases 1 and 3, the GMPs of the pointed V-notch specimen with notch depth, *DP*, notch angle, 60°, are calculated by using the model of notch depth effect on k^* proposed in this study. By using the obtained GMPs of the pointed V-notch specimen with notch depth, *DP*, notch angle, 60°, K_{vc} = 26.3 MPa$m^{1-\lambda}$ (λ = 0.5122) and σ_0 = 358.7 MPa, then fracture analysis for case 2 is performed by means of the local stress field failure model proposed in this study.

Fracture analysis for case 4 listed in Table 5.44 is performed in the following method:

Here, the notch specimen of case 4 can be considered the U-notch specimen with notch depth, 1.95 mm, notch radius, 4 mm. By using K_{Ic} = 24.03 MPa$m^{0.5}$, σ_0 = 358.7 MPa and the values of the GMPs of the U-notch specimen, k^* = 2.97, L_c^* = 0.847 mm; ρ^* = 0.8 mm, thus, the fracture analysis of case 4 is performed by means of the local stress field failure model proposed in this study. The obtained results are shown in Table 5.44, with relative error 10.7%.

TABLE 5.45
Mechanical Properties of Al 6061-T6 and Al 7075-T6 (Data from Torabi *et al.* [51])

Material Property	Al 6061-T6	Al 7075-T6
Elastic modulus (GPa)	67	71
Poisson's ratio	0.33	0.33
Ultimate tensile strength (MPa)	292	583
Tensile yield strength (MPa)	276	521
Fracture toughness (MPa $m^{0.5}$)	38	50
Elongation at break (%)	11	5.8
True fracture stress (MPa)	299	610
Strain hardening coefficient (MPa)	314	698
Strain hardening exponent	0.021	0.046

5.12 FRACTURE ANALYSIS OF CENTER NOTCH PLATES MADE OF METAL MATERIALS

Here, an attempt is made to show that the local stress field failure model proposed in this study is used to perform fracture analysis of center notch plate made of metallic materials. From the viewpoint of Elastic Plastic Fracture Mechanics, no doubt, the attempt is not feasible. From the study [64] on the unified fatigue lifetime equation, in low/medium/high cycle fatigue regime of metallic materials, by using the *linear elastic stress quantity,* however, the author considers that the attempt will be successful. In the following, three examples are given to illustrate that the attempt is successful.

Example 1: Fracture Analysis of Center U-Notch Specimens Made of Al 7075-T6 and Al 6061-T6

In 2015, fracture of center U-notch specimens were investigated experimentally by Torabi et al. [51]. The materials selected were two aluminum alloys AL 7075-T6 and Al 6061-T6 which exhibit moderate and large plastic deformations under tension, respectively. The materials were provided in the forms of plates of 2 mm and 4 mm thick, respectively. The engineering and true stress-strain curves for the tested materials are shown in Figure 5.18. The measured mechanical properties are listed in Table 5.45.

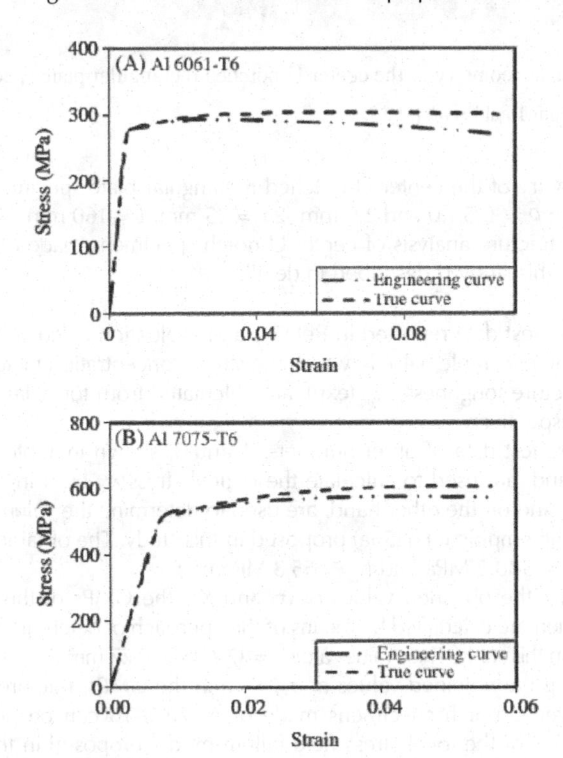

FIGURE 5.18 The true and engineering stress-strain curves for the tested aluminum alloys: A, Al 6061-T6; B, Al 7075-T6.

Source: (Data from Torabi *et al.* [51])

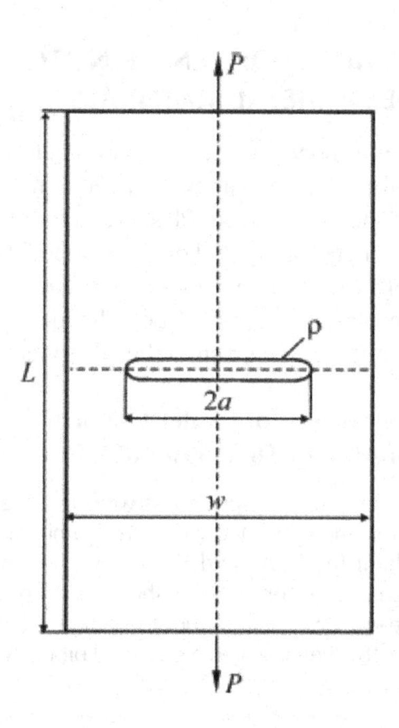

FIGURE 5.19 The geometry of the center U-notched rectangular plate specimens.

Source: (Data from Torabi *et al.* [51])

The geometry of the center U-notched rectangular plate specimens is shown in Figure 5.19: $\rho = 0.5,1.0$ and 2.0 mm, $2a = 25$ mm, $L = 160$ mm, $W = 50$ mm.

Here, the fracture analysis of center U-notch specimens made of Al 7075-T6 performed in this study is described in detail.

(1) Fracture test data reported in Ref.[51] and evaluation calculation results are shown in Table 5.46, in which the stress concentration factor, k, and the fracture toughness, $K_{\rho c}$ (exp), are calculated from formulas (27) and (13), respectively;

(2) Fracture test data of order numbers, 1 and 3, shown in Table 5.46, on one hand, are used to calculate the critical stress σ_0 by using Eqs. (14) to (16), and on the other hand, are used to determine the value of K_{Ic} by using the empirical Eq.(56a) proposed in this study. The obtained results are $\sigma_0 = 580.8$ MPa and $K_{Ic} = 55.3$ MP$am^{0.5}$;

(3) By using the obtained values of σ_0 and K_{Ic}, the GMPs of the U-notch specimen are calculated by means of the approach of determining k^* proposed in this study. Their values are $k^* = 17.2$, $\rho^* = 0.27$ mm, $L_c^* = 1.55$ mm;

(4) By using the obtained values of σ_0, K_{Ic} and the GMPs, fracture analysis of center U-notch specimens made of Al 7075-T6 can be performed by means of the local stress field failure model proposed in this study. The obtained results are given in Table 5.46, with the calculation error indexes: M.E. = −0.4%, M.A.E. = 1.2%, E.R. = [−2.4,0.8]%, from which it can be seen that the predicted fracture toughness is in excellent agreement with that measured experimentally.

TABLE 5.46
Fracture Test Data Reported in Ref.[51] and Evaluation Calculation Results (Center U-Notch Specimens Made of Al 7075-T6)

No.	Notch Radius (mm)	Failure Load Fc(N)	SCFs k	σ_{max} (MPa)	$K_{\rho c}$ (exp) (MPam$^{0.5}$)	$K_{\rho c}$ (cal) (MPam$^{0.5}$)	E.R. (%)
1	0.5	25,369	12.89	3270.06	64.8	63.26	–2.4
2	1	27,598	9.41	2596.28	72.76	73.08	0.4
3	2	30,250	6.95	2100.86	83.26	83.91	0.8

Note: (a) $K_{Ic} = 55.3\ MPam^{0.5}$, $\sigma_0 = 580.8$ MPa; (b) $k^* = 17.2$, $\rho^* = 0.27$ mm, $L_c^* = 1.55$ mm; (c) M.E. = –0.4%, M.A.E. = 1.2%, E.R. = [–2.4,0.8]%

TABLE 5.47
Fracture Test Data Reported in Ref.[51] and Evaluation Calculation Results (Center U-Notch Specimens Made of Al 6061-T6)

No.	Notch Radius (mm)	Failure Load Fc(N)	SCFs k	σ_{max} (MPa)	$K_{\rho c}$ (exp) (MPam$^{0.5}$)	$K_{\rho c}$ (cal) (MPam$^{0.5}$)	E.R. (%)
1	0.5	29,329	12.89	1890.25	37.46	35.5	–5.2
2	1	30,822	9.41	1449.79	40.63	41.01	0.9
3	2	33,245	6.95	1154.43	45.75	47.09	2.9

Note: (a) $K_{Ic} = 33.1\ MPam^{0.5}$, $\sigma_0 = 327.6$ MPa; (a) $k^* = 15.0$, $\rho^* = 0.36$ mm, $L_c^* = 1.76$ mm; (b) M.E. = 0.4%, M.A.E. = 3.0%, E.R. = [–5.2,2.9]%

By the way, it is pointed out that the fracture toughness K_{Ic} of Al 7075-T6 reported in Ref.[46] is 50 MP$am^{0.5}$ (see Table 5.45). Its value estimated by using Eq.(56a) is 55.3 MP$am^{0.5}$. Here, the author considers that, for the failure assessment of center U-notch plate specimens, using the fracture toughness estimated by using Eq.(56a) is more suitable than that reported in Ref.[51].

Fracture analysis of center U-notch plate specimens made of Al 6061-T6, as done for center U-notch plate specimens made of Al 7075-T6, can be performed. The obtained results and fracture test data are shown in Table 5.47, with the calculated error indexes: M.E. = 0.4%, M.A.E. = 3.0%, E.R. = [–5.2,2.9]%.

Example 2: Fracture Analysis of Center Circular Hole Specimens Made of Pure Copper-T2

In 2016, tensile experiments [52] were conducted on a series of pure copper (commercially pure copper-T2) plate specimens with a center circular hole. The main geometric sizes of rectangular plate specimens are length, 180 mm, width, 20 mm, with different thicknesses, 4 mm, 6 mm and 8 mm and with different circular hole diameters, 3 mm, 6 mm and 12 mm.

For center circular notch plate specimens under tensile loading, here, the stress concentration factor, k, is calculated by using the following equation reported in Ref.[53].

$$k = 2 + 0.15\left(\frac{D}{W}\right)^2 + \left(1 - \frac{D}{W}\right)^3 \tag{58}$$

where D and W are the center circular hole diameter and the plate width, respectively.

Here, a center circular hole specimen is considered to be a U-notch specimen with a notch depth and a notch radius that both are equal to the circular hole radius. Thus fracture toughness of the center circular hole specimen, $K_{\rho c}$, can be calculated by using formula (13).

For plate thickness = 8 mm, fracture test data and evaluation calculation results are shown in Table 5.48, in which the values of the failure loads are taken from Figure 8 in Ref [52]. Fracture analysis performed here is described as follows:

(1) By using fracture test data of notch radii equal to 3 mm and 6 mm listed in Table 5.48, the fracture toughness, K_{Ic}, and the critical stress, σ_0, are estimated by means of the empirical equation (56a). The obtained results are: K_{Ic} = 12 Mpam$^{0.5}$, σ_0 = 440 MPa.
(2) By using fracture test data of notch radius equal to 3 mm listed in Table 5.48 and the obtained values of K_{Ic} and σ_0, the GMPs corresponding to the U-notch specimen with notch depth, 3 mm, are calculated by means of the approach of determining k^* proposed in this study. Their values are k^* = 5.662, ρ^* = 0.52 mm, L_c^* = 0.093 mm.
(3) For center notch specimens, as done for DEN specimens, the effect of notch depth on k^* is assumed to be

$$F \cdot L_c^* = c \tag{59}$$

where F is the geometry shape function of the center crack with the crack length equal to notch depth; c is a constant. Here, $c = 1.059 \times 0.093mm = 0.09849mm$.
(4) By using the obtained value of c in Eq.(59), the values of L_c^* of the U-notch specimens with notch depths equal to 1.5 mm and 6 mm are calculated. Their values are 0.098 mm and 0.076 mm, respectively.

TABLE 5.48
Fracture Test Data Reported in Ref.[52] and Evaluation Calculation Results (Thickness = 8 mm)

No.	Notch Depth (mm)	Notch Radius (mm)	SCFs k	Failure Load F_c (N)	σ_{max} (MPa)	$K_{\rho c}$ (exp) (MPam$^{0.5}$)	$K_{\rho c}$ (cal) (MPam$^{0.5}$)	E.R. (%)
1	1.5	1.5	2.62	29000	558.1	19.2	18.44	−3.8
2	3.0	3.0	2.36	24000	505.0	24.5	24.48	−0.1
3	6.0	6.0	2.12	14000	463.3	31.8	32.55	2.3

TABLE 5.49
Variation of the GMPs with Notch Depths (Thickness = 8 mm)

No.	Notch Depth (mm)	k^*	ρ^*	L_c^*
1	1.5	4.53	0.50	0.098
2	3.0	5.66	0.52	0.093
3	6.0	6.59	0.62	0.076

(5) By using the obtained values of L_c^*, K_{Ic} and σ_0, the value of ρ^* is calculated from the following failure equation:

$$F(K_{Ic}, L_c^*, \rho^*, \alpha, \sigma_0) = 0 \tag{60}$$

where α is the notch angle, for a U-notch specimen, $\alpha = 0$. Here, the obtained values of ρ^* of the U-notches with notch depths equal to 1.5 mm and 6 mm are 0.50 mm and 0.62 mm, respectively.

(6) From the geometry of notched specimens and the obtained value of ρ^*, the value of k^* can be obtained by using the Finite Element Method or related references (e.g., Refs [19–22,24–27]). Here, the obtained values of k^* of the U-notches with notch depths equal to 1.5 mm and 6 mm are 4.53 and 6.59, respectively.

Here, variation of the GMPs with notch depths is listed in Table 5.49.

By using the obtained values of the GMPs, fracture analysis of notched specimens is easily performed by means of the local stress field failure model proposed in this study. The obtained results are shown in Table 5.48, from which it can be seen that the predicted fracture toughness is in excellent agreement with that measured experimentally.

Here, an attempt is made to study the plate thickness effect on the fracture behavior of pure copper studied. From the variation of the failure loads with the plate thickness shown in Figure 8 in Ref.[52], it is found that the ratio of the failure loads to the plate thickness is not changed with notch depths. Thus it is assured that the fracture toughness of notched specimens, $K_{\rho c}$ (exp), is not varied with the plate thickness in the studied range of the plate thickness.

Example 3: Fracture Analysis of Center Circular Notch Specimens Made of Q345

In 2008, tensile experiments [54] were conducted on a series of structure steel (Q345) plate specimens with a center circular hole. The main geometric sizes of rectangular plate specimens are: length, 600 mm, width, 50 mm, with different thicknesses, 4 mm and 10 mm and with different circular hole diameters, 10 mm,20 mm and 30 mm.

As done in Example 2 of this section, here, fracture analysis of center circular hole plates made of structure steel (Q345) can be performed. The obtained results and fracture test data reported in Ref.[54] are shown in Table 5.50. Variation of the GMPs with notch depths is listed in Table 5.51.

TABLE 5.50
Fracture Test Data Reported in Ref.[54] and Evaluation Calculation Results

No.	W (mm)	T (mm)	D (mm)	Failure Load F_c (KN)	SCFs k	σ_{max} (MPa)	K_{pc} (exp) (MPam$^{0.5}$)	K_{pc} (cal) (MPam$^{0.5}$)	E.R. (%)
1	51.2	3.5	10.0	61.58	2.53	1074.05	67.62	80.92	19.7
2	50.1	3.5	20.5	46.65	2.23	1004.75	90.15	90.11	0.1
3	49.7	3.7	30.1	31.53	2.12	920.15	100.03	96.01	−3.0

Note: For thickness = 4 mm, K_{Ic} = 65 pam$^{0.5}$, σ_0 = 700 MPa, which are estimated by using the empirical equation (56a).

No.	W (mm)	T (mm)	D (mm)	Failure Load F_c (KN)	SCFs k	σ_{max} (MPa)	K_{pc} (exp) (MPam$^{0.5}$)	K_{pc} (cal) (MPam$^{0.5}$)	E.R. (%)
4	53.8	9.4	10.2	200.30	2.54	1240.2	77.80	84.95	9.2
5	50.3	9.9	20.3	146.10	2.24	1100.2	98.23	98.45	0.2
6	51.1	9.6	30.3	99.80	2.12	1059.7	115.59	106.79	−7.6

Note: For thickness = 10 mm, K_{Ic} = 70 pam$^{0.5}$, σ_0 = 780 MPa, which are estimated by using the empirical equation (56a).

No.	W (mm)	T (mm)	D (mm)	Failure Load F_c (KN)	SCFs k	σ_{max} (MPa)	K_{pc} (exp) (MPam$^{0.5}$)	K_{pc} (cal) (MPam$^{0.5}$)	E.R. (%)
1'	50.0	3.5	10.0	61.58	2.52	1107.6	69.41	81.04	16.8
4'	50.0	9.4	10.2	200.30	2.51	1344.2	85.30	84.32	1.2

TABLE 5.51
Variation of the GMPs with Notch Depths

No.	k^*	ρ^* (mm)	L_c^* (mm)
1	3.73	2.30	1.608
2	3.57	4.00	1.474
3	3.61	5.16	1.257
4	3.52	2.60	1.468
5	3.66	3.80	1.346
6	3.69	5.00	1.148

From Table 5.50, it is found that the specimen that has the most large error in specimens with plate thickness = 4 mm has the largest width. Specimens with plate thickness = 10 mm are also true. For it, an attempt is made here to adjust the width of the specimen with the largest width to be the mean value, 50 mm. The obtained results are listed also in Table 5.50, with better predicted ones than the original ones.

Here, plate thickness has some effect on the fracture toughness. For example, for circular hole diameter = 30 mm, the fracture toughness of the specimen with plate thickness = 10 mm is 15% larger than that with plate thickness = 4 mm.

5.13 CONCLUDING REMARKS

On the basis of the linear elastic notch tip stress fields, the local stress field failure model for fracture analysis of notched specimens is proposed in this study. In the local stress field failure model, an important concept, the stress concentration factor eigenvalue (denoted by k^*), is introduced. For any notch (V- or U-notch), the GMPs corresponding to the notch are calculated by using an approach of determining k^* and models such as notch angle effect on k^*, notch depth effect on k^* and different materials effect on k^* proposed in this study. It can be seen from this study that fracture analysis for the notch performed by using the obtained GMPs and by means of the local stress field failure model has high accuracy within an error interval of a few percent.

Taking into account that the fracture toughness K_{Ic} is unknown at some circumstances, an empirical equation of determining K_{Ic} is proposed in this study. By using the empirical equation, the local stress field failure model proposed in this study has a wide range of applications such as fracture analysis of the RV-BD specimens and fracture analysis of center circular hole plates made of metallic materials.

In this study, an attempt was made to show that the local stress field failure model proposed in this study was used to perform fracture analysis of center notch plates made of metallic materials. From the viewpoint of Elastic Plastic Fracture Mechanics, no doubt, the attempt is not feasible. From the study [46] on the unified fatigue life-time equation, in low/medium/high cycle fatigue regime of metallic materials, by using the *linear elastic stress quantity,* however, the attempt was successful from the experimental verification examples. More explanations were given in Ref.[46].

Here, particular *emphasis* is placed on the following three points:

(1) For a **blunt notch**, as is well known to us, the peak stress criterion can be usually applied to its fracture analysis. From this study, a criterion according to which a notch is considered to be a blunt notch is as follows: When fracture analysis for a notch is performed by means of the local stress field failure model, if the solution of L_c^* for the studied notch cannot be found from Eq.(12-A), the notch is considered to be a blunt notch and its failure analysis is carried out by using the peak stress criterion.

(2) For a notch, if its stress concentration factor k is more than k^* or if its notch radius ρ is less than , (note here: k^* and ρ^* are the GMPs corresponding to the notch), the notch is considered to be the **super-sharp notch** and its fracture analysis is performed by means of the local stress field failure model for sharply notched specimens proposed in Ref.[46].

(3) When fracture analysis for a notch is performed, if the notch depth is less than a_0 calculated by (13-A), its fracture analysis is performed by using the **short crack approach**.

REFERENCES

1 Neuber, H. Zur Theorie der technischen Formzahl. *Forschg Ing-Wes* 7, 271–281 (1936).
2 Neuber, H. *Theory of notch stresses: Principles for exact calculation of strength with reference to structural form and material.* 2nd ed. Berlin: Springer Verlag; 1958.

3 Peterson, R.E. Notch sensitivity. In: Sines, G., Waisman, J.L., editors. *Metal fatigue.* New York, USA: McGraw Hill; 1959. p. 293–306.
4 Taylor, D. *The theory of critical distances: A new perspective in fracture mechanics.* Oxford, UK: Elsevier; 2007.
5 Susmel, L. *Multiaxial notch fatigue: From nominal to local stress/strain quantities.* Cambridge: Woodhead & CRC; 2009.
6 Berto, F., Lazzarin, P. Recent developments in brittle and quasi-brittle failure assessment of engineering materials by means of local approaches. *Materials Science and Engineering R* 75, 1–48 (2014).
7 Lazzarin, P., Zambardi, R. A finite-volume-energy based approach to predict the static and fatigue behaviour of components with sharp V-shaped notches. *International Journal of Fracture* 112, 275–298 (2001).
8 Lazzarin, P., Lazzarin, F.X. Some expressions for the strain energy in a finite volume surrounding the root of blunt V-notches. *International Journal of Fracture* 135, 161–185 (2005).
9 Taylor, D. Predicting the fracture strength of ceramic materials using the theory of critical distances. *Engng Frac Mech* 71, 2407–2416 (2004).
10 Taylor, D., Merlo, M., Pegley, R., Cavatorta, M.P. The effect of stress concentrations on the fracture strength of polymethylmethacrylate. *Material Science and Engineering* A382, 288–294 (2004).
11 Taylor, D., Cornetti, P., Pugno, N. The fracture mechanics of finite crack extension. *Engng Frac Mech* 72, 1021–1038 (2005).
12 Susmel, L., Taylor, D. An elasto-plastic reformulation of the theory of critical distances to estimate lifetime of notched components failing in the low/medium-cycle fatigue regime. *J Eng Technol* 132, 210021–210028 (2010).
13 Susmel, L., Taylor, D. A novel formulation of the theory of critical distances to estimate lifetime of notched components in the medium-cycle fatigue regime. *Fatigue Fract Eng Mater Struct* 30(7), 567–581 (2007).
14 Susmel, L., Taylor, D. On the use of the theory of critical distances to estimate fatigue strength of notched components in the medium-cycle fatigue regime. In: *Proceedings of FATIGUE 2006.* Atlanta, USA; 2006.
15 Yang, X.G., Wang, J.K., Liu, J.L. High temperature LCF life prediction of notched DS Ni-based superalloy using critical distance concept. *International Journal of Fatigue* 33, 1470–1476 (2011).
16 Filippi, S., Lazzarin, P., Tovo, R. Developments of some explicit formulas useful to describe elastic stress field ahead of the notches. *International Journal of Solids and Structures* 39, 4543–4565 (2002).
17 Lazzarin, P., Filippi, S. A generalized stress intensity factor to be applied to rounded V-shaped notches. *Int. J. Solids Struct.* 43, 2461–2478 (2006).
18 Gómez, F.J., Elices, M. A fracture criterion for blunted V-notched samples. *International Journal of Fracture* 127, 239–264 (2004).
19 Noda, N.A., Sera, M., Takase, Y. Stress concentration factors for round and flat test specimens with notches. *Int. J. Fatigue* 17(3), 163-178 (1995).
20 Noda, N.A., Takase, Y. Stress concentration formula useful for all notch shape in a round bar (comparison between torsion, tension and bending). *International Journal of Fatigue* 28, 151–163 (2006).
21 Noda, N.A., Takase, Y. Stress concentration formulas useful for any shape of notch in a round test specimen under tension and under bending. *Fatigue Fract Eng Mater Struct* 22, 1071–1082 (1999).
22 Noda, N.A., Takase, Y. Stress concentration factor formulas useful for all notch shapes in a flat test specimen under tension and bending. *J Test Eval* 30(5), 369–381 (2002).

23 Mykhaylo, P.S., Kazberuk, A. Two-dimensional fracture mechanics problems for solids with sharp and rounded V-notches. *Int J Fract* 161, 79–95 (2010).
24 Noda, N.A., Takase, Y. Generalized stress intensity factors of V-shaped notch in a round bar under torsion, tension, and bending. *Engineering Fracture Mechanics* 70, 1447–1466 (2003).
25 Liu, Y., Liu, X.M., Hu, L., Xie, Y.D. Numerical analysis of stress concentration factors for U-shaped notch with four-point bending (in Chinese). Journal of Shanghai University of Engineering Science 28(2), 1009-444X(2014)02-0141-04
26 Wang, Q.Z., Wang, K. Stress concentration factors for three-point bending beam with single-edge U-shaped notch in Chinese). *Journal of Sichuan University* 39(3), 1009-3087(2007)03-0001-06.
27 Barati, E., Alizadeh, Y. A notch root radius to attain minimum fracture loads in plates weakened by U-notches under Mode I loading. *Scientia Iranica B* 19(3), 491–502 (2012).
28 Shin, C.S., Man, K.C., Wang, C.M. A practical method to estimate the stress concentration of notches. *Fatigue* 16, 242–256 (1994).
29 Lee, B.W., Jang, J., Shin, D.X. Evaluation of fracture toughness using small notched specimens. *Materials Science and Engineering* A334, 207–214 (2002).
30 BS7448, Fracture Mechanics Toughness Tests, British Standards Institution, London; 1997.
31 Wilshaw, T.R., Rau, C.A., Tetelman, A.S. Effect of temperature and strain-rate on deformation and fracture of mild-steel charpy specimens. *Eng. Frac. Mech.* 1, 191 (1968).
32 Ritche, R.O., Francis, B., Server, W.L. *Metall. Mater. Trans. A* 7, 831 (1976).
33 Swanson, R.E., Thompson, A.W., Bernstein, I.M. *Metall. Mater. Trans. A* 17, 1633 (1986).
34 Frost, N.E. Non-propagating cracks in V-notched specimens subjected to fatigue loading. *Aeronautical Quarterly* VIII, 1–20 (1957).
35 Frost, N.E. A relation between the critical alternating propagation stress and crack length for mild steel. *Proceedings of the Institution of Mechanical Engineers* 173, 811–834 (1959).
36 Smith, R.A., Miller K.J. Prediction of fatigue regimes in notched components. *International Journal of Mechanical Sciences* 20, 201–206 (1978).
37 Atzori, B., Lazzarin, P., Meneghetti, G. A unified treatment of the mode I fatigue limit of components containing notches or defects. *International Journal of Fracture* 133, 61–87 (2005).
38 Lazzarin, P., Filippi, S. A generalized stress intensity factor to be applied to rounded V-shaped notches. *International Journal of Solids and Structures* 43, 2461–2478 (2006).
39 Gómez, F.J., Elices, M., Planas, J. The cohesive crack concept: Application to PMMA at –60°. *Engineering Fracture Mechanics* 72, 1268–1285 (2005).
40 Savruk, M.P., Kazberuk, A. Two-dimensional fracture mechanics problems for solids with sharp and rounded V-notches. Int J Fract 161, 79–95 (2010).
41 Gross, B., Mendelson, A. Plane elastostatic analysis of V-notched plates. *Int. J. of Fracture Mech.* 8, 267–276 (1972).
42 Gómez, F.J., Elices, M., Valiente, A. Cracking in PMMA containing U-shaped notches. *Fatigue Fract. Eng. Mater. Struct.* 23, 795–803 (2000).
43 Gogotsi, G.A. Fracture toughness of ceramics and ceramic composites. *Ceram Int* 7, 777–784 (2003).
44 Berto, F., Barati, E. Fracture assessment of U-notches under three point bending by means of local energy density. *Materials and Design* 32, 822–830 (2011).

45 Berto, F., Lazzarin, P., Marangon. C. Brittle fracture of U-notched graphite plates under mixed mode loading. *Materials and Design* 41, 421–432 (2012).
46 Yan, X.Q. A local approach for fracture analysis of V-notch specimens under I loading. To be published in *Engineering Fracture Mechanics.*
47 Susmel, L., Taylor, D. The theory of critical distances to predict static strength of notched brittle components subjected to mixed-mode loading. *Engineering Fracture Mechanics* 75, 534–550 (2008).
48 Ayatollahi, M.R., Torabi, A.R. Investigation of mixed mode brittle fracture in rounded-tip. *Engineering Fracture Mechanics* 77, 3087–3104 (2010).
49 Susmel, L., Taylor, D. The theory of critical distances to estimate the static strength of notched samples of Al6082 loaded in combined tension and torsion. Part I: Material cracking behavior. *Engineering Fracture Mechanics* 77, 452–469 (2010).
50 Susmel, L., Taylor, D. The theory of critical distances to estimate the static strength of notched samples of Al6082 loaded in combined tension and torsion. Part II: Multiaxial static assessment. *Engineering Fracture Mechanics* 77, 470–478 (2010).
51 Torabi, A.R., Habibi, R., Hosseini, B.M. On the ability of the equivalent material concept in predicting ductile failure of U-notches under moderate-and large-scale yielding conditions. *Phys Mesomech* 18, 337–347 (2015).
52 Cai, D.L., Qin, S.H., Gao, L., Liu, G.L., Zhang, K.S. Study on tensile fracture of pure copper plate with a center hole (in Chinese). *Journal of Guangxi University* 41(4), 1178–1186 (2016).
53 Yao, W.X., Gu, Y. Influence of finite wide and elliptical aspect on notched laminate strength predictions (in Chinese). *Acta Materiae Compositae Sinica* 6, 21–29 (1989).
54 Wang, W.Z., Zhang, J.J., Sun, Y.P., Zhang, D.L. An experimental study on structural steel fracture (in Chinese). *Journal of Gansu Sciences* 19(4), 112—115 (2007).
55 Zappalorto, M.,·Lazzarin, P., Filippi, S. Stress field equations for U and blunt V-shaped notches in axisymmetric shafts under torsion. *Int J Fract* 164, 253–269 (2010).
56 Zappalorto, M., Lazzarin, P., Yates, J.R. Elastic stress distributions for hyperbolic and parabolic notches in round shafts under torsion and uniform antiplane shear loadings. *International Journal of Solids and Structures* 45 4879 (2008).
57 Zappalorto, M., Lazzarin, P., Berto, F. Elastic notch stress intensity factors for sharply V-notched rounded bars under torsion. *Engineering Fracture Mechanics* 76, 439–453 (2009).
58 Coulomb, M. Recherches théoriques et expérimentales sur la force de torsion et sur l'élasticité des fils de metal. *Histoire de l'Académie Royale des Sciences*, 229–269 (1784).
59 Zappalorto, M., Bertoand, F., Lazzarin, P. Practical expressions for the notch stress concentration factors of round bars under torsion. *International Journal of Fatigue* 33, 382–395 (2011).
60 Berto, F., Elicesb, M., Lazzarin, P., Zappalorto, M. Fracture behaviour of notched round bars made of PMMA subjected to torsion at room temperature. *Engineering Fracture Mechanics* 90, 143–160 (2012).
61 Berto, F., Cendon, D.A., Lazzarin, P., Elices, M. Fracture behaviour of notched round bars made of PMMA subjected to torsion **at** -60°C. *Engineering Fracture Mechanics* 102, 271–287 (2013).
62 El Haddad, M.H., Topper, T.H., Smith, K.N. Prediction of non propagating cracks. *Engineering Fracture Mechanics* 11, 573–584 (1979).
63 Berto, F., Lazzarin, P., Ayatollahi, M.R. Brittle fracture of sharp and blunt V-notches in isostatic graphite under torsion loading. *CARBON* 50, 1942–1952 (2012).
64 Frost, N.E. Non-propagating cracks in vee-notched specimens subjected to fatigue loadings. *Aeronaut. Q.* 8, 1–20 (1957).

65 Murakami, Y. *Stress intensity factors handbook*. Beijing: Pergamon Press World Publishing Corporation; 1989.
66 Weiss, B., Stickler, R., Lukas, P., Kunz, L. Non-damaging notches in fatigue: A short crack problem? In: 2nd intl. workshop on small fatigue cracks, Santa Barbara, Calif, 1986, 5–10 January.
67 DuQuesnay, D.L., Yu, M., Topper, T.H. An analysis of notch size effect on the fatigue limit. *J. Test. Eval.* 4, 375–385 (1988).
68 Tada, H. A note on the finite width corrections to the stress intensity factor. *Engng Frac. Mech.* 3(3), 345–347 (1971).
69 Lukas, P., Kunz, L. Influence of notches on high cycle fatigue life. *Materials Science and Engineering* 47, 93–91 (1981).
70 Liu, B.W., Yan, X.Q. A new model of multiaxial fatigue life prediction with influence of different mean stresses. *Int. J. Damage Mech.* 28(9), 1323–1343 (2019).
71 Susmel, L., Taylor, D. The modified Wöhler curve method applied along with the theory of critical distances to estimate finite life of notched components subjected to complex multiaxial loading paths. *Fatigue and Fracture of Engineering Materials and Structures* 3112, 1047–1064 (2008).

APPENDIX A: A LOCAL STRESS FIELD FAILURE MODEL OF SHARP NOTCHES UNDER III LOADING

Here, an attempt is made to propose a local stress field failure model of sharp notches (including U- and V-notches) under III (torsion) loading. Under the assumption that the stress concentration factor eigenvalue k^* for V-notches under I loading proposed in the main body of this study is valid for V-notches under III loading, in fact, the local stress field failure model of V-notches under III loading is parallel to that under I loading. For brevity, thus, the linear stress field of V-notches under III loading on the model descriptions is given only because of the difference between the linear stress fields of I and III V-notches. Because the model descriptions on the local stress field failure model and the approach of determining k^* for V-notches under III loading are the same as those for V-notches under I loading, the model descriptions are not given here, which is in fact the reason why this Appendix A occurs here.

Thus the contents of this Appendix are outlined as follows:

A-1 A brief description of local stress field ahead of rounded V-notches under III loading;
A-2 Experimental verifications;
A-3 Effect of notch depth on k^*;
A-4 Effect of notch angle on effect k^*;
A-5 Comments on the Appendix

A-1 A BRIEF DESCRIPTION OF LOCAL STRESS FIELD AHEAD OF ROUNDED V-NOTCHES UNDER III LOADING

In 2010, analytical solutions were developed by Zappalorto et al. [55] for the stress fields induced by circumferential U- and V-shaped notches in axisymmetric shafts under torsion, with a finite value of the notch root radius. Reference system chosen in

Ref.[55] was shown in Figure 1-A. The stress τ_{zy} along the notch bisector is expressed to be

$$\tau_{zy} = \frac{\tau_{max}}{\omega_3}\left(\frac{r}{r_0}\right)^{\lambda_3-1}\left[1+\left(\frac{r}{r_3}\right)^{\mu_3-\lambda_3}\right]\left(1-\frac{r-r_0}{R}\right) \tag{1-A}$$

where the geometrical parameters

$$q = \frac{2\pi-2\alpha}{\pi} = \frac{1}{\lambda_3}, \qquad r_0 = \frac{q-1}{q}\rho, \qquad r_3 \cong (1-\mu_3)\times\rho \tag{2-A}$$

and R is the radius of the net section of the shaft. Values of coefficients, λ_3, μ_3, ω_3 in (1-A) were given by Zappalorto et al. [55], see Table 1-A.

By means of the following relation [56]:

$$K_{v\rho} = \tau_{max}\sqrt{2\pi r_0^{1-\lambda_3}} \tag{3-A}$$

The stress τ_{zy} along the notch bisector is also expressed in terms of the generalized N-SIFs $K_{v\rho}$ defined in (3-A).

$$\tau_{zy} = \frac{K_{v\rho}}{\omega_3\sqrt{2\pi r_0^{1-\lambda_3}}}\left(\frac{r}{r_0}\right)^{\lambda_3-1}\left[1+\left(\frac{r}{r_3}\right)^{\mu_3-\lambda_3}\right]\left(1-\frac{r-r_0}{R}\right) \tag{4-A}$$

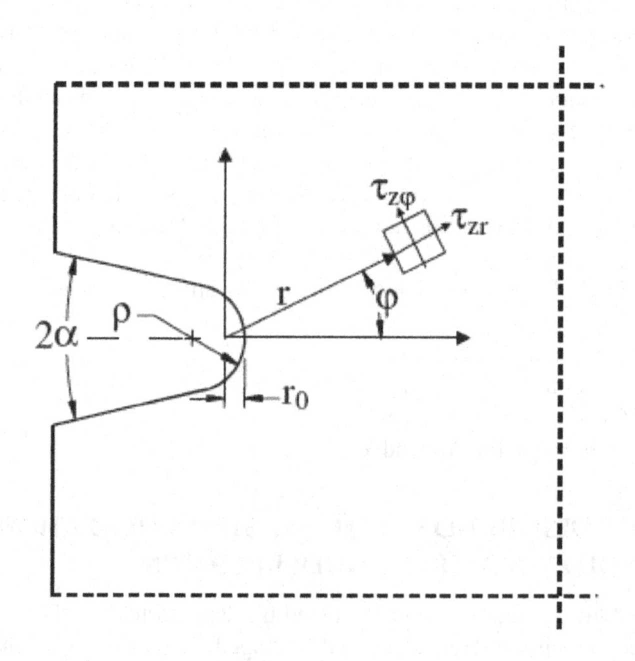

FIGURE 1-A Reference system for U- or V-notches.

TABLE 1-A
Values of Coefficients in Eq.(1-A) (from Ref. [55])

$2\alpha / \pi$	q	λ_3	μ_3	ω_3
0	2.0000	0.5000	0.40978	2.0155
1/6	1.8333	0.5454	0.45082	2.0181
1/4	1.7500	0.5714	0.47473	2.0196
1/3	1.6667	0.6000	0.50146	2.0213
5/12	1.5833	0.6315	0.53160	2.0232
1/2	1.5000	0.6667	0.56594	2.0253
7/12	1.4167	0.7058	0.60554	2.0275
2/3	1.3333	0.7500	0.65196	2.0297
3/4	1.2500	0.8000	0.70756	2.0315
5/6	1.1667	0.8571	0.77634	2.0317

By the way, it is pointed out here that the solution of the elastic notch stress intensity factors for sharply V-notched rounded bars under torsion by Zappalorto et al. [57] sometimes is very useful for the fracture analysis of blunt V-notch specimens:

$$K_3 = \tau_{nom,n} \cdot F_3 \cdot R^{1-\lambda_3} \tag{4-A}$$

where $\tau_{nom,n}$ is the net section nominal stress evaluated according to Coulomb's expression as [58]

$$\tau_n = \frac{16M_T}{\pi(\phi - 2D)^3} \tag{5-A}$$

and the nondimensional coefficient F_3 was given in Ref.[57,59].

A-2 EXPERIMENTAL VERIFICATIONS

Experimental verifications given here are parallel to those in Section 5 of the main body of this study.

For the sake of clear discussions, the experimental verifications given here are described in the forms of examples, respectively.

Example 1: Fracture Analysis of Notched Round Bars Made of PMMA Subjected to Torsion at Room Temperature

In 2012, fracture behavior of notched round bars made of PMMA subjected to torsion at room temperature was investigated by Berto et al. [60]. The geometry size of the specimens studied was shown in Figure 2-A.

For U-notched specimens (Figure 2(b)-A), notches with four different notch root radii, 0.3, 0.5, 1.0, 2.0 mm, were tested. Two values of the notch depth were used, d = 2 mm and d = 5 mm, with a constant gross diameter equal to 20 mm.

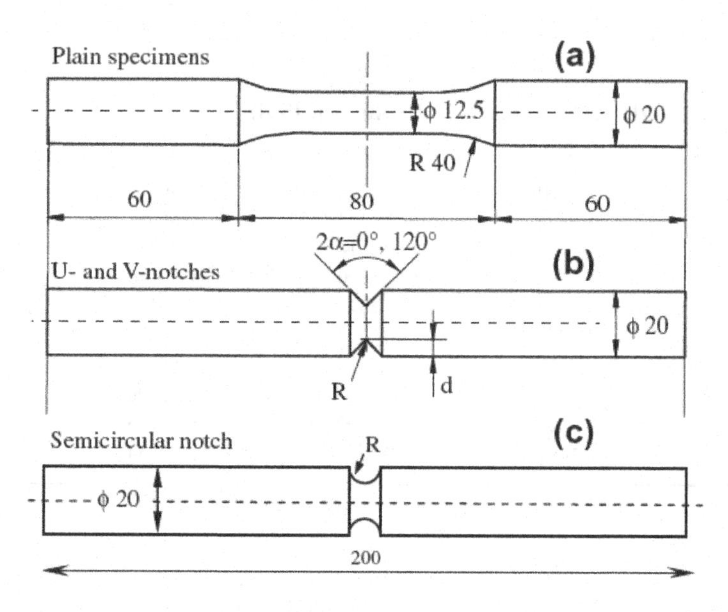

FIGURE 2-A Geometry of plain and notched specimens (a–c).

Source: (Data from Berto *et al.* [60]).

For semicircular notches (Figure 2(c)-A), notches with seven different notch root radii, 0.5, 1.0, 2.0, 4.0, 5.0, 6.0 and 7.0 mm, were tested.

At room temperature, the torsion strength and the torsion fracture toughness of PMMA reported in Ref.[6] are τ_{tor} = 67 MPa and K_{IIIc} = 3.35 MPam$^{0.5}$, respectively.

For notch depth = 5 mm, fracture test data reported in Ref.[60] were shown in Table 2-A. The fracture analysis details performed here are described as follows:

(1) Fracture test data of cases, 1 and 4, shown in Table 2-A, are used to calculate the critical stress τ_0 by using the model equations (14) to (16). Its value is τ_0 = 58.6 MPa;

By the way, it is pointed out here that $K_{\rho c}$ (exp) shown in Table 2-A is calculated by using Eq.(3-A).

(2) In order to efficiently implement the model calculations, a simple method of estimating the stress concentration factor is proposed and is illustrated through an example. For cases shown in Table 2-A, fracture test data of case 4 are taken as reference data: $\tau_{nom,n0}$ = 90.03 Mpa, ρ_0 = 2 mm, k_0 = 1.21, $K_{\rho c0}$ = 8.64 Mpam$^{0.5}$. For any notch with notch radius ρ, τ_{max} is calculated by using Eq.(3-A) by letting $K_{3\rho} = K_{3\rho 0}$ and the stress concentration factor of the notch is defined to be a ratio τ_{max} to $\tau_{nom,n0}$. For notch radius ρ = 2.0 mm, 1.0 mm, 0.5 mm, 0.3 mm and 0.047 mm, the calculated stress concentration factors are given in Table 3-A;

(3) By using the obtained τ_0 and K_{IIIc} = 3.35 MPam$^{0.5}$, the GMPs of the U-notch are calculated by means of the approach of determining k^*. Their values are k^* = 7.89, ρ^* = 0.047 mm, L_c^* = 0.373 mm;

TABLE 2-A
Fracture Test Data reported in Ref.[60] and Some Calculated Results (Notch Depth d = 5 mm)

No.	Notch Radius ρ (mm)	$\tau_{nom,n}$ (MPa)	K	τ_{max} (Mpa)	$K_{\rho c}$ (exp)
1	0.3	81.63	2.3	187.75	5.76
2	0.5	84.9	1.88	159.61	6.33
3	1.0	88.27	1.49	131.52	7.37
4	2.0	90.03	1.21	108.94	8.64

TABLE 3-A
The Stress Concentration Factors Calculated by Using a Simple Method Proposed in This Study

Notch Radius (mm)	2.0	1.0	0.5	0.3	0.047
SCFs k	1.21	1.71	2.42	3.12	7.89

TABLE 4-A
Comparison of the Predicted Notch Fracture Toughness with That Measured Experimentally (Notch Depth = 5 mm) (PMMA at Room Temperature)

Notch Radius ρ (mm)	SCFs k	L_c (mm)	$K_{\rho c}$ (pre) (MPa)	$K_{\rho c}$ (exp) (MPa)	R.E.(%)
0.3	3.12	0.942	5.38	5.76	−6.7
0.5	2.42	1.2161	6.18	6.33	−2.4
1	1.71	1.7198	7.51	7.37	1.9
2	1.21	2.4322	9.24	8.64	7

Note: (a) $k^* = 7.89$, $\rho^* = 0.047$ mm, $L_c^* = 0.373$ mm; (b) M.E. = −0.04%, M.A.E. = 4.5%, E.R. = [−6.7,7]%

TABLE 5-A
Comparison of the Predicted Notch Fracture Toughness with That Measured Experimentally (Notch Depth = 2 mm) (PMMA at Room Temperature)

Notch Radius ρ (mm)	SCFs k	L_c (mm)	$K_{\rho c}$ (pre) (MPa)	$K_{\rho c}$ (exp) (MPa)	R.E.(%)	Notch Radius ρ (mm)
0.3	81.3	3.2	1.2048	6.04	6.44	−6.2
0.5	81.2	2.47	1.5553	6.91	6.89	0.4
1.0	82.57	1.75	2.1996	8.35	8.1	3.1

Note: (a) $k^* = 10.46$, $\rho^* = 0.028$ mm, $L_c^* = 0.368$ mm; (b) M.E. = −0.9%, M.A.E. = 3.2%, E.R. = [−6.2,3.1]%

(4) By using the obtained values of the GMPs, the fracture analysis of the U-notched round bars is performed by means of the local stress field failure model. The obtained results are listed in Table 4-A, with the calculation error indexes: M.E. = −0.04%, M.A.E. = 4.5%, E.R. = [−6.7, 7.0]%.

For notch depth = 2 mm, the GMPs of the U-notch calculated by using the obtained τ_0 and K_{IIIc} = 3.35 MPam$^{0.5}$ and by means of the approach of determining k^* are k^* = 10.46, ρ^* = 0.028 mm, L_c^* = 0.368 mm and the obtained fracture analysis results are shown in Table 5-A, with the calculation error indexes: M.E. = −0.9%, M.A.E. = 3.2%, E.R. = [−6.2,3.1]%.

Example 2: Fracture Analysis of Notched Round Bars Made of PMMA Subjected to Torsion at −60°C

In 2013, fracture behavior of notched round bars made of PMMA subjected to torsion at −60°C was investigated by Berto et al. [61]. Taking into account that the study reported in Ref.[61] is parallel to that in Ref.[60], the model calculation details are not described here and the calculation results are given only (see Table 6-A and Table 7-A), with the calculation error indexes: M.E. = 0.2%, M.A.E. = 6.2%, E.R. = [−8.7,6.8]% and M.E. = 0.7%, M.A.E. = 2.7%, E.R. = [−2.9,3.5]%, for notch depths, 5 mm and 2 mm, respectively. Obviously, the calculation results are very satisfactory.

TABLE 6-A

Comparison of the Predicted Notch Fracture Toughness with That Measured Experimentally (Notch Depth = 5 mm) (PMMA at −60°C)

Notch Radius ρ (mm)	SCFs k	L_c (mm)	$K_{\rho c}$ (pre) (MPam$^{0.5}$)	$K_{\rho c}$ (exp) (MPam$^{0.5}$)	R.E.(%)
0.3	81.3	3.2	6.04	6.44	−6.2
0.5	81.2	2.47	6.91	6.89	0.4
1.0	82.57	1.75	8.35	8.1	3.1

Note: (a) $k^* = 10.46$, $\rho^* = 0.028$ mm, $L_c^* = 0.368$ mm; (b) M.E. = −0.9%, M.A.E. = 3.2%, E.R. = [−6.2,3.1]%

TABLE 7-A

Comparison of the Predicted Notch Fracture Toughness with That Measured Experimentally (Notch Depth = 2 mm) (PMMA at −60°C)

Notch Radius ρ (mm)	SCFs k	L_c (mm)	$K_{\rho c}$ (pre) (MPa)	$K_{\rho c}$ (exp) (MPa)	R.E.(%)	Notch Radius ρ (mm)
0.3	131.75	3.2	0.3312	10.13	10.44	−2.9
0.5	138.53	2.47	0.4275	11.94	11.75	1.6
1	148.8	1.75	0.6046	15.11	14.6	3.5

Note: (a) $k^* = 7.83$, $\rho^* = 0.05$ mm, $L_c^* = 0.135$ mm; (b) M.E. = 0.7%, M.A.E. = 2.7%, E.R. = [−2.9,3.5]%

A-3 EFFECT OF NOTCH DEPTH ON k^*

The GMPs of the U-notch made of PMMA at room temperature and at -60°C, obtained in Section 2-A, are listed in Table 8-A. From the values of the GMPs, here, an attempt is made to propose a model of predicting notch depth effect on k^* of different notched materials

From Table 8-A, it can be seen that k^* is varied with notch depth. It is assumed here that the variation of k^* with notch depth is linear. In order to determine the linear variation of k^* with notch depth, the following parameter, *ratio*, is defined:

$$ratio = \frac{k^*(2d/\phi = 0.2)}{k^*(2d/\phi = 0.5)} \tag{6-A}$$

For PAMM at -60°C, by letting $ratio = ratio_1$, its value is $ratio_1 = 1.77$. For PAMM at room temperature, similarly, by letting $ratio = ratio_2$, its value is $ratio_2 = 1.32$.

From Table 8-A, it can be seen that k^* is varied with notched materials. In order to determine the effect of notched materials on k^*, the torsion fracture toughness of PMMA at -60°C and at room temperature are denoted by K_{IIIc1} and K_{IIIc2}, respectively, and at the same time, the following parameter, ra, is defined:

$$ra = \left(\frac{K_{IIIc1}}{K_{IIIc2}}\right)^{0.5} \tag{7-A}$$

Here, $ra = 1.34$. By observing the values of $ratio_1$, $ratio_2$ and ra shown in Table 8-A, we find that the following relation is valid:

$$ratio_2 \approx \frac{ratio_1}{ra} \tag{8-A}$$

According to (8-A), for any material 3 (whose fracture toughness K_{IIIc} and *ratio* are denoted by K_{IIIc3} and *ratio*3, respectively), a model of predicting the variation of k^* with notch depth is proposed as follows:

If $K_{IIIc3} \le K_{IIIc1}$, *ratio*3 is

$$ratio_3 = \frac{ratio_1}{\left(\frac{K_{IIIc1}}{K_{IIIc3}}\right)^{0.5}} \tag{9-A}$$

IF $K_{IIIc3} \ge K_{IIIc2}$, *ratio*3 is

$$ratio_3 = ratio_2 \cdot \left(\frac{K_{IIIc3}}{K_{IIIc2}}\right)^{0.5} \tag{10-A}$$

For any material 3, if $k^*(2d/\phi = 0.5)$ is known, $k^*(2d/\phi = 0.2)$ is calculated by the following equation:

$$k^*(2d/\phi = 0.2) = k^*(2d/\phi = 0.5) \cdot ratio_3 \tag{11-A}$$

TABLE 8-A
The GMPs of U-Notch (PMMA at Room Temperature and at $-60^{\circ}C$)

Materials	Notch Depth d (mm)	$\frac{2d}{\phi}$	k^*	ρ^* (mm)	L_c^* (mm)	Note
PAMM at -60°C	5	0.5	4.42	0.15	0.099	K_{IIIc} = 6.02 MPam$^{0.5}$,
	2	0.2	7.83	0.05	0.135	τ_0 = 172 MPa
PAMM at room	5	0.5	7.89	0.047	0.373	K_{IIIc} = 3.35 MPam$^{0.5}$,
temperature	2	0.2	10.46	0.028	0.368	τ_0 = 58.6 MPa

In the following, the experimental verifications described in the forms of examples are given to verify the notch depth effect on k^* proposed in this section

Example 1: U-Notch Made of PMMA at Room Temperature

Here, material 3 is taken PMMA at room temperature:
$k^*(2d/\phi = 0.5)$ = 7.89, $ratio_3$ is calculated from (9-A):

$$ratio_3 = \frac{ratio_1}{\left(\frac{K_{IIIc1}}{K_{IIIc3}}\right)^{0.5}} = \frac{1.77}{\left(\frac{6.02}{3.35}\right)^{0.5}} = \frac{1.77}{1.34} = 1.32$$

From (11-A), thus, we have:

$$k^*(2d/\phi = 0.2) = k^*(2d/\phi = 0.5)\cdot ratio3$$
$$= 7.89 \times 1.32 = 10.43$$

which is in excellent agreement with that ($k^*(2d/\phi = 0.2)$ = 10.46) determined by using fracture test data.

Here, fracture analysis for semicircular notches reported in Ref.[60] is performed by using the obtained $k^*(2d/\phi = 0.5)$ = 7.89 and $k^*(2d/\phi = 0.2)$ = 10.43 and by linear interpolation or linear extrapolation. The obtained values of k^* for various semicircular notches are shown in Table 9-A. By using the obtained values of k^* and the geometry of semicircular notches, the values of ρ^* for various semicircular notches can be obtained (see Table 9-A). The values of L_c^* can be calculated further from the following failure equation:

$$F(K_{IIIc}, L_c^*, \rho^*, \alpha, \tau_0) = 0 \tag{12-A}$$

where α is notch angle, for U-notch, α = 0 (note here: Eq. (12-A) is similar to Eq.(60)). The obtained values of L_c^* for various semicircular notches are shown in Table 9-A. By using the obtained values of the GMPs shown in Table 9-A, the fracture analysis for semicircular notches reported in Ref.[60] is easily performed by means of the local stress field failure model. The obtained results and fracture test data reported in Ref.[60] are given in Table 9-A, from which it can be seen that the obtained results are very satisfactory, excepting semicircular notches with small radii.

For semicircular notches with small radii, an attempt is made to show that fracture analysis is carried out by using the short crack approach [62]:

$$\tau_{nom} = \frac{K_{IIIc}}{\sqrt{\pi(a+a_0)}} \tag{13-A}$$

where

$$a_0 = \frac{1}{\pi}\left(\frac{K_{IIIc}}{\tau_0}\right)^2 \tag{14-A}$$

Here, $a_0 = 1.041$ mm. The obtained results from (13-A) are shown in Table 9-A with larger errors.

By the way, it is pointed out that when fracture analysis for a notch specimen is performed by means of the local stress field failure model, if notch depth $a < a_0$, fracture analysis for the notch specimen can be carried out by using the short crack approach [62].

Taking into account that the results predicted by formula (13-A) have larger errors (see Table 9-A), a coefficient cof in (13-A) is introduced as follows:

$$\tau_{nom} = \frac{K_{IIIc}}{\sqrt{\pi(a+cof \cdot a_0)}} \tag{15-A}$$

The variation of the calculation errors with cof is shown in Table 10-A, from which it can be seen that, if cof is taken as 0.25, the mean error is about 0.2%.

TABLE 9-A
Fracture Test Data of Semicircular Notches Reported in Ref.[60] and Evaluation Calculation Results (PMMA at Room Temperature)

Semicircular Notches (mm)	**0.5**	**1.0**	**2.0**	**4.0**	**5.0**	**6.0**
$\frac{2d}{\phi}$	0.05	0.1	0.2	0.4	0.5	0.6
k^*	11.29	11.08	10.43	8.78	7.89	7.05
ρ^* (mm)	0.0119	0.0206	0.036	0.076	0.105	0.146
L_c^* (mm)	0.354	0.364	0.371	0.373	0.368	0.358
SCFs k	1.74	1.59	1.4	1.21	1.14	1.1
$\tau_{nom,n}$ (MPa)	73.1	80.26	83.33	86	77.87	73.76
$K_{\rho c}$ (exp) (MPam$^{0.5}$)	5.04	7.15	9.25	11.67	11.13	11.14
$K_{\rho c}$ (pre) (MPam$^{0.5}$)	8.32	8.90	9.72	10.65	10.95	11.16
R.E.(%)	65	24.4	5.1	−8.7	−1.6	0.14
R.E.(%) (short crack)	−23.2	−28.5				

Note: (a) The values of k and $\tau_{nom,n}$ are taken from Ref.[22]; (b) $K_{\rho c}$ (exp) is calculated from (7); (d) τ_0 = 58.56 MPa, τ_{tor} = 67 MPa, a_0 = 1.041 mm.

TABLE 10-A
The Variation of the Calculation Errors with *cof*

cof	1.0	0.7	0.5	0.4	0.3	0.25	0.0	Note
R.E. (%)	−23.2	−14.0	−5.6	−0.4	5.8	9.4	34.9	Semicircular notches (0.5 mm)
R.E. (%)	−28.5	−22.3	−17.2	−14.2	−10.8	-9.0	2.2	Semicircular notches (1.0 mm)

Example 2: U-Notch Made of PMMA at $-60^{\circ}C$

As another example, here, material 3 is taken as PMMA at $-60^{\circ}C$: $k^{*}(2d/\phi=0.5)$ = 4.42, $ratio_3$ is calculated from (10-A):

$$ratio_3 = ratio_2 \cdot \left(\frac{K_{IIIc3}}{K_{IIIc2}}\right)^{0.5} = 1.32 \times \left(\frac{6.02}{3.35}\right)^{0.5} = 1.77$$

From (11-A), thus, we have:

$$\begin{aligned} k^{*}(2d/\phi=0.2) &= k^{*}(2d/\phi=0.5) \cdot ratio3 \\ &= 4.42 \times 1.77 = 7.82 \end{aligned}$$

which is in excellent agreement with ($k^{*}(2d/\phi=0.2)$ = 7.83), determined by using fracture test data.

As done in Example 1 of this section, here, fracture analysis for semicircular notches reported in Ref.[61] is performed by using the obtained $k^{*}(2d/\phi=0.5)$ = 4.42 and $k^{*}(2d/\phi=0.2)$ = 7.82 and by linear interpolation or linear extrapolation. The obtained results are shown in Table 11-A. For semicircular notches with radii equal to 5 mm, the predicted fracture toughness ($K_{\rho c}$ (pre)) is in good agreement with ($K_{\rho c}$ (exp)) measured experimentally, with relative error 5.9%. But it is necessary to point out here that because the solutions of L_c^{*} for semicircular notches with radii equal to 6 mm and 7 mm cannot be found from Eq.(12-A), fracture analysis of the two semicircular notches cannot be performed by means of the local stress field failure model. For that, an attempt is made to show that the maximum stress approach is used to analyze the fracture behavior of semicircular notches reported in Ref.[61]. The obtained results are given in Table 11-A, from which it can be seen that the calculation results obtained by taking the reference stress to be τ_0 = 172 MPa are better than those by taking the reference stress to be τ_{tor} = 153.1 Mpa, with the calculation error indexes: M.E. = −0.05%, A.M.E. = 2.36%, E.R. = [−2.24,3.47]%, and M.E. = 12.29%, A.M.E. = 12.29%, E.R. = [9.83,16.25]%.

From the fracture analysis for semicircular notches reported in Ref.[61] (see Table 11-A), it is concluded perhaps that the condition that the peak stress criterion is applied to the fracture analysis for a notched specimen is that the solution of L_c^{*} for the studied notched specimen cannot be found from Eq.(12-A). The details are explained as follows:

By substituting $r = r_0 + L_c^{*}$ and $K_{v\rho} = K_{v\rho c}$ into (4-A), the stress $\tau_{zy}(L_c^{*})$ at the critical distance L_c^{*} can be expressed as:

$$\tau_{zy}(L_c^{*}) = H(K_{v\rho c}, \rho, \alpha, L_c^{*}) \tag{16-A}$$

TABLE 11-A

Fracture Test Data of Semicircular Notches Reported in Ref.[61] and Evaluation Calculation Results (PMMA at $-60^\circ C$)

Semicircular Notches (mm)	**5**	**6**	**7**
$2d/\phi$	0.5	0.6	0.7
ρ^* (mm)	**0.332**	**0.675**	**1.72**
L_c^* (mm)	0.02	no solution	no solution
k	1.14	1.1	1.06
$\tau_{nom,n}$ (MPa)	**147.50**	**154.20**	167.89
$K_{\rho c}$ (exp)	21.07	23.29	26.39
$K_{\rho c}$ (pre)	22.32		
R.E.(%)	5.9		
τ_{max} (MPa)	168.15	**169.62**	177.97
$(\tau_{max}-\tau_0)\times 100/\tau_0$	-2.24	-1.38	3.47
$(\tau_{max}-\tau_{tor})\times 100/\tau_{tor}$	9.83	10.79	16.25

Note: (a) τ_{tor} = 153.1 MPa, τ_0 = 172 MPa;

When $L_c^* \to 0$, $\tau_{zy}(L_c^*) < \tau_0$ occurs, the solution of L_c^* for the studied notched specimen cannot be found from Eq.(12-A) occurs. At this time, the peak stress criterion is applied to the fracture analysis for a notched specimen. Perhaps this can be considered a criterion: when fracture analysis for a notched specimen is performed by means of the local stress field failure model, if the solution of L_c^* for the studied notched specimen cannot be found from Eq.(12-A), the peak stress criterion is applied to the fracture analysis for the notched specimen.

Example 3: Brittle Fracture of Sharp and Blunt V-Notches in Isostatic Graphite Under Torsion Loading

In 2012, fracture behavior of notched round bars made of isostatic graphite subjected to torsion at room temperature was investigated by Berto et al. [63]. The geometry size of the specimens studied was shown in Figure 2-A.

For U-notched specimens in (Figure 2(b)-A), notches with two different notch root radii were tested: ρ = 1 and 2.0 mm. The effect of the net section area was studied by changing the notch depth *d*. Two values were used, *d* = 2 and 5 mm, while keeping the gross diameter constant (20 mm).

For V-notched specimens with a notch angle $2\alpha = 30^\circ$ (Figure 2(b)-A), three different notch root radii were used in the experiments: ρ = 0.1, 0.3 and 0.5 mm. Moreover, a larger notch angle ($2\alpha = 120^\circ$) was also considered, combined with five notch root radii, ρ = 0.1, 0.3, 0.5, 1.0, 2.0 mm. With a constant gross diameter (20 mm), the net section area was varied in each specimen by changing the notch depth, *d* = 2 and 5 mm.

For semicircular notches (Figure 2(c)-A), notches with four different notch root radii were tested: ρ = 0.5, 1.0, 2.0 and 4.0 mm.

The ultimate torsion strength of isostatic graphite reported in Ref.[63] was τ_{tor} = 30 MPa. But the fracture toughness K_{IIIc} was not reported in Ref.[63] and in the reviewed Ref.[6]. Here, the value of K_{IIIc} is determined by using the empirical Eq.(11–1a).

For U-notched specimens, fracture test data reported in Ref.[63] are listed in Table 12-A, where $K_{\rho c}$ is calculated from (3-A). The test data of two cases shown in Table 12-A, on one hand, are used to calculate the critical stress τ_0 by using Eqs. 14 to 16, and on the other hand, are used to determine the value of K_{IIIc} by using the empirical Eq.56a. The obtained results are τ_0 = 35.76 MPa and K_{IIIc} = 1.94 MPam$^{0.5}$. By using the obtained values of τ_0 and K_{IIIc}, the GMPs of the U-notch are calculated by means of the approach of determining k^*. The obtained results are k^* = 3.23, ρ^* = 0.34 mm, L_c^* = 0.17 mm. By using the obtained values of the GMPs, fracture analysis for U-notched specimens shown in Table 12-A is performed by means of the local stress field failure model. The obtained results are very satisfactory (see Table 13-A).

In the following, the model of notch depth effect on k^* is verified further.

Here, material 3 is taken as isostatic graphite:

$k^*(2d/\phi = 0.5) = 3.23$, and $ratio_3$ is calculated from (9-A):

$$ratio_3 = ratio_1 / \left(\frac{K_{IIIc1}}{K_{IIIc3}}\right)^{0.5} = 1.77 / \left(\frac{6.02}{1.94}\right)^{0.5} = 1.0055$$

From this result, it is assumed that k^* is not varied with notch depth. Based on this assumption, fracture analysis of U-notch with depth d = 2 mm, notch radius ρ = 1 mm is performed and the obtained results are shown in Table 14-A, with relative error 5.4%. Based on this assumption, in addition, fracture analysis of semicircular notches is performed and the obtained results are shown in Table 15-A. Obviously the obtained results shown in Table 15-A are satisfactory.

By the way, it is pointed out here that fracture analysis of semicircular notches is carried out by the maximum stress approach and the obtained results are given in Table 16-A, with larger errors.

By the way, it is pointed out that, for the semicircular notch with radius = 0.5 mm< a_0 = 0.94 mm, by τ_{nom} = 28.9 MPa predicted by using the short crack approach (13-A) with τ_{nom} = 28.54 PMa measured experimentally, relative error is 1.3%.

TABLE 12-A

Fracture Test Data of U-Notched Specimens Made of Isostatic Graphite Reported in Ref.[63]

Notch Depth d (mm)	Notch Radius (mm)	SCFs k	M_T (NM)	$\tau_{nom,n}$ (MPa)	τ_{max} (MPa)	$K_{\rho c}$ (exp)
5	1	1.57	6.516	33.19	52.1	2.92
5	2	1.33	6.827	34.77	46.24	3.67

TABLE 13-A
Comparison of the Predicted Fracture Toughness with That Measured Experimentally (d = 5 mm)

Notch Radius (mm)	SCFs k	L_c (mm)	$K_{\rho c}$ (pre) (MPam$^{0.5}$)	$K_{\rho c}$ (exp) (MPam$^{0.5}$)	R.E.(%)
1	1.88	0.4253	2.86	2.92	–2.2
2	1.33	0.6015	3.73	3.67	1.9

Note: (a) $k^* = 3.23$, $\rho^* = 0.34$ mm, $L_c^* = 0.248$ mm; (b) M.E. = –0.16%, M.A.E. = 2.03%, E.R. = [–2.2.1.9]%

TABLE 14-A
Fracture Test Data Reported in Ref.[63] and Evaluation Calculation Results (d = 2 mm)

Notch Radius (mm)	$\tau_{nom,n}$ (MPa)	SCFs k	L_c (mm)	$K_{\rho c}$ (pre) (MPam$^{0.5}$)	$K_{\rho c}$ (exp) (MPam$^{0.5}$)	R.E.(%)
1	29.33	1.72	0.5014	2.98	2.83	5.4

$k^* = 3.23$, $\rho^* = 0.29$ mm, $L_c^* = 0.27$ mm

TABLE 15-A
Fracture Test Data of Semicircular Notches Reported in Ref.[63] and Evaluation Calculation Results

Semicircular Notches (mm)	0.5	1.0	2.0	4.0
k^*	3.23	3.23	3.23	3.23
ρ^* (mm)	**0.156**	**0.262**	**0.405**	0.57
L_c^* (mm)	0.32	0.28	0.22	0.15
k	1.79	1.64	1.44	1.21
$\tau_{nom,n}$ (MPa)	**33.28**	31.45	**32.49**	37.19
$K_{\rho c}$ (exp)	2.59	3.05	3.59	4.51
$K_{\rho c}$ (pre)	2.36	2.89	**3.71**	5.04
R.E.(%)	9.7	5.7	–3.2	–10.6

TABLE 16-A
Fracture Analysis Results of Semicircular Notches Reported in Ref.[63] Obtained by the Maximum Stress Approach

Semicircular Notch (mm)	$\tau_{nom,n}$ (MPa)	k	τ_{max}	$(\tau_{max} - \tau_0) \times 100 / \tau_0$
0.5	33.28	1.89	**59.58**	66.4
1	**31.45**	**1.64**	**51.58**	44.1
2	**32.49**	**1.44**	**46.79**	30.1
4	37.19	1.21	45.00	25.7

$\tau_0 = 35.8$ MPa

A-4 EFFECT OF NOTCH ANGLES ON EFFECT k^*

It this section, fracture analysis for V-notched specimens made of isostatic graphite reported in Ref.[63] is performed first. By using the obtained GMPs, a model of predicting an effect of notch angles on k^* is proposed.

For V-notched specimens with notch angle $2\alpha = 30^\circ$ and notch depth d = 5 mm, the details of fracture analysis are described as follows:

(1) By using the obtained values of the critical stress $\tau_0 = 35.8$ MPa and the fracture toughness $K_{IIIc} = 1.94$ MPam$^{0.5}$, the fracture toughness of point V-notch with notch angle $2\alpha = 30^\circ$ and notch depth $d = 5$ mm is calculated by using the prediction model of the fracture toughness of point V-notches proposed in [46]. Its value is $K_{vc} = 2.72\ MPam^{1-\lambda_3}$ ($\lambda_3 = 0.5455$).
(2) By using the obtained value of K_{vc} and fracture test data of V-notched specimens with notch angle $2\alpha = 30^\circ$ and notch depth $d = 5$ mm (see Table 17-A), the GMPs are calculated by means of the approach of determining k^*. The obtained results are: $k^* = 4.03$, $\rho^* = 0.1$ mm, $L_c^* = 0.3$.
(3) By using the obtained values of the GMPs, fracture analysis for V-notched specimens with notch angle $2\alpha = 30^\circ$ and notch depth $d = 5$ mm can be performed by means of the local stress field failure model. The obtained results are given in Table 17-A, with the calculation error indexes: M.E. = −2.4%, M.A.E. = 9.4.8%, E.R. = [−17.6,9.3]%.

For V-notched specimens with notch angle $2\alpha = 120^\circ$ and notch depth $d = 5$ mm, the similar results to those shown in Table 17-A can be obtained (see Table 18-A), with the calculation error indexes: M.E. = −1.7%, M.A.E. = 1.8%, E.R. = [−4.1,0.4]%.

For U- and V-notches made of isostatic graphite, the GMPs obtained here are shown in Table 19-A. At the same time, the geometric parameters λ_3, μ_3, F_1 and F_3, are listed in Table 19-A, in which λ_3 and μ_3 are Williams' eigenvalues, F_1 and F_3 are the pointed V-notch shape functions given by Gross and Mendelson [41] and by by Zappalorto et al. [57], respectively.

TABLE 17-A
Fracture Test Data of V-Notched Specimens with Notch Angle $2\alpha = 30^\circ$ and Notch Depth $d = 5$ mm and Evaluation Calculation Results

M_T (NM)	k	Notch Radius (mm)	$\tau_{nom,n}$ (MPa)	K_{vpc} (Predicted)	K_{vpc} (exp)	R.E.(%)
6.778	3.57	0.1	34.52	2.71	3.28	−17.6
6.622	2.32	0.3	33.73	3.48	3.43	1.3
6.600	1.94	0.5	33.61	3.94	3.61	9.3

Note: (a) Unit of K_{vpc} is $MPam^{1-\lambda_3}$ ($\lambda_3 = 0.5455$); (b) K_{vpc} is calculated by using (3-A); (c) $K_{vc} = 2.72$ $MPam^{1-\lambda_3}$ ($\lambda_3 = 0.5455$); (d) $k^* = 4.03$, $\rho^* = 0.1$ mm, $L_c^* = 0.3$; (e) M.E. = −2.4%, M.A.E. = 9.4.8%, E.R. = [−17.6,9.3]%

TABLE 18-A
V-Notched Specimens with Notch Angle $2\alpha = 120^{\circ}$ and Notch Depth d = 5 mm and Evaluation Calculation Results

T(NM)	k	Notch Radius (mm)	$\tau_{nom,n}$ (MPa)	$K_{\rho c}$ (Predicted)	$K_{\rho c}$	R.E.(%)
6.699	2.43	0.1	34.12	14.09	14.69	−4.1
6.632	1.93	0.3	33.78	14.95	15.2	−1.7
6.699	1.72	0.5	34.12	15.48	15.55	−0.4
6.888	1.48	1	35.08	16.43	16.36	0.4
7.335	1.3	2	37.36	17.71	18.2	−2.7

Note: (a) Unit of $K_{v\rho c}$ is $MPam^{1-\lambda_3}$(λ_3 = 0.75); $K_{v\rho c}$ is calculated by using (3-A); (c) K_{vc} = 10.14 $MPam^{1-\lambda_3}$ (λ_3 = 0.75); (d) k^* = 23.11, ρ^* = 0.00002 mm, L_c^* = 0.038 mm; (e) M.E. = −1.7%, M.A.E. = 1.8%, E.R. = [−4.1,0.4]%

TABLE 19-A
Summary of Parameters Used in the Notch Angle Model

Notch Angle (°)	0	30	60	90	120
K_{vc} ($MPam^{1-\lambda}$)	1.94	2.72	4.00	6.18	10.14
k^*	3.23	4.03			23.11
ρ^* (mm)	0.34	0.10			0.00002
L_c^* (mm)	0.248	0.30			0.038
F_3	0.665	0.782	0.943	1.163	1.456
F_1	1.767	1.775	1.857	2.118	2.77
λ_3	0.5000	0.5455	0.6000	0.6667	0.7500
μ_3	0.40978	0.45082	0.50146	0.56594	0.65196
kro	13.41	11.84			11.72

In order to propose a model of predicting the notch angle effect on k^*, the following parameter is defined:

$$kro = \frac{k^*}{F_1 \cdot F_3 \cdot \lambda_3 \cdot \mu_3} \tag{17-A}$$

According to the definition of *kro*, *kro* can be considered to be a parameter that depends both on notch angles and on notched materials. According to the variation of *kro* with notch angles (see Table 19-A), it is assumed here that *kro* is not varied with notch angles, i.e.,

$$kro=c \tag{18-A}$$

where *c* is a constant. For isostatic graphite, *c* = 13.41.

After constant c in (18-A) is determined by using fracture test data of U-notches and the approach of determining k^*, thus, k^* at any notch angle can be calculated using the condition (18-A). By using the obtained value of k^* and notched geometry, further, ρ^* can be determined. Finally, L_c^* at any notch angle is calculated by the following failure equation:

$$F(K_{vc}, L_c^*, \rho^*, \alpha, \tau_0) = 0 \tag{19-A}$$

where K_{vc} is the fracture toughness of pointed V-notch with notch angle α under III loading.

In the following, two examples of V-notches made of isostatic graphite are given to illustrate the notch angle effect model to be valid.

For V-notches with notch angle $\mathbf{2\alpha = 30^\circ}$ and notch depth $d = 5$ mm, fracture test data reported in Ref.[63] and the calculation results given here are shown in Table 20-A. The GMPs obtained by using the model of predicting notch angle effect on k^* proposed in this section are: $k^* = 4.57$, $\rho^* = 0.076$ mm, $L_c^* = 0.317$ mm. By using the obtained values of the GMPs, further, fracture analysis for the V-notches is performed by means of the local stress field failure model and the obtained results are shown in Table 20-A, with the calculation error indexes: M.E. = −0.4%, M.A.E. = 10.1%, E.R. = [−15.7,11.3]%.

For V-notches with notch angle $2\alpha = 120^\circ$ and notch depth $d = 5$ mm, a similar analysis can be performed as done for V-notches with notch angle $2\alpha = 30^\circ$ and notch depth $d = 5$ mm and the obtained results are given in Table 21-A, with the calculation error indexes: M.E. = 0.16%, M.A.E. = 1.4%, E.R. = [−1.9,2.1]%.

According to the study on notch depth effect on k^* in Section 3-A, k^* for U-notches made of isostatic graphite is not varied with notch depth. Here, this assumption is extended to V-notches. Thus k^* obtained for V-notches with notch depth $d = 5$ mm can be used to deal with V-notches with notch depth $d = 2$ mm, as follows:

(1) For V-notches with notch angle $2\alpha = 30^\circ$ and notch depth $d = 2$ mm, the obtained results and fracture test data reported in Ref.[63] are shown in Table 22-A. The GMPs obtained are $k^* = 4.57$, $\rho^* = 0.094$ mm, $L_c^* = 0.313$ mm. The calculation error indexes are: M.E. = 3.6%, M.A.E. = 12.7%, E.R. = [−13.7,15.2]%.
(2) For V-notches with notch angle $2\alpha = 120^\circ$ and notch depth $d = 2$ mm, the obtained results and fracture test data reported in Ref.[63] are shown in Table 23-A. The GMPs obtained are $k^* = 26.49$, $\rho^* = 0.000014$ mm, $L_c^* = 0.037$ mm. The calculation error indexes are: M.E. = 0.5%, M.A.E. = 4.4%, E.R. = [−8.1,5.4]%.

By comparing the calculation error indexes shown in Table 20-A with those shown in Table 22-A for V-notches with notch angle $2\alpha = 30^\circ$ (see also Table 24-A), and by comparing the calculation error indexes shown in Table 21-A with those shown in Table 23-A for V-notches with notch angle $2\alpha = 120^\circ$ (see also

TABLE 20-A
Fracture Test Data Reported in Ref.[63] and the Calculation Results (V-Notches with Notch Angle $2\alpha = 30^{\circ}$ and Notch Depth d = 5 mm)

Notch Radius	k	L_c	K_{vpc} (pre)	K_{vpc} (exp)	R.E.(%)
0.1	4.03	0.326	2.77	3.28	−15.7
0.3	2.45	0.5372	3.55	3.43	3.3
0.5	1.94	0.6776	4.02	3.61	11.3

Note: (a) Unit of K_{vpc} is $MPam^{1-\lambda_3}$ ($\lambda_3 = 0.5455$); (b) $k^* = 4.57$, $\rho^* = 0.076$ mm, $L_c^* = 0.317$ mm; (c) M.E. = −0.4%, M.A.E. = 10.1%, E.R. = [−15.7,11.3]%

TABLE 21-A
Fracture Test Data Reported in Ref.[63] and the Calculation Results (V-Notches with Notch Angle $2\alpha = 120^{\circ}$ and Notch Depth = 5 mm)

Notch Radius	k	L_c	K_{vpc} (pre)	K_{vpc} (exp)	R.E.(%)
0.1	2.75	0.3469	14.41	14.69	−1.9
0.3	2.09	0.4565	15.26	15.2	0.4
0.5	1.84	0.5187	15.79	15.55	1.5
1	1.55	0.6169	16.71	16.36	2.1
2	1.3	0.7336	17.97	18.2	−1.3

Note: (a) Unit of K_{vpc} is $MPam^{1-\lambda_3}$ ($\lambda_3 = 0.75$); (b) $k^* = 26.49$, $\rho^* = 0.0000116$ mm, $L_c^* = 0.036$ mm; (c) M.E. = 0.16%, M.A.E. = 1.4%, E.R. = [−1.9,2.1]%

TABLE 22-A
Fracture Test Data Reported in Ref.[63] and the Calculation Results (V-Notches with Notch Angle $2\alpha = 30^{\circ}$ and Notch Depth = 2 mm)

Notch Radius	k	NM	$\tau_{nom,n}$	K_{pc}^{V} (cal)	K_{pc}^{V} (exp)	R.E.(%)
0.1	4	24.07	29.93	2.75	3.19	−13.7
0.3	2.58	22.926	28.51	3.53	3.23	9.4
0.5	2.14	23.585	29.33	4	3.47	15.2

Note: (a) Unit of K_{vpc} is $MPam^{1-\lambda_3}$ ($\lambda_3 = 0.5455$); (b) $k^* = 4.57$, $\rho^* = 0.094$ mm, $L_c^* = 0.313$ mm; (c) M.E. = 3.6%, M.A.E. = 12.7%, E.R. = [−13.7,15.2]%

Table 24-A), at the same time, by comparing the calculation error indexes shown in Table 22-A and Table 23-A (see also Table 24-A), it can be concluded that the variation law of k^* with notch depths obtained for U-notches can be extended to V-notches.

TABLE 23-A

Fracture Test Data Reported in Ref.[63] and the Calculation Results (V-Notches with Notch Angle $2\alpha = 120^{\circ}$ and Notch Depth = 2 mm)

Notch Radius	k	NM	$\tau_{nom,n}$	K^V_{pc} (cal)	K^V_{pc} (exp)	R.E.(%)
0.1	2.76	25.625	31.86	14.33	15.58	−8.1
0.3	2.13	25.032	31.12	15.19	15.46	−1.8
0.5	1.89	24.61	30.6	15.71	15.33	2.5
1	1.62	24.863	30.91	16.64	15.78	5.4

Note: (a) Unit of K_{vpc} is $MPam^{1-\lambda_3}$ ($\lambda_3 = 0.75$); (b) $k^* = 26.49$, $\rho^* = 0.000014$ mm, $L^*_c = 0.037$ mm; (c) M.E. = 0.5%, M.A.E. = 4.4%, E.R. = [−8.1,5.4]%

TABLE 24-A

Summary of GMPs and Error Indexes for U- and V-Notches Under Different Notch Depths

GMPs	$k^* = 3.23$, $\rho^* = 0.34$ mm, $L^*_c = 0.248$ mm	$2\alpha = 0^{\circ}$, d = 5 mm
	$k^* = 3.23$, $\rho^* = 0.29$mm, $L^*_c = 0.27$mm	$2\alpha = 0^{\circ}$, d = 2 mm
Error indexes	M.E. = −0.16%, M.A.E. = 2.03%, E.R. = [−2.2.1.9]%	$2\alpha = 0^{\circ}$, d = 5 mm
	Single specimen, R.E. = 5.4%	$2\alpha = 0^{\circ}$, d = 2 mm
GMPs	$k^* = 4.57$, $\rho^* = 0.076$mm, $L^*_c = 0.317$mm	$2\alpha = 30^{\circ}$, d = 5 mm
	$k^* = 4.57$, $\rho^* = 0.094$mm, $L^*_c = 0.313$mm	$2\alpha = 30^{\circ}$, d = 2 mm
Error indexes	M.E. = −0.4%, M.A.E. = 10.1%, E.R. = [−15.7,11.3]%	$2\alpha = 30^{\circ}$, d = 5 mm
	M.E. = 3.6%, M.A.E. = 12.7%, E.R. = [−13.7,15.2]%	$2\alpha = 30^{\circ}$, d = 2 mm
GMPs	$k^* = 26.49$, $\rho^* = 0.0000116$mm, $L^*_c = 0.036$mm	$2\alpha = 120^{\circ}$, d = 5 mm
	$k^* = 26.49$, $\rho^* = 0.000014$ mm, $L^*_c = 0.037$ mm	$2\alpha = 120^{\circ}$, d = 2 mm
Error indexes	M.E. = 0.16%, M.A.E. = 1.4%, E.R. = [−1.9,2.1]%	$2\alpha = 120^{\circ}$, d = 5 mm
	M.E. = 0.5%, M.A.E. = 4.4%, E.R. = [−8.1,5.4]%	$2\alpha = 120^{\circ}$, d = 2 mm

A-5 COMMENTS ON THE APPENDIX

From the study of this Appendix, an attempt was successfully made to propose a local stress field failure model of sharp notches (including U- and V-notches) under III (torsion) loading. Here, particular emphasis is placed on the following three points:

(1) The variation tendency of k^* with notch depths obtained for U-notches is valid for V-notches.
(2) When fracture analysis for a notched specimen is performed by means of the local stress field failure model, if the solution of L^*_c for the studied notched specimen cannot be found from Eq.(12-A), the peak stress criterion is applied to the fracture analysis for the notched specimen.

(3) When fracture analysis for a notched specimen is performed by means of the local stress field failure model, if the notch depth is less than a_0 calculated by (14-A), the short crack approach is applied to the fracture analysis for the notched specimen.

APPENDIX B: FATIGUE LIMIT ANALYSIS OF NOTCHED COMPONENTS

B-1 A BRIEF DESCRIPTION OF A LOCAL APPROACH

Here, the local approach of fracture analysis for notched components under static loading proposed in the text is used to perform fatigue limit analysis of notched components. Naturally, the fatigue limit analysis is limited to the load ratio R = –1, which is the basis of taking into account the mean stress effect and the nonproportional loading effect.

In fatigue limit analysis of notched components performed by using the local approach, two assumptions made here are as follows:

(1) The critical stress σ_c is taken to be the fatigue limit σ_0 of plain components.
(2) Under fatigue limit loading, fracture behavior of notched components behaves like static loading.

Figure 5.B-1 shows the normalized values of the generalized stress intensity factor to failure of U-notched components of brittle materials reported in Ref.[6]. According to the second assumption, an empirical equation of determining K_{Ic0} is proposed here:

$$\frac{K_{\rho c0}}{K_{Ic0}} = a_0 x^{0.5} + a_1 x + a_2 x^2 + a_3 x^3 + a_4 x^4 \tag{B-1a}$$

where $x = \rho / L_{ch}$, $a_0 = 0.130479$, $a_1 = 0.213032$, $a_2 = -0.00576$, $a_3 = -7.844\text{E-}07$, $a_4 = 2.59275\text{E-}06$, $K_{\rho c0}$ and K_{Ic0} are the fracture toughness of U-notched components and the corresponding crack under fatigue limit loading, respectively, and L_{ch} is defined as:

$$L_{ch} = \left(\frac{K_{Ic0}}{\sigma_0}\right)^2 \tag{B-2a}$$

For V-notched components, Eqs. (B-1a) and (B-2a) are replaced by

$$\frac{K_{v\rho c0}}{K_{vc0}} = a_0 x^{0.5} + a_1 x + a_2 x^2 + a_3 x^3 + a_4 x^4 \tag{B-1b}$$

and

$$L_{ch} = \left(\frac{K_{vc0}}{\sigma_0}\right)^{\frac{1}{1-\lambda}} \tag{B-2b}$$

where $K_{v\rho c0}$ and K_{vc0} are the fracture toughness of V-notched components with notch radius ρ and the corresponding pointed V-notched component under fatigue limit loading, respectively, and λ is the Williams' eigenvalue corresponding to the notch angle, α.

For V-notched components with notch angle, $\alpha = 0°$, obviously, (B-1b) and (B-2b) become (B-1a) and (B-2a), respectively.

By using the fatigue limit σ_0 of plain components and fatigue limit test data of a notched component (called a *sample specimen*) with a notch radius ρ, according to the two assumptions made above, fracture analysis of any notched components under fatigue limit loading can be performed by using the local approach for fracture analysis of notch components under static loading proposed in the text. For a U-notched component (***sample specimen***), for example, its generalized notch fracture toughness, $K_{\rho c0}$, is calculated by using its fatigue limit test data (including its notch fatigue limit σ_{0n}, its notch radius, ρ, its notch angle, α, and its stress concentration factor, k), and by the following formula by Lazzarin and Filippi [17]:

$$K_{v\rho} = \sqrt{2\pi}\sigma_{\max}\frac{r_0^{1-\lambda}}{1+\omega_1} = \sigma_{\max}\frac{\sqrt{2\pi}}{1+\omega_1}\left(\frac{q-1}{q}\rho\right)^{1-\lambda} \tag{B-3}$$

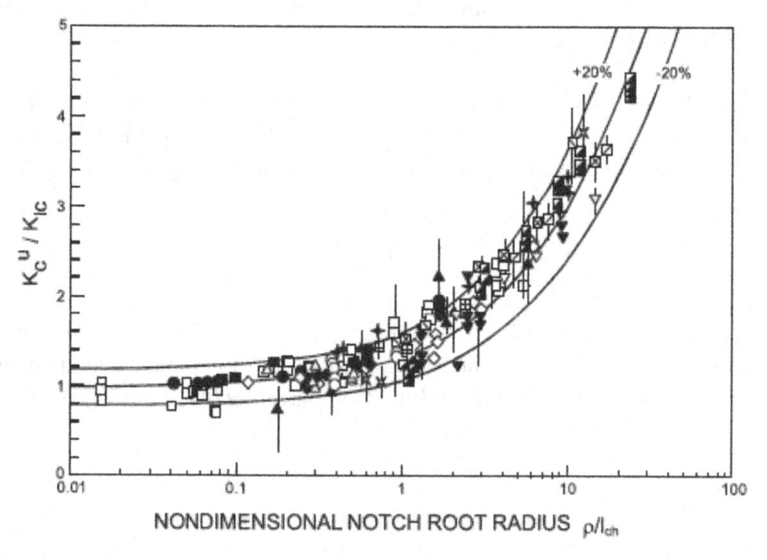

FIGURE 5.B-1 Normalized values of the generalized stress intensity factor to failure

Source: (Data from Berto and Lazzarin Ref.[6])

which is the formula (13b) in the text. By using the obtained generalized notch fracture toughness, $K_{\rho c0}$, notch radius, ρ, and fatigue limit, σ_0, K_{Ic0} can be found by solving the empirical equations (B-1a) and (B-2a). By using the approach of determining k^* proposed in the text, further, the GMPs (k^*, ρ^*, L_c^*) of the notched component (***sample specimen***) can be calculated by using $K_{\rho c0}$, σ_0 and K_{Ic0}. Thus, fracture analysis of any notch components under fatigue limit loading can be performed by using the notch depth effect model, the notch angle effect model, the different material effect model proposed in the text and the local approach for fracture analysis of the pointed V-notch components reported in Ref.[46].

B-2 EXPERIMENTAL VERIFICATIONS

By using the fatigue limit test data of V-notched components from the literature, here, the local approach for fatigue limit analysis of notched components proposed in this Appendix will be verified. For the sake of clear discussions, the experimental verifications given here are described in the forms of examples, respectively.

Example 1: Fatigue Limit Analysis of V-Notched Cylindrical Bars Made of Ni–Cr Steel

In 1957, reversed direct stress fatigue tests were carried out by Frost [64] on V-notched and unnotched cylindrical bars made of Ni–Cr Steel. The fatigue test data of the notched components were shown in Table 5.B-1 and the fatigue limit of the material was $\sigma_0 = 500$ MPa.

Here, the local approach for fatigue limit analysis of notched components proposed in this Appendix is used to perform fatigue limit analysis for the V-notched cylindrical bars made of Ni–Cr Steel. The details are as follows:

(1) By using the fatigue limit test data of case 1 (***sample specimen***) and the empirical equations (B-1b) and (B-2b), K_{vc0} is calculated: here, $K_{v\rho c0}$ = 5.8755MPam$^{1-\lambda}$ ($\lambda = 0.512$) is calculated by using formula (B-3), $\sigma_0 = 500$ MPa, and the obtained K_{vc0} is equal to 1.5 MPam$^{1-\lambda}$ ($\lambda = 0.512$).

(2) By using the approach of determining k^* proposed in the text, the GMPs of case 1 are calculated by using the fatigue limit test data of case 1 shown in Table 5.B-1 and the obtained $K_{v\rho c0}$ = 5.8755 MPam$^{1-\lambda}$ ($\lambda = 0.512$), K_{vc0} = 1.5 MPam$^{1-\lambda}$ ($\lambda = 0.512$) and σ_0 = 500 MPa. The obtained results are k^* = 24.56, ρ^* = 0.0042 mm, L_c^* = 0.00097 mm.

(3) Taking into account that the geometric size d_g of case 2 shown in Table 5.B-1 is different from that of case 1 (see Table 5.B-1), it is assumed here that there is a notched component, called a **fictitious notched component**, denoted by case 1′ (see Table 5.B-2). The fictitious notched component has the same geometric size d_g as that of case 2 and its geometric sizes are proportional to those of case 1 (see Table 5.B-2). Then from ρ^* = 0.0042 mm of case 1, ρ^* = 0.0059 mm of case 1′ can be obtained by letting k^* of case 1 be equal to that of

case 1′. Further, L_c^* of case 1′ can be calculated by solving the following failure equation:

$$F(K_{vc0}, L_c^*, \rho^*, \alpha, \sigma_0) = 0 \tag{B-4}$$

where $\rho^* = 0.0059$ mm, $K_{vc0} = 1.5$ MPam$^{1-\lambda}$ ($\lambda = 0.512$), $\sigma_0 = 500$ MPa, $\alpha = 55°$. The obtained L_c^* is equal to 0.00069 mm. The comparison of the GMPs of case 1 with those of case 1′ is given in Table 5.B-3.

The specific form of equation (B-4) is referred to (14a) in the text.

(4) From the notch depth model for RNT notch specimens proposed in the text, i.e., the following equation:

$$\frac{F_{DEN}}{L_c^*} = C \tag{B-5}$$

$L_c^* = 0.000692$ mm of case 2 can be obtained from $L_c^* = 0.00069$ mm of case 1′.

From the notch depth model for RNT notch specimens proposed in the text, data processing calculating L_c^* of case 2 are shown in Table 5.B-4.

(5) Then by using $L_c^* = 0.000692$ mm of case 2, and $K_{vc0} = 1.5$ MPam$^{1-\lambda}$ ($\lambda = 0.512$), $\sigma_0 = 500$ MPa, $\alpha = 55°$, $\rho^* = 0.0059$ mm of case 2 can be calculated from the failure equation (B-4).
(6) By using the generalized notch fracture toughness $K_{v\rho c0} = 5.8755$ MPam$^{1-\lambda}$ ($\lambda = 0.512$) (calculated by using formula (B-3)) of case 2 and the obtained $\rho^* = 0.0059$ mm of case 2, $k^* = 31.16$ of case 2 is calculated by using formula (B-3). Here, the method of calculating the stress concentration factor, k^*, by using the σ_{max} calculated by using the generalized notch fracture toughness, $K_{v\rho c0}$, and formula (B-3), is called a Lazzarin & Filippi method, (called a $K_{v\rho c}$ method for short) (see in detail data processing schematic Figure 5.B-2).
(7) By using the obtained GMPs of case 2 and case 1, fatigue limit analysis of case 2 and case 1 shown in Table 5.B-1 is performed by using the local approach for fracture analysis of notched components proposed in the text. The obtained results are given in Table 5.B-5, from which it is found that the predicted generalized notch fracture toughnesses are in excellent agreement with those measured experimentally.

$\sigma_{0n} = 71.10\ MPa$, $k = 8$, $\rho = 0.00013m \Rightarrow$ (*formula* (*B*–3)) $\rightarrow K_{v\rho c0} = 5.8755\ MPam^{1-\lambda}$

$K_{v\rho c0} = 5.8755\ MPam^{1-\lambda}$, $\rho^* = 0.0059$ mm $\Rightarrow$ (*formula* (*B*–3)) $\rightarrow \sigma_{max} \Rightarrow k^* = \sigma_{max}/\sigma_{0n} = 31.16$

FIGURE 5.B-2 Schematic of calculating k^* by using the $K_{v\rho c}$ method.

TABLE 5.B-1
Fatigue Limit Test Data of V-Notched Cylindrical Bars Made of Ni–Cr Steel [64]

No.	Notch Depth T (mm)	Notch Radius ρ (mm)	d_n (mm)	Notch Angle (degree)	d_g (mm)	SCFs k	σ_{0n} (MPa)
1	0.508	0.13	21.59	55	22.6	4.6	123.65
2	5.08	0.13	21.64	55	**31.8**	8.0	71.10

TABLE 5.B-2
Geometric Sizes of the Fictitious Notched Component and Its Stress Concentration Factor

No.	Notch Depth T (mm)	Notch Radius ρ (mm)	d_n (mm)	d_g (mm)	SCFs k
1′	0.7146	0.1829	30.3706	**31.8**	4.6

TABLE 5.B-3
The GMPs of Case 1 and Case 1′

No.	SCFs k	k^*	ρ^* (mm)	L_c^* (mm)
1	4.6	24.56	0.0042	0.00097
1′	4.6	24.56	0.0059	0.00069

TABLE 5.B-4
Data Processing of Calculating L_c^* of Case 2 from the Notch Depth Model for RNT Notch Specimens Proposed in the Text

No.	Notch Depth T (mm)	Notch Radius ρ (mm)	d_n (mm)	d_g (mm)	SCFs k	$2 \times T/d_g$	F_{DEN} [65]	L_c^* (mm)
1′	0.7146	0.1829	30.3706	**31.8**	4.6	0.045	1.118	0.000690
2	5.08	0.13	21.64	**31.8**	8.0	0.319	1.121	0.000692

Example 2: Fatigue Limit Analysis of Semicircular Notched Cylindrical Bars Made of 99.98%Gu

Fatigue limit test data of semicircular notched cylindrical bars made of 99.98%Gu reported by Weiss et al.[66] were shown in Table 5.B-6. The fatigue limit of the material was σ_0 = 73 MPa. Here, the local approach for fatigue limit analysis of notched components proposed in this Appendix is used to perform fatigue limit analysis of the notched components shown in Table 5.B-6. The details are as follows:

(1) By using the fatigue limit test data ($K_{\rho c0}$ = 1.4233 Mpam$^{0.5}$, ρ = 0.10 mm) of case 2 (***sample specimen***) shown in Table 5.B-6, and σ_0 = 73 MPa, K_{Ic0} = 1.25 MPam$^{0.5}$ is calculated by using the empirical Eqs. (B-1a) and (B-2a).
(2) By using the approach of determining k^* proposed in the text, the GMPs of case 2 are calculated by using the fatigue limit test data ($K_{\rho c0}$ = 1.4233 Mpam$^{0.5}$, ρ = 0.10 mm, k = 2.86) of case 2 shown in Table 5.B-6, σ_0 = 73 Mpa and K_{Ic0} = 1.25 Mpam$^{0.5}$. The obtained results are k^* = 3.63, ρ^* = 0.062 mm, L_c^* = 0.05512 mm.
(3) From the notch depth model for RNT notch specimens (see Eq. (B-5)), the values of L_c^* for the other cases are calculated by using L_c^* = 0.05512 mm of case 2. The obtained results are given in Table 5.B-7.
(4) For case 1, ρ^* = 0.062 mm is calculated by using the obtained L_c^* = 0.05512 mm (see Table 5.B-7), σ_0 = 73 MPa, K_{Ic0} = 1.25 MPam$^{0.5}$ and α = 0° and by the failure equation (B-4). Further, k^* = 3.63 of case 1 is calculated by using the $K_{v\rho c}$ method. By using the local approach for fracture analysis of notched components proposed in the text, thus fatigue limit analysis of case 1 can be performed by using the obtained the GMPs of case 1 (k^* = 3.63, ρ^* = 0.062 mm, L_c^* = 0.05512 mm).

Here, note that ρ = 0.04 mm < ρ^* = 0.062 mm. Thus, a failure analysis is carried out by comparing $K_{\rho c0}$ = 1.1951 MPam$^{0.5}$ with K_{Ic0} = 1.25 MPam$^{0.5}$, with relative error −4.4% (see Table 5.B-8).

(5) For case 2, fatigue limit analysis is carried out by using the obtained GMPs (k^* = 3.63, ρ^* = 0.062 mm, L_c^* = 0.05512 mm) of case 2 and the local approach for fracture analysis of notched components proposed in the text. The obtained results are shown in Table 5.B-9, with a relative error −0.6%.
(6) For case 3, ρ^* = 0.062 mm is calculated by using the obtained L_c^* = 0.05512 mm (see Table 5.B-7), σ_0 = 73 MPa, K_{Ic0} = 1.25 MPam$^{0.5}$ and α = 0° and by the failure equation (B-4). Further, k^* = 4.32 of case 3 is calculated by using the $K_{v\rho c}$ method. By using the local approach for fracture analysis of notched components proposed in the text, thus fatigue limit analysis of case 3 can be performed by using the obtained GMPs of case 3 (k^* = 4.32, ρ^* = 0.062 mm, L_c^* = 0.05512 mm). The obtained results are shown in Table 5.B-10, with a relative error 10.6%.
(7) For cases 4 to 7, as done for case 3, fatigue limit analysis can be performed. The obtained results are given in Table 5.B-10, from which it is found that the predicted generalized notch fracture toughness is in good agreement with that measured experimentally.

TABLE 5.B-5
Comparison of the Predicted Generalized Notch Fracture Toughness with Those Measured Experimentally

No.	$K_{v\rho c0}$ (pre) ($MPam^{1-\lambda}$)	$K_{v\rho c0}$ (exp) ($MPam^{1-\lambda}$)	R.E. (%)
1	5.8747	5.8755	−0.01
2	5.6830	5.8755	−3.28

Note: (a) For case 1, $k^* = 24.56$, $\rho^* = 0.0042$ mm, $L_c^* = 0.00097$; (b) For case 2, $k^* = 31.16$, $\rho^* = 0.0059$ mm, $L_c^* = 0.000692$ mm

TABLE 5.B-6
Fatigue Limit Test Data of Semicircular Notched Cylindrical Bars Made of 99.98%Gu [66]

No.	Notch Radius ρ (mm)	d_g (mm)	SCFs k	σ_{0n} (MPa)	σ_{max} (MPa)	$K_{\rho c0}$ $MPam^{0.5}$
1	0.04	5	2.95	72.28	213.22	1.1951
2	0.1	5	2.86	56.15	160.6	1.4233
3	0.15	5	2.78	47.4	131.78	1.4303
4	0.2	5	2.71	49.32	133.67	1.6753
5	0.3	5	2.56	47.4	121.35	1.8627
6	0.5	5	2.27	45.63	103.57	2.0524
7	0.8	5	1.83	48.99	89.66	2.2474

Note: (a) notch depth = notch radius; (b) $K_{\rho c0}$ is calculated by using formula (B-3)

TABLE 5.B-7
Variation of L_c^* with Notch Depth

No.	Notch Depth T (mm)	d_g (mm)	$2 \times T/d_g$	CSFs F_{DEN} [65]	L_c^* (mm)
1	0.04	5	0.016	1.118	0.05512
2	0.1	5	0.04	1.118	0.05512
3	0.15	5	0.06	1.118	0.05512
4	0.2	5	0.08	1.118	0.05512
5	0.3	5	0.12	1.118	0.05512
6	0.5	5	0.2	1.118	0.05512
7	0.8	5	0.32	1.121	0.05527

TABLE 5.B-8
Fatigue Limit Analysis Results of Case 1

No.	Notch Radius ρ (mm)	$K_{\rho c0}$ (MPam$^{0.5}$)	K_{Ic0} (Mpam$^{0.5}$)	R.E. (%)
1	0.04	1.1951	1.25	–4.4

Note: (a) $k^* = 3.63$, $\rho^* = 0.062$ mm, $L_c^* = 0.05512$ mm; (b) Because $\rho = 0.04$ mm $< \rho^* = 0.062$ mm. Thus, a failure analysis is carried out by comparing $K_{\rho c0}$ with K_{Ic0}, with relative error –4.4%

TABLE 5.B-9
Fatigue Limit Analysis Results of Case 2

No.	Notch Radius ρ (mm)	SCFs k	$K_{\rho c0}$(exp) (MPam$^{0.5}$)	$K_{\rho c0}$(pre) (MPam$^{0.5}$)	R.E. (%)
2	0.1	2.86	1.423	1.415	–0.6

Note: $k^* = 3.63$, $\rho^* = 0.062$ mm, $L_c^* = 0.05512$ mm

TABLE 5.B-10
Fatigue Limit Analysis Results of Cases 3 to 7

No.	Notch Radius ρ (mm)	SCFs k	$K_{\rho c0}$(exp) (MPam$^{0.5}$)	$K_{\rho c0}$(pre) (MPam$^{0.5}$)	R.E. (%)
3	0.15	2.78	1.430	1.581	10.6
4	0.2	2.71	1.675	1.718	2.5
5	0.3	2.56	1.863	1.940	4.2
6	0.5	2.27	2.052	2.285	11.3
7	0.8	1.83	2.247	2.659	18.3

Note: (a) for case 3: $k^* = 4.32$, $\rho^* = 0.062$ mm, $L_c^* = 0.05512$ mm;
(b) for case 4: $k^* = 4.87$, $\rho^* = 0.062$ mm, $L_c^* = 0.05512$ mm;
(c) for case 5: $k^* = 5.63$, $\rho^* = 0.062$ mm, $L_c^* = 0.05512$ mm;
(d) for case 6: $k^* = 6.45$, $\rho^* = 0.062$ mm, $L_c^* = 0.05512$ mm;
(e) for case 6: $k^* = 6.37$, $\rho^* = 0.066$ mm, $L_c^* = 0.05527$ mm.

Example 3: Fatigue Limit Analysis of V-Notched Cylindrical Bars Made of AL-Alloy BS L65

The fatigue limit test data on V-notched cylindrical bars made of AL-Alloy BS L65 reported by Frost [64] were shown in Table 5.B-11. The fatigue limit of the material is σ_0 = 150 MPa. Here, the local approach for fatigue limit analysis of notched components proposed in this Appendix is used to perform fatigue limit analysis of the notched components shown in Table 5.B-11. The details are as follows:

(1) By using the fatigue limit test data (K_{vpc0} = 2.17 $MPam^{1-\lambda}$ (= 0.512), ρ = 0.203 mm) of case 3 (***sample specimen***) shown in Table 5.B-11, and σ_0 = 150 MPa, K_{vc0} = 0.55 $MPam^{1-\lambda}$(= 0.512) is calculated by using the empirical Eqs. (B-1b) and (B-2b).

(2) By using the approach of determining k^* proposed in the text, the GMPs of case 3 are calculated by using the fatigue limit test data (K_{vpc0} = 2.17 $MPam^{1-\lambda}$ (= 0.512), ρ = 0.203 mm, k = 7.3) of case 3 shown in Table 5.B-11, σ_0 = 150 MPa and K_{vc0} = 0.5517 $MPam^{1-\lambda}$(= 0.512). The obtained results are k^* = 37.8, ρ^* = 0.007 mm, L_c^* = 0.00135 mm.

(3) By using the local approach for fracture analysis of notched components proposed in the text, fatigue limit analysis for cases 1 to 3 is performed by using the obtained GMPs. The obtained results are shown in Table 5.B-12, with error indexes: M.E. = 4.7%, A.M.E. = 4.8%, E.R. = [−0.2,9.3]%, from which it can be seen that the predicted results are in good agreement with those measured experimentally.

TABLE 5.B-11
Fatigue Limit Test Data of V-Notched Cylindrical Bars Made of AL-Alloy BS L65 [64] and the Generalized Notch Fracture Toughness

No.	Notch Radius ρ (mm)	Notch Depth D (mm)	d_n (mm)	d_g (mm)	SCFs k	σ_{0n} (MPa)	K_{vpc0} ($MPam^{1-\lambda}$)
1	1.27	5.08	33.02	43.18	3.3	46.368	4.81
2	0.508	5.08	33.02	43.18	4.85	30.912	3.01
3	0.203	5.08	33.02	43.18	7.3	23.184	2.17

Note: (a) notch angle α = 55°, λ = 0.512; (b) K_{vpc0} is calculated by using formula (B-3)

TABLE 5.B-12
Comparison of the Predicted Generalized Notch Fracture Toughness with That Measured Experimentally

No.	K_{vpc0} (exp) ($MPam^{1-\lambda}$)	K_{vpc0} (pre) ($MPam^{1-\lambda}$)	R.E. (%)
1	4.81	5.05	5.1
2	3.01	3.29	9.3
3	2.17	2.17	−0.2

Note: (a) k^* = 37.8, ρ^* = 0.007 mm, L_c^* = 0.00135 mm; (b) M.E. = 4.7%, A.M.E. = 4.8%, E.R. = [−0.2,9.3]%

Example 4: Fatigue Limit Analysis of Center Circular Hole Components Made of SAE 1045 Steel

The fatigue limit test data on center circular hole components made of SAE 1045 steel reported by DuQuesnay et al. [67] were shown in Table 5.B-13. The fatigue limit of the material was σ_0 = 304 MPa. Here, the local approach for fatigue limit analysis of notched components proposed in this Appendix is used to perform fatigue limit analysis of the notched components shown in Table 5.B-13. The details are as follows:

(1) By using the fatigue limit test data of case 4 (***sample specimen***) shown in Table 5.B-13, including the generalized notch fracture toughness, $K_{\rho c0}$ = 12.01 MPam$^{0.5}$, notch radius, ρ = 1.5 mm, and fatigue limit of unnotched specimens, σ_0 = 304 MPa, K_{Ic0} = 6 MPam$^{0.5}$ is calculated by using the empirical Eqs. (B-1a) and (B-2a).

(2) By using the approach of determining k^* proposed in the text, the GMPs of case 4 are calculated by using the fatigue limit test data ($K_{\rho c0}$ = 12.01 MPam$^{0.5}$, ρ = 0.0015 m, k = 2.82) of case 4 shown in Table 5.B-13, σ_0 = 304 MPa and K_{Ic0} = 6 MPam$^{0.5}$. The obtained results are k^* = 6.65, ρ^* = 0.27 mm, L_c^* = 0.0488 mm.

(3) For center notched plates, the notch depth model for DEN notch specimens proposed in the text is employed here:

$$F_{cc}(a_1)\cdot L_{c1}^* = F_{cc}(a_2)\cdot L_{c2}^* \tag{B-6}$$

where the crack shape function F_{cc} is taken to be [68]:

$$F_{cc} = \left[1-0.025\left(\frac{2a}{W_g}\right)^2+0.06\left(\frac{2a}{W_g}\right)^4\right]\cdot\sqrt{\sec\left(\frac{\pi a}{W_g}\right)} \tag{B-7}$$

According to formula (B-7), the values of F_{cc} are almost equal to 1 within the range of 2 × a/W_g from 0.0054 to 0.1125. At the range [0.0054,0.1125], thus, the values L_c^* are not varied according to Eq. (B-6) (see Table 5.B-14). At the range [0.0054,0.1125], further, the vales of ρ^* are not varied either according to the failure equation (B-4).

(3) By using the $K_{v\rho c}$ method, k^* = 1.987 of case 1 can be obtained by using ρ^* = 0.00027 m, $K_{\rho c0}$ = 5.2 MPam$^{0.5}$, ρ = 0.00012 m, and σ_{0n} = 179.88 MPa.

(4) For cases 2, 3 and 5, as done for case 1, k^* can be calculated. The obtained results are shown in Table 5.B-14.

(5) By using the obtained GMPs, fatigue limit analysis for various cases can be performed by using the local approach for fracture analysis of notched components proposed in the text. The obtained results are shown in Table 5.B-15.

For case 1, here, note that ρ = 0.12 mm < ρ^* = 0.27 mm. thus, a failure analysis is carried out by comparing $K_{\rho c0}$ = 5.2 Mpam$^{0.5}$ with K_{Ic0} = 6 Mpam$^{0.5}$, with relative error −13.3% (see Table 5.B-15).

For case 2, here, note that ρ = 0.25 mm < ρ^* = 0.27 mm. Thus, a failure analysis is carried out by comparing $K_{\rho c0}$ = 6.43 Mpam$^{0.5}$ with K_{Ic0} = 6 Mpam$^{0.5}$, with relative error 6.8% (see Table 5.B-15).

From Tables 5.B-15 and 5.B-16 it is found that the predicted results are in good agreement with those measured experimentally.

TABLE 5.B-13
Fatigue Limit Test Data of Center Notched Plate Made of SAE 1045 Steel [67]

No.	W_n (mm)	W_g (mm)	Notch Radius ρ (mm)	SCFs k	σ_{0n} (MPa)	$K_{\rho c0}$ (exp) (MPam$^{0.5}$)
1	44.21	44.45	0.12	2.98	179.88	5.2
2	43.95	44.45	0.25	2.96	155.1	6.43
3	43.45	44.45	0.5	2.94	140.09	8.16
4	41.45	44.45	1.5	2.82	124.08	12.01
5	39.45	44.45	2.5	2.7	131.03	15.68

Note: $K_{\rho c0}$ is calculated by using formula (B-3)

TABLE 5.B-14
Variation of the GMPs with Notch Depth

No.	W_g (mm)	Notch Depth a (mm)	$2 \times a / W_g$	CSFs F_{cc}	L_c^* (mm)	ρ^* (mm)	k^*
1	44.45	0.12	0.0054	1.0	0.0488	0.27	1.987
2	44.45	0.25	0.0112	1.0	0.0488	0.27	1.848
3	44.45	0.5	0.0225	1.0	0.0488	0.27	4.000
4	44.45	1.5	0.0675	1.0	0.0488	0.27	6.647
5	44.45	2.5	0.1125	1.0	0.0488	0.27	8.216

Note: Crack shape functions are denoted by CSFs

TABLE 5.B-15
Comparison of the Predicted Results with Those Measured Experimentally

No.	Notch Radius ρ (mm)	SCFs k	σ_{0n} (MPa)	$K_{\rho c0}$ (exp) (MPam$^{0.5}$)	$K_{\rho c0}$ (pre) (MPam$^{0.5}$)	R.E. (%)
3	0.5	2.94	140.09	8.16	7.61	−6.8
4	1.5	2.82	124.08	12.01	12.04	0.27
5	2.5	2.7	131.03	15.68	15.09	−3.8

TABLE 5.B-16
Fatigue Limit Analysis Results of Cases 1 and 2

No.	Notch Radius ρ (mm)	$K_{\rho c0}$ (exp) (MPam$^{0.5}$)	K_{Ic0} (pre) (Mpam$^{0.5}$)	R.E. (%)
1	0.12	5.2	6	−13.3
2	0.25	6.43	6	6.7

Example 5: Fatigue Limit Analysis of Center Circular Hole Components Made of 12010 Steel

The fatigue limit test data on center circular hole components made of 12010 steel reported by Lukas and Kunz [69] were shown in Table 5.B-17. The fatigue limit of the material is σ_0 = 200 MPa. Here, the local approach for fatigue limit analysis of notched components proposed in this Appendix is used to perform fatigue limit analysis of the notched components shown in Table 5.B-17. The details are as follows:

(1) By using the fatigue limit test data of case 1 (***sample specimen***) shown in Table 5.B-17, including the generalized notch fracture toughness, $K_{\rho c0}$ = 5.36 MPam$^{0.5}$, notch radius, ρ = 0.5 mmm, and fatigue limit of unnotched specimens, σ_0 = 200 MPa, K_{Ic0} = 3.6 MPam$^{0.5}$ is calculated by using the empirical Eqs (B-1a) and (B-2a).

(2) By using the approach of determining k^* proposed in the text, the GMPs of case 1 are determined by using the fatigue limit test data ($K_{\rho c0}$ = 5.36 MPam$^{0.5}$, ρ = 0.0005 m, k = 2.9) of case 1 shown in Table 5.B-17, σ_0 = 200 MPa and K_{Ic0} = 3.6 MPam$^{0.5}$. The obtained results are k^* = 5.13, ρ^* = 0.16 mm, L_c^* = 0.052 mm.

(3) According to the notch depth model (B-6), L_c^* is calculated by using the obtained value of L_c^* for case 1 and the values of the crack shape function F_{CC}. The obtained results are given in Table 5.B-18.

(4) For case 2, ρ^* = 0.177 mm is calculated by using the obtained L_c^* = 0.050 mm (see Table 5.B-18), σ_0 = 200 MPa, K_{Ic0} = 3.6 MPam$^{0.5}$ and α = 0° and by the failure equation (B-4). By using the $K_{v\rho c}$ method, further, k^* = 11.14 of case 2 is calculated by using the fatigue limit test data (ρ = 0.00375 m, k = 2.42, $\sigma_0 n$ = 85.2 MPa, $K_{\rho c0}$ = 11.19 MPam$^{0.5}$) of case 2, and ρ^* = 0.177 mm.

(5) For case 3, as done for case 2, the GMPs are calculated. The obtained results are given in Table 5.B-18.

(6) By using the obtained GMPs, fatigue limit analysis for various cases can be performed by using the local approach for fracture analysis of notched components proposed in the text. The obtained results are shown in Table 5.B-19, from which it is found that the predicted results are in good agreement with those measured experimentally.

TABLE 5.B-17
Fatigue Limit Test Data of Center Notched Plate

No.	W_n (mm)	W_g (mm)	Notch Radius ρ (mm)	SCFs k	σ_{0n} (MPa)	$K_{\rho c0}$ (exp) (MPam$^{0.5}$)
1	29	30	0.5	2.9	93.3	5.36
2	22.5	30	3.75	2.42	85.2	11.19
3	15	30	7.5	2.16	88.1	14.61

Note: $K_{\rho c0}$ is calculated by using formula (B-3)

TABLE 5.B-18
Variation of the GMPs with Notch Depth

No.	W_g (mm)	Notch Depth *a* (mm)	$2 \times a/W_g$	CSFs F_{cc}	L_c^* (mm)	ρ^* (mm)	k^*
1	30	0.5	0.033333	1.000	0.052	0.160	5.13
2	30	3.75	0.25	1.040	0.050	0.177	11.14
3	30	7.5	0.5	1.189	0.0437	0.215	12.76

TABLE 5.B-19
Comparison of the Predicted Results with Those Measured Experimentally

No.	Notch Radius ρ (mm)	SCFs k	σ_{0n} (MPa)	$K_{\rho c0}$ (exp) (MPam$^{0.5}$)	$K_{\rho c0}$ (pre) (MPam$^{0.5}$)	R.E. (%)
1	0.5	2.9	93.3	5.36	5.36	−0.4
2	3.75	2.42	85.2	11.19	12.17	8.7
3	7.5	2.16	88.1	14.61	16.40	12.3

APPENDIX C: A LOCAL APPROACH FOR FATIGUE LIFE ANALYSIS OF NOTCHED COMPONENTS

C-1 A BRIEF DESCRIPTION OF A LOCAL APPROACH

Here, the local approach for fracture analysis for notched components under static loading proposed in the text is used to perform fatigue life analysis of notched components. Naturally, the fatigue life analysis is limited to the load ratio R = −1, which is the basis of taking into account the mean stress effect and the nonproportional loading effect.

According to the local approach for fatigue life analysis of notched components, the basic fatigue test data which are required to be known are the fatigue life test data of plain material studied and the fatigue life test data of a notched material (***sample notched component***).

For the plain material studied, it is assumed that fatigue life equation under multiaxial loading can be expressed as [70]:

$$\log \sigma_{e,a} = A_\rho \log N + C_\rho \tag{C-1}$$

where σ_e is a mechanical quantity which is a measure of stress states under multiaxial loading, and here, the von Mises equivalent stress is adopted; $\sigma_{e,a}$ is the amplitude of the mechanical quantity, and ρ is a multiaxial parameter, defined as follows:

$$\rho = \frac{\sigma_{11,a}}{\sigma_{e,a}} \tag{C-2}$$

where $\sigma_{11,a}$ is the amplitude of the first invariant of stress tensor. It is evident that for the axial and pure shear fatigue conditions, the values of multiaxial parameter ρ are equal to 1 and 0, respectively A_ρ, C_ρ in Eq. (C-1) are material parameters, which appear to be constants. In fact, they are varied with the multiaxial parameter ρ.

Under the axial and pure shear loadings, the Eq.(C-1) is simplified as

$$\log \sigma_a = A_1 \log N + C_1 \tag{C-3}$$

and

$$\log\left(\sqrt{3}\tau_a\right) = A_0 \log N + C_0 \tag{C-4}$$

In view of the complexity of fatigue life analysis in the multiaxial stress states, and at the same time, taking into account that the existing literature has accumulated a large number of fatigue experimental data under the axial and pure shear loadings, from the viewpoint of application, it is assumed that the material parameters in Eq. (C-1) can be obtained by interpolating the material parameters in Eqs. (C-3) and (C-4), i.e.:

$$A_\rho = A_1 \cdot \rho + A_0 \cdot (1-\rho) \tag{C-5}$$

$$C_\rho = C_1 \cdot \rho + C_0 \cdot (1-\rho) \tag{C-6}$$

For the notched material (***sample notched component***), it is assumed that there are similar equations as (C-1) to (C-6). For brevity, the following multiaxial fatigue life equation is listed here:

$$\log \sigma_{an} = A_{1n} \log N + C_{1n} \tag{C-7}$$

where A_{1n}, C_{1n}, A_{0n} and C_{0n} are material constants similar to A_1, C_1, A_0 and C_0, respectively.

For the ***sample notched component,*** it is assumed that there are n sets of fatigue life test data, σ_{ani}, N_i ($i = 1, 2, \ldots, n$). For any set of fatigue life test data, σ_{ani}, N_i, the local approach for fracture analysis of notched components proposed in the text is used to predict σ_{ani}, N_i $(i = 1, 2, \cdots, n)$ of any notched component. The details are as follows:

Two assumption made here are:

(1) By substituting N_i into Eq.(C-3), the obtained σ_a is denoted by σ_0, which is taken to be the critical stress σ_c in the local approach.
(2) Under fatigue life N_i, fracture behavior of notched components behaves like static loading.

Figure 5.C-1 shows the normalized values of the generalized stress intensity factor to failure of U-notched components of brittle materials reported in Ref.[6]. According to the second assumption, an empirical equation of determining K_{Ic0} is proposed here

$$\frac{K_{\rho c0}}{K_{Ic0}} = a_0 x^{0.5} + a_1 x + a_2 x^2 + a_3 x^3 + a_4 x^4 \tag{C-8a}$$

where $x = \rho/L_{ch}$, $a_0 = 0.130479$, $a_1 = 0.213032$, $a_2 = -0.00576$, $a_3 = -7.844\text{E-}07$, $a_4 = 2.59275\text{E-}06$, $K_{\rho c0}$ and K_{IC0} are the fracture toughness of U-notched components with notch radius ρ and the corresponding crack under fatigue loading **corresponding to fatigue life** N_i, respectively, and L_{ch} is defined as:

$$L_{ch} = \left(\frac{K_{Ic0}}{\sigma_0}\right)^2 \tag{C-9a}$$

For V-notched components, Eqs (C-8a) and (C-9a) are replaced by

$$\frac{K_{v\rho c0}}{K_{vc0}} = a_0 x^{0.5} + a_1 x + a_2 x^2 + a_3 x^3 + a_4 x^4 \tag{C-8b}$$

and

$$L_{ch} = \left(\frac{K_{vc0}}{\sigma_0}\right)^{\frac{1}{1-\lambda}} \tag{C-9b}$$

where $K_{v\rho c0}$ and K_{vc0} are the fracture toughness of V-notched components with notch radius ρ and the corresponding pointed V-notched component under fatigue loading **corresponding to fatigue life** N_i, respectively, and λ is the Williams' eigenvalue corresponding to the notch angle, α.

For V-notched components with notch angle, $\alpha = 0°$, obviously, (C-8b) and (C-9b) become (C-8a) and (C-9a), respectively.

At the fatigue life N_i, by using the critical stress σ_0 and the fatigue test data σ_{ani} of the *sample specimen* with a notch radius ρ, according to the two assumptions made above, fracture analysis of any notched components can be performed by using the local approach for fracture analysis of notch components under static loading proposed in the text. For a U-notched component (***sample specimen***), for example, its generalized notch fracture toughness, $K_{\rho c0}$, is calculated by using its fatigue test data (including its notch fatigue load σ_{ani}, its notch radius, ρ, and its stress concentration factor, k), and by the following formula by Lazzarin and Filippi [17]:

$$K_{v\rho} = \sqrt{2\pi}\sigma_{\max}\frac{r_0^{1-\lambda}}{1+\omega_1} = \sigma_{\max}\frac{\sqrt{2\pi}}{1+\omega_1}\left(\frac{q-1}{q}\rho\right)^{1-\lambda} \tag{C-10}$$

which is the formula (13b) in the text. By using the obtained generalized notch fracture toughness, $K_{\rho c0}$, notch radius, ρ, and the critical stress, σ_0, K_{Ic0} can be found by solving the empirical equations (C-8a) and (C-9a). By using the approach of determining k^* proposed in the text, further, the GMPs (k^*, ρ^*, L_c^*) of the notched component (***sample notch***) can be calculated by using $K_{\rho c0}$, σ_0 and K_{Ic0}. Thus, fracture analysis of any notch components under fatigue loading **corresponding to fatigue life** N_i can be performed by using the notch depth effect model, the notch angle effect model, the different material effect model proposed in the text and the local approach for fracture analysis of the pointed V-notch components reported in Ref.[46].

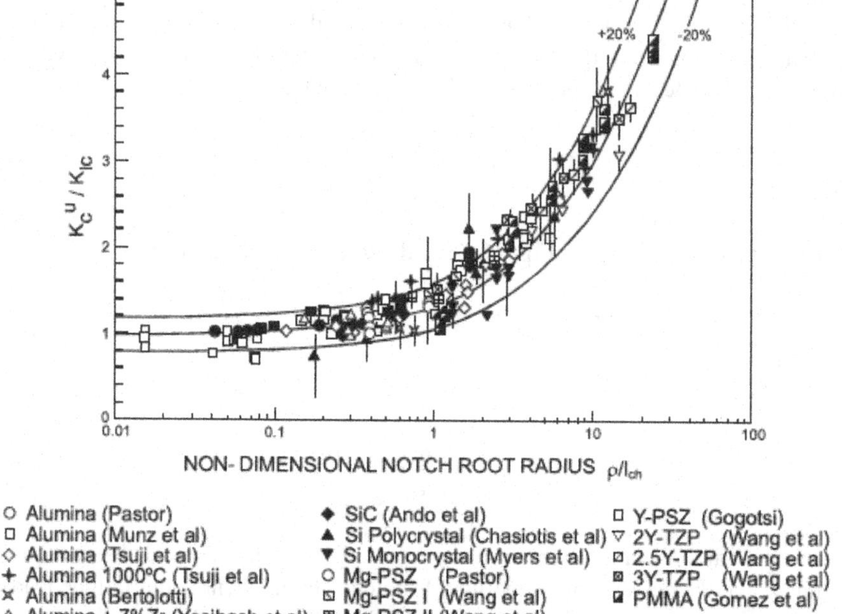

FIGURE 5.C-1 Normalized values of the generalized stress intensity factor to failure.

Source: (Data from Berto and Lazzarin Ref.[6])

After the fracture load, σ'_{ani}, of any notched component at the **fatigue life** N_i is calculated by using the local approach described above, the fatigue life prediction equation of the notched component can be obtained from σ'_{ani}, N_i $(i = 1, 2, \cdots, n)$ and expressed as:

$$\log \sigma'_{an} = A'_{1n} \log N + C'_{1n} \tag{C-11}$$

which is similar to Eq. (C-7).

Here, Eq.(C-7) and Eq.(C-11) are called the inherent and predicted notch *S-N* equations, respectively.

C-2 EXPERIMENTAL VERIFICATIONS

By using the fatigue test data of notched cylindrical bars made of En3B (a commercial cold-rolled low-carbon steel) reported by Susmel and Taylor [71] (see Figure 5.C-2) (the investigated V-notches had a notch angle equal to 60°, net diameter, d_n, equal to 5 mm and notch root radii, ρ, equal to 0.2 mm, 1.25 mm; the investigated U-notch had a notch radius $\rho = 4$ mm. All specimens had the same gross diameter $d_g = 8$ mm and

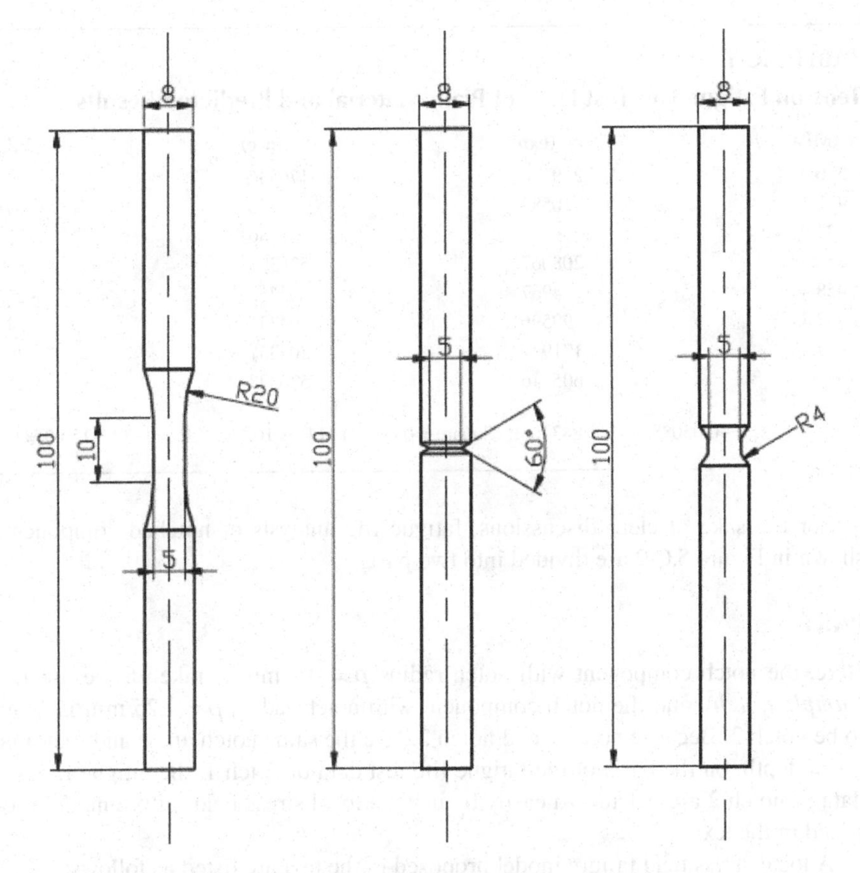

FIGURE 5.C-2 Geometric sizes of plain and notched components made of En3B.

Source: (Data from Susmel and Taylor [71])

the same notch depth $T = 1.5$ mm), here, the local approach for fatigue life analysis of notched components proposed in this Appendix will be verified. For the sake of clear discussions, the experimental verifications given here are divided into three parts: tension fatigue life analysis, torsion fatigue life analysis and tension-torsion fatigue life analysis.

Tension Fatigue Life Analysis

Tension fatigue life test data of the **plain material** (En3B) studied are shown in Table 5.C-1. As the test data are expressed in the *S-N* equation, i.e., Eq.(C-3), $A_1 = -0.05088$, $C_1 = 2.84316$ in Eq.(C-3) are obtained. The comparison of the fatigue life test data with those predicted by the *S-N* are shown in Table 5.C-1, with M.E. = 10.5%, E.R. = [−39.3,153.0]%.

TABLE 5.C-1
Tension Fatigue Life Test Data of Plain Material and Predicted Results

σ_a (MPa)	N_f (exp)	N_f (pre)	R.E.
376.9	249199	176336	−29.2
432.9	10583	11586	9.5
432.9	9473	11586	22.3
356.5	208067	526398	153
458.4	4967	3762	−24.3
387.1	99596	104331	4.8
387.1	171937	104331	−39.3
356.5	605146	526398	−13

Note: (1) $A_1 = -0.05088$, $C_1 = 2.84316$, R- Square = 0.93; (2) M.E. = 10.5%, E.R. = [−39.3,153.0]%

For the sake of clear discussions, fatigue life analysis of notched components shown in Figure 5.C-2 are divided into two parts:

PART 1

Here, the notch component with notch radius $\rho = 0.2$ mm is taken to be notch 1 (***sample notch***), and the notch component with notch radius $\rho = 1.25$ mm is taken to be notch 2. Because notch 1 and notch 2 have the same notch angle and the same notch depth, on the basis of the fatigue life test data of notch 1, the fatigue life test data of notch 2 are calculated easily by using a local stress field failure model proposed in the text.

A local stress field failure model proposed in the text are listed as follows:

For notch 1, the failure equation is written to be:

$$F(K_{v\rho c1}, L_{c1}, \rho_1, \alpha, \sigma_0) = 0 \tag{C-12}$$

For notch 2, a similar equation to Eq. (C-12) is written to be

$$F(K_{v\rho c2}, L_{c2}, \rho_2, \alpha, \sigma_0) = 0 \tag{C-13}$$

According to Yang's finding [15], the following condition is valid:

$$k_2 \cdot L_{c2} = k_1 \cdot L_{c1} \tag{C-14}$$

Fatigue life test data of notch 1 are shown in Table 5.C-2, including 12 sets of test data.

For any set of fatigue life test data (σ_{an}, N_f) of notch 1 (***sample notch***), according to the local approach proposed in this Appendix, the critical stress σ_0 is calculated by the following equation:

$$\log \sigma_0 = A_1 \log N_f + C_1 \tag{C-15}$$

By substituting σ_0 from Eq.(C-15) into Eqs. (C-12) and (C-13), at the same time, using Eq.(C-14), the fatigue strength of notch 2 **corresponding to fatigue life** N_f can be calculated.

The calculated fatigue strength of notch 2 (denoted by σ'_{an}) at the various fatigue lives N_f are shown in Table 5.C-2, in which some process data such as the critical stress σ_0, the critical distance L_{c1} and the critical distance L_{c2}, which all correspond to N_f, are listed also.

As the fatigue life test data (σ_{an}, N_f) of notch 1 shown in Table 5.C-2 are expressed in the inherent notch *S-N* equation, i.e., Eq.(C-7), $A_{1n} = -0.24764$, $C_{1n} = 3.53879$ in Eq.(C-7) is obtained.

As the fatigue life test data (σ'_{an}, N_f) of notch 2 shown in Table 5.C-2 are expressed in the predicted notch *S-N* equation, i.e., Eq.(C-11), $A'_{1n} = -0.16557$, $C'_{1n} = 3.30173$ in Eq.(C-11) is obtained.

PART 2

Here, the notch component with notch radius $\rho = 0.2$ mm, notch depth $T = 1.5$ mm, notch angle $\alpha = 60°$ shown in Figure 5.C-2 is taken to be notch 1 (***sample notch***), and the notch component with notch radius $\rho = 4$ mm, notch depth $T = 1.5$ mm, notch angle $\alpha = 0°$ shown in Figure 5.C-2 is taken to be notch 2. Because notch 1 and notch 2 have different notch angles, on the basis of the fatigue life test data of notch 1, the fatigue life test data of notch 2 can be calculated by using the local approach for fracture analysis of notched components proposed in the text.

TABLE 5.C-2
Tension Fatigue Life Test Data of Notch 1 (*Sample Notch*) and Predicted Fatigue Strength of Notch 2

σ_{an} (exp) (MPa)	N_f (exp)	σ_0 (MPa)	L_{c1} (mm)	L_{c2} (mm)	σ'_{an} (pre) (MPa)
331	10852	434.3434	1.96E-01	4.06E-01	4.21E+02
203.7	96785	388.5815	8.72E-02	1.81E-01	2.96E+02
137.5	236230	371.334	3.00E-02	6.23E-02	2.38E+02
137.5	314269	365.98	3.19E-02	6.62E-02	2.36E+02
331	18762	422.4113	2.07E-01	4.29E-01	4.18E+02
331	22505	418.5198	2.11E-01	4.37E-01	4.17E+02
203.7	148533	380.2048	9.19E-02	1.91E-01	2.93E+02
300.5	26164	415.3242	1.76E-01	3.66E-01	3.88E+02
137.5	1313155	340.2985	4.18E-02	8.66E-02	2.27E+02
142.6	224361	372.3092	3.45E-02	7.14E-02	2.42E+02
142.6	237312	371.2477	3.48E-02	7.22E-02	2.42E+02
203.7	52908	400.7073	8.09E-02	1.68E-01	3.00E+02

Note: (1) Fatigue test data (σ_{an}, N_f) are from Ref.[71]; (2) the SCFs of notch 1 and notch 2 are 3.607 and 1.74, respectively; (3) $A_{1n} = -0.24764$, $C_{1n} = 3.53879$, R-square = 0.86; (4) $A'_{1n} = -0.16557$, $C'_{1n} = 3.30173$, R-square = 0.89.

Tension fatigue life test data (σ_{an}, N_f) of notch 1 are shown in Table 5.C-3. In Table 5.C-3, some process data are listed also:

(1) The critical stress σ_0 is calculated from Eq.(C-15);
(2) The generalized notch fracture toughness K_{vpc0} ($MPam^{1-\lambda}$, $\lambda = 0.512$) is calculated by using formula (C-10), in which the notch radius $\rho = 0.0002$ m, the SCFs $k = 3.607$, the notch angle, $\alpha = 60°$ and $\sigma_{max} = k \cdot \sigma_{an}$;
(3) The critical distance L_{c1} is determined from failure Eq. (C-12), in which $K_{vpc1} = K_{vpc0}$, $\alpha = 60°$, $\rho_1 = 0.0002$ m;
(4) By using obtained K_{vpc0}, σ_0 and $\rho = 0.2$ mm, then the corresponding pointed V-notch fracture toughness K_{vc0} is calculated from the empirical equations (C-8b) and (C-9b).

By using the approach of determining k^* proposed in the text, thus, the GMPs of notch 1 are calculated. The obtained results are given in Table 5.C-4.

By using the local approach for fracture analysis of the pointed V-notch specimens reported in Ref.[46], the fracture toughness K_{Ic0} is calculated by using the pointed V-notch fracture toughness K_{vc0} of notch 1, the critical stress σ_0 and the V-notch shape function reported by [41]. The obtained results are given in Table 5.C-3.

TABLE 5.C-3
Tension Fatigue Life Test Data of Notch 1 (the *Sample Notched Component*) and Some Process Data

No.	σ_{an} (exp) (MPa)	N_f (exp)	σ_0 (MPa)	L_{c1} (mm)	K_{vpc0} ($MPam^{1-\lambda}$)	K_{vc0} ($MPam^{1-\lambda}$)	K_{Ic0} ($MPam^{0.5}$)
1	331	10852	434.3434	1.96E-01	15.22	13.8	12.35
2	203.7	96785	388.5815	8.72E-02	9.37	7.6	6.72
3	137.5	236230	371.334	3.00E-02	6.32	3.8	3.31
4	137.5	314269	365.98	3.19E-02	6.32	3.8	3.31
5	331	18762	422.4113	2.07E-01	15.22	13.8	12.36
6	331	22505	418.5198	2.11E-01	15.22	13.8	12.36
7	203.7	148533	380.2048	9.19E-02	9.37	7.7	6.81
8	300.5	26164	415.3242	1.76E-01	13.82	12.2	10.9
9	137.5	1313155	340.2985	4.18E-02	6.32	4.2	3.67
10	142.6	224361	372.3092	3.45E-02	6.56	4	3.48
11	142.6	237312	371.2477	3.48E-02	6.56	4	3.48
12	203.7	52908	400.7073	8.09E-02	9.37	7.5	6.62

Note: (1) the critical stress σ_0 is calculated from Eq.(C-15); (2) the generalized notch fracture toughness K_{vpc0} is calculated by using formula (C-10), in which notch radius $\rho = 0.0002$ m, the SCFs $k = 3.607$, notch angle, $\alpha = 60°$ and.$\sigma_{max} = k \cdot \sigma_{an}$

By using the notch angle model proposed in the text, the **characteristic critical distance** L_c^* of notch 2 is calculated by using the **characteristic critical distance** L_c^* of notch 1 and the V-notch shape function reported by [41]. The obtained results are given in Table 5.C-5.

TABLE 5.C-4
Variation of the GMPs of Notch 1 with Fatigue Lives

No.	N_f (exp)	σ_0 (MPa)	K_{vc0} ($MPam^{1-\lambda}$)	k^*	ρ^* (mm)	L_c^* (mm)
1	10852	434.3434	13.8	4.45	0.13	0.1524
2	96785	388.5815	7.6	5.24	0.093	0.056
3	236230	371.334	3.8	7.31	0.047	0.0126
4	314269	365.98	3.8	7.47	0.045	0.0133
5	18762	422.4113	13.8	4.45	0.13	0.1613
6	22505	418.5198	13.8	4.45	0.13	0.1639
7	148533	380.2048	7.7	5.13	0.097	0.0601
8	26164	415.3242	12.2	4.67	0.118	0.1305
9	1313155	340.2985	4.2	6.71	0.056	0.0199
10	224361	372.3092	4	7.39	0.046	0.015
11	237312	371.2477	4	7.39	0.046	0.015
12	52908	400.7073	7.5	5.33	0.09	0.051

Note: Notch angle $\alpha = 60°$, notch depth $T = 1.5$ mm, diameter of cylindrical bars = 8 mm

TABLE 5.C-5
Variation of the GMPs of Notch 2 with Fatigue Lives and Predicted Fatigue Strength of Notch 2

No.	N_f (exp)	σ_0 (MPa)	K_{Ic0} ($MPam^{0.5}$)	L_c^* (mm)	ρ^* (mm	k^*	σ'_{an} (pre) (MPa)
1	10852	434.3434	12.35	0.1601	0.16	6.397	468.62
2	96785	388.5815	6.72	0.0588	0.07	9.672	369.34
3	236230	371.334	3.31	0.0132	0.037	13.303	309.97
4	314269	365.98	3.31	0.014	0.035	13.678	307.25
5	18762	422.4113	12.36	0.1695	0.14	6.839	471.07
6	22505	418.5198	12.36	0.1722	0.14	6.839	469.07
7	148533	380.2048	6.81	0.0631	0.08	9.047	361.63
8	26164	415.3242	10.9	0.1371	0.136	6.939	439.71
9	1313155	340.2985	3.67	0.0209	0.041	12.637	293.07
10	224361	372.3092	3.48	0.015	0.034	13.877	315.05
11	237312	371.2477	3.48	0.015	0.042	12.486	314.15
12	52908	400.7073	6.62	0.053	0.034	13.877	400.65

Note: $A'_{1n} = -0.12888$, $C'_{1n} = 3.21097$, R-square = 0.94

By using the obtained **characteristic critical distance** L_c^* of notch 2, the **characteristic notch radius** ρ^* of notch 2 is calculated by using the following failure equation:

$$F(K_{Ic0}, L_c^*, \rho^*, \alpha, \sigma_0) = 0 \tag{C-16}$$

where $\alpha = 0°$. The obtained results are shown in Table 5.C-5.

By using the notch radius $\rho = 4$ mm and the SCFs $k = 1.28$ of notch 2 and by letting σ_{an} be equal to some value (for example, 100 MPa), the generalized notch stress intensity factor $K_{\rho c}$ is calculated by formula (C-10). Then the **characteristic** SCFs k^* of notch 2 is defined as $k^* = \sigma_{max}/\sigma_{an}$, where σ_{max} is calculated by (C-10) by using the calculated $K_{\rho c}$ and the **characteristic notch radius** ρ^* of notch 2 (see Figure 5.C-3). The obtained results are shown in Table 5.C-5.

By using the local approach for fracture analysis of notched components proposed in this text, the generalized notch fracture toughness $K_{\rho c0}$ of notch 2 is calculated by using the obtained GMPs of notch 2. Thus fatigue strength (denoted by) σ'_{an} of notch 2 is calculated by using the obtained $K_{\rho c0}$ and the notch radius $\rho = 4$ mm and the SCFs $k = 1.28$ of notch 2. The obtained results are shown in Table 5.C-5.

According to the variation of fatigue strength of σ'_{an} of notch 2 with fatigue lives N_f shown in Table 5.C-5, $A'_{1n} = -0.12888$, $C'_{1n} = 3.21097$ in the predicted notch *S-N* equation are calculated.

Torsion Fatigue Life Analysis

Torsion fatigue life test data of the **plain material** (En3B) studied are shown in Table 5.C-6. As the test data are expressed in the *S-N* equation, i.e., Eq.(C-4), $A_0 = -0.03961$, $C_0 = 2.91508$ in Eq.(C-4) are obtained. The comparison of the fatigue life test data with those predicted by the *S-N* equation are shown in Table 5.C-6, with M.E. = 25.7%, E.R. = [−65.5,148.3]%.

For the sake of clear discussions, fatigue life analysis of notched components shown in Figure 5.C-2 are also divided into two parts:

$$\sigma_{an}(\text{ssigned value}), k = 1.28, \rho = 4mm \Rightarrow (formula(C-10)) \rightarrow K_{\rho c}$$
$$K_{\rho c}, \rho^* \Rightarrow (formula(C-10)) \rightarrow \sigma_{max} \Rightarrow k^* = \sigma_{max} / \sigma_{an}$$

FIGURE 5.C-3 Schematic of Calculating k^* by Using the $K_{\rho c}$ Method.

PART 1

This part is parallel to Part 1 in Tension fatigue life analysis.

Here, the notch component with notch radius $\rho = 0.2$ mm is taken to be notch 1 (***sample notch***), and the notch component with notch radius $\rho = 1.25$ mm is taken to be notch 2. The obtained results are shown in Table 5.C-7.

As the fatigue life test data (τ_{an}, N_f) of notch 1 shown in Table 5.C-7 are expressed in the *S-N* equation, $A_{0n} = -0.11637$, $C_{0n} = 3.17584$ in the **inherent** notch *S-N* equation are obtained.

TABLE 5.C-6
Torsion Fatigue Life Test Data of Plain Material (En3B) and Predicted Results by the *S-N* Curve

τ_a (MPa)	N_f (exp)	N_f (pre)	R.E.
285.2	707338	387915	−45.2
317.8	15765	25238	60.1
285.2	276550	387915	40.3
317.8	72802	25238	−65.3
301.5	67062	95356	42.2
273	613038	1169720	90.8
301.5	276412	95356	−65.5
273	471108	1169720	148.3

Note: (1) $A_0 = -0.03961$, $C_0 = 2.91508$, R- Square = 0.70; (2) M.E. = 25.7%, E.R. = [−65.5,148.3]%

TABLE 5.C-7
Torsion Fatigue Life Test Data of Notch 1 (*Sample Notch*) and Predicted Fatigue Strength of Notch 2

τ_{an} (exp) (MPa)	N_f (exp)	τ_0 (MPa)	L_{c1} (mm)	L_{c2} (mm)	τ'_{an} (pre) (MPa)
171.1	1025331	274.4284	5.06E-02	8.09E-02	2.38E+02
264.8	31705	314.9413	1.52E-01	2.43E-01	3.27E+02
264.8	30028	315.62	1.51E-01	2.41E-01	3.27E+02
171.1	306633	287.8686	3.87E-02	6.19E-02	2.44E+02
171.1	904657	275.7928	4.94E-02	7.90E-02	2.38E+02
203.7	344380	286.5479	8.89E-02	1.42E-01	2.67E+02
203.7	614536	280.0496	9.66E-02	1.55E-01	2.64E+02
244.5	70232	305.1744	1.32E-01	2.11E-01	3.06E+02
297.4	7700	333.1006	1.78E-01	2.85E-01	3.61E+02

Note: (1) notch radii of notch 1 and notch 2 are 0.2 mm and 1.25 mm, respectively; (2) The SCFs of notch 1 and notch 2 are 2.093, 1.275, respectively; (3) (τ_{an}, N_f) are the torsion fatigue life test data of notch 1; (4) $A_{0n} = -0.11637$, $C_{0n} = 3.17584$, R-square = 0.88; (5) $A'_{0n} = -0.08704$, $C'_{0n} = 3.13939$, R-square = 0.94

As the fatigue life test data (σ'_{an}, N_f) of notch 2 shown in Table 5.C-7 are expressed in the *S-N* equation, $A'_{0n} = -0.08704$, $C'_{0n} = 3.13939$ in the **predicted** notch *S-N* equation are obtained.

PART 2

This part is parallel to Part 2 in Tension fatigue life analysis.

Here, the notch component with notch radius $\rho = 0.2$ mm, notch depth $T = 1.5$ mm, notch angle $\alpha = 60°$ shown in Figure 5.C-2 is taken to be notch 1 (***sample notch***), and the notch component with notch radius $\rho = 4$ mm, notch depth $T = 1.5$ mm, notch angle $\alpha = 0°$ shown in Figure 5.C-2 is taken to be notch 2. The obtained results are shown in Tables 5.C-8, 5.C-9 and 5.C-10.

TABLE 5.C-8
Torsion Fatigue Life Test Data of Notch 1 (The *Sample Notched Component*) and Some Process Data

No.	τ_{an} (exp) (MPa)	N_f (exp)	τ_0 (MPa)	L_{c1} (mm)	K_{vpc0} ($MPam^{1-\lambda}$)	K_{vc0} ($MPam^{1-\lambda}$)	K_{IIIc0} ($MPam^{0.5}$)
1	171.1	1025331	274.4284	5.06E-02	20.097	18.4	7.821
2	264.8	31705	314.9413	1.52E-01	31.103	29.5	13.632
3	264.8	30028	315.62	1.51E-01	31.103	29.5	13.625
4	171.1	306633	287.8686	3.87E-02	20.097	18.4	7.728
5	171.1	904657	275.7928	4.94E-02	20.097	18.4	7.811
6	203.7	344380	286.5479	8.89E-02	23.926	22.5	9.949
7	203.7	614536	280.0496	9.66E-02	23.926	22.5	10.006
8	244.5	70232	305.1744	1.32E-01	28.719	27	12.3
9	297.4	7700	333.1006	1.78E-01	34.932	33.5	15.758

TABLE 5.C-9
Variation of the GMPs of Notch 1 with Fatigue Lives

No.	N_f (exp)	τ_0 (MPa)	K_{vc0} ($MPam^{1-\lambda}$)	k^*	ρ^* (mm)	L_c^* (mm)
1	1025331	274.4284	18.4	2.32	0.145	0.0449
2	31705	314.9413	29.5	2.29	0.15	0.1578
3	30028	315.62	29.5	2.27	0.153	0.1572
4	306633	287.8686	18.4	2.29	0.149	0.0338
5	904657	275.7928	18.4	2.31	0.146	0.0441
6	344380	286.5479	22.5	2.26	0.155	0.0867
7	614536	280.0496	22.5	2.26	0.154	0.0946
8	70232	305.1744	27	2.29	0.15	0.1337
9	7700	333.1006	33.5	2.22	0.161	0.1953

Note: Notch angle, notch depth $T = 1.5$mm, gross diameter of cylindrical bars = 8 mm

TABLE 5.C-10
Variation of the GMPs of Notch 2 with Fatigue Lives and Predicted Fatigue Strength of Notch 2

No.	N_f (exp)	τ_0 (MPa)	K_{IIIc0} (MPam$^{0.5}$)	k^* (mm)	ρ^* (mm	L_c^*	τ'_{an} (MPa)
1	1025331	274.4284	7.821	1.061	4.25	No solution	251.08
2	31705	314.9413	13.632	1.046	4.36	No solution	288.14
3	30028	315.62	13.625	1.038	4.44	No solution	288.76
4	306633	287.8686	7.728	1.049	4.34	No solution	263.37
5	904657	275.7928	7.811	1.058	4.28	No solution	252.33
6	344380	286.5479	9.949	1.033	4.48	No solution	262.17
7	614536	280.0496	10.006	1.035	4.46	No solution	256.22
8	70232	305.1744	12.3	1.046	4.36	No solution	279.21
9	7700	333.1006	15.758	1.017	4.62	No solution	304.76

Note: (1) The fatigue strength (τ'_{an}) of notch 2 is calculated by using the peek stress criterion: $\tau'_{an}=\tau_0/k$, where $k = 1.093$ is the SCFs of notch 2; $A'_{0n} = -0.03961$, $C'_{0n} = 2.87645$, R-square = 1.

Table 5.C-8 shows the torsion fatigue life test data of notch 1 (***sample notch***) and some process data.

Table 5.C-9 shows the variation of the GMPs of notch 1 with fatigue lives;

Table 5.C-10 shows the variation of the GMPs of notch 2 with fatigue lives and predicted fatigue strength of notch 2.

According to the variation of fatigue strength of τ'_{an} of notch 2 with fatigue lives N_f shown in Table 5.C-10, $A'_{0n} = -0.03961$, $C'_{0n} = 2.87645$ in the **predicted** notch *S-N* equation are calculated.

Here, it is worth mentioning that the fatigue strength (τ'_{an}) of notch 2 shown in Table 5.C-10 is calculated by using the peek stress criterion because the solution of L_c^* for the notched specimen studied cannot be found from the following failure equation:

$$F(K_{IIIc}, L_c^*, \rho^*, \alpha, \tau_0) = 0 \qquad \text{(C-17)}$$

which is the equation (12-A) in Appendix A. Eq. (C-17) is similar to Eq.(C-16).

Tension-Torsion Fatigue Life Analysis

Based on the studies on the tension and torsion fatigue life analysis performed by using the local approach proposed in this Appendix, the obtained constants in the *S-N* equations are shown in Table 5.C-11.

For the **sample notch** with notch angle $\alpha = 60°$, notch radius $\rho = 0.2$ mm, and notch depth $T = 1.5$ mm, the comparison of experimental multiaxial fatigue lives with those predicted by using the inherent notch *S-N* equation is given in Table 5.C-12 (see also Figure 5.C-3), with M.E. = −19.4%, E.R. = [−67.0,34.1]%.

TABLE 5.C-11
Summery of Constants in the *S-N* Equations for Plain and Notched Specimens

Type of Specimens	Tension Fatigue			Torsion Fatigue		
Plain	A_1	C_1	R- Square	A_0	C_0	R- Square
	−0.05088	2.84316	0.93	−0.03961	2.91508	0.70
sample notch	A_{1n}	C_{1n}	R- Square	A_{0n}	C_{0n}	R- Square
α = 60°, ρ = 0.2 mm	−0.24764	3.53879	0.86	−0.11637	3.17584	0.88
V-notch α = 60°	A'_{1n}	C'_{1n}	R- Square	A'_{0n}	C'_{0n}	R- Square
ρ = 1.25 mm	−0.16557	3.30173	0.89	−0.08704	3.13939	0.94
U-notch	A'_{1n}	C'_{1n}	R- Square	A'_{0n}	C'_{0n}	R- Square
ρ = 4 mm	−0.12888	3.21097	0.94	−0.03961	2.87645	1

Note: All specimens have the same gross diameter d_g = 8 mm, the same notch depth T = 1.5 mm

TABLE 5.C-12
Comparison of Experimental Multiaxial Fatigue Lives with Those Predicted by the Inherent Notch *S-N* Equation for the V-Notched Component with Notch Radius = 0.2 mm, Notch Angle 60° Under Tension-Torsion Loading

σ_{an} (exp) (Mpa)	τ_{an} (exp) (Mpa)	N_f (exp)	N_f (pre)	R.E. (%)
259.6	155.9	14743	13120	−11
216.3	129.9	30837	31632	2.6
183.9	110.4	87177	69224	−20.6
146	87.7	460400	210525	−54.3
135.2	81.2	227391	305037	34.1
135.2	81.2	924890	305037	−67

Note: (1) A_{1n} = −0.24764, C_{1n} = 3.53879, A_{0n} = −0.11637, C_{1n} = 3.17584; (2) M.E. = −19.4%, E.R. = [−67.0,34.1]%

For the V-notch with notch angle α = 60°, notch radius ρ = 1.25 mm, notch depth T = 1.5 mm, the comparison of experimental multiaxial fatigue lives with those calculated by using the predicted notch *S-N* equation is given in Table 5.C-13 (see also Figure 5.C-4), with M.E. = −26.5%, E.R. = [−58.9,−1.3]%.

For the U-notch with notch radius ρ = 4 mm, notch depth T = 1.5 mm, the comparison of experimental multiaxial fatigue lives with those calculated by using the predicted notch *S-N* equation is given in Table 5.C-14 (see also Figure 5.C-5), with M.E. = −14.5%, E.R. = [−73.9,79.2]%.

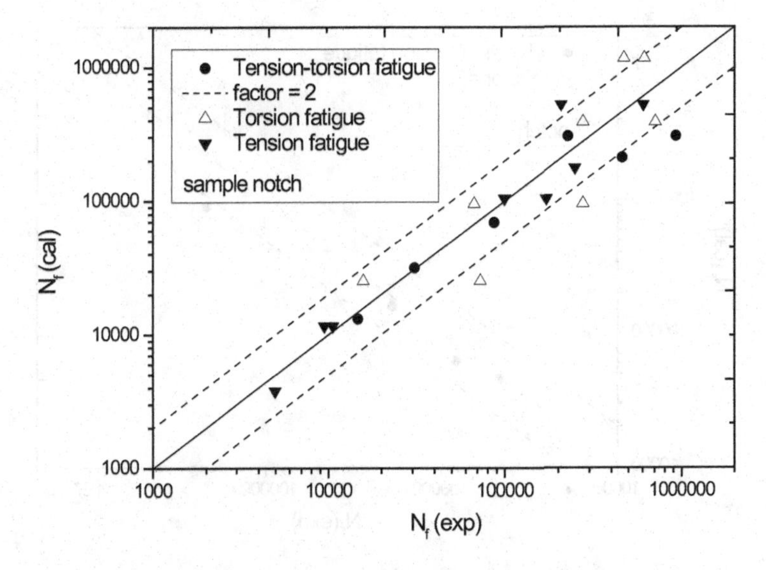

FIGURE 5.C-3 Comparison of experimental multiaxial fatigue lives with those predicted by the inherent notch *S-N* equation for the V-notched component with notch radius = 0.2 mm, notch angle = 60°.

TABLE 5.C-13

Comparison of Experimental Fatigue Lives with Those by the Predicted Notch *S-N* Equation for the V-Notched Component with Notch Radius = 1.25 mm, Notch Angle = 60° Under Tension-Torsion Loading

σ_{an} (exp) (Mpa)	τ_{an} (exp) (Mpa)	N_f (exp)	N_f (pre)	R.E. (%)
259.6	155.9	82952	62409	−24.8
200	115.5	437907	426180	−2.7
180	103.9	2174897	893158	−58.9
275	158.8	46254	45661	−1.3
230	132.8	188480	159956	−15.1
190	109.7	1400006	610989	−56.4

Note: (1) A'_{1n} = −0.16557, C'_{1n} = 3.30173, A'_{0n} = −0.08704, C'_{0n} = 3.13939; (2) M.E. = −26.5%, E.R. = [−58.9,−1.3]%

From Tables 5.C-12, 5.C-13 and 5.C-14, it is seen that the experimental multiaxial fatigue lives are in good agreement with those predicted by the notch S-N equation, which verifies the accuracy and reliability of the local approach for fatigue life analysis of notched components proposed in this Appendix.

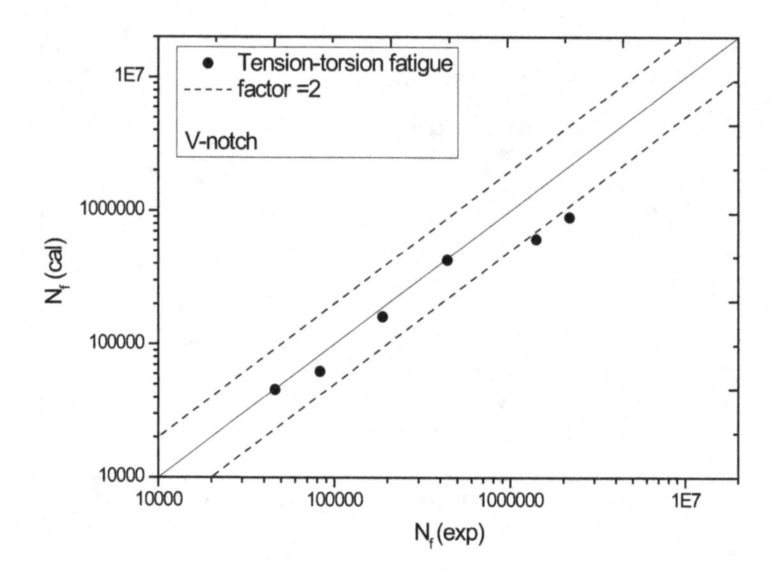

FIGURE 5.C-4 Comparison of experimental fatigue lives with those calculated by the predicted notch *S-N* equation for the V-notched component with notch radius = 1.25 mm, notch angle = 60° under tension-torsion loading.

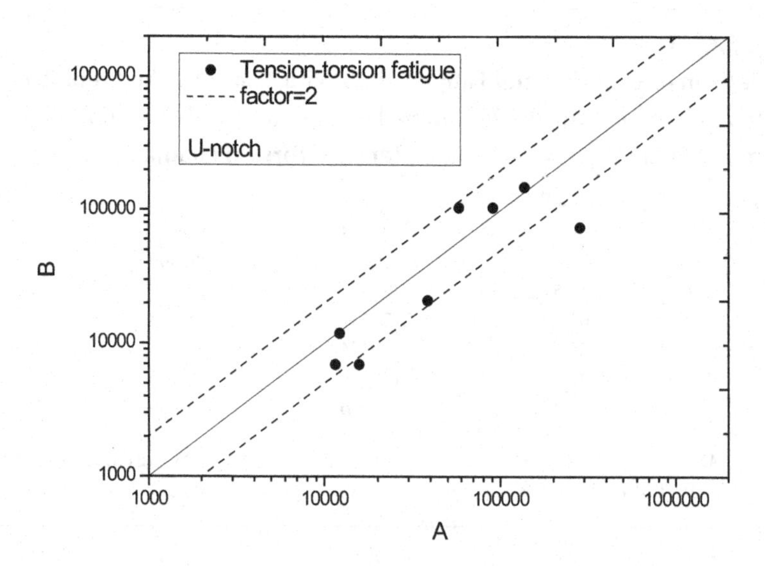

FIGURE 5.C-5 Comparison of experimental fatigue lives with those calculated by the predicted notch *S-N* equation for the U-notched component with notch radius = 4 mm under tension-torsion loading.

TABLE 5.C-14

Comparison of Experimental Fatigue Lives with Those Calculated by the Predicted Notch *S-N* Equation for the U-Notched Component with Notch Radius = 4 mm Under Tension-Torsion Loading

σ_{an} (exp) (Mpa)	τ_{an} (exp) (Mpa)	N_f (exp)	N_f (pre)	R.E. (%)
290	167.4	282833	73801	−73.9
270	155.9	136165	147791	8.5
330	190.5	38446	20978	−45.4
370	213.6	11543	6887	−40.3
370	213.6	15717	6887	−56.2
350	202.1	12200	11816	−3.1
280	161.7	57847	103664	79.2
280	161.7	90140	103664	15

Note: (1) $A'_{1n} = -0.12888$, $C'_{1n} = 3.21097$, $A'_{0n} = -0.03961$, $C'_{0n} = 2.87645$; (2) M.E. = −14.5%, E.R. = [−73.9,79.2]%

6 An Empirical Fracture Equation of Mixed Mode Cracks

6.1 INTRODUCTION

In the special case of Mode I loading, fracture toughness methodology is well established. However, cracks in real structures can often be subjected to some combination of the three loading modes first identified by Irwin [1], and it is important to develop an understanding of cracking behavior in these "mixed mode" situations.

Up to now, a number of theories based on continuum mechanics have been advanced. These theories can be roughly classed into three categories. The first category maintains that the tangential tensile stress component, $\sigma_{\theta\theta}$, of the crack tip stress field controls the cracking behavior. This is the so-called "maximum $\sigma_{\theta\theta}$" hypothesis, originally proposed by Erdogan and Sih [2], with modified versions advanced byWilliams and Ewing [3], and Maiti and Smith [4]. The second category attempts to extend to mixed mode situations the energy release rate (i.e., G) concept of Griffith [5]. Workers who have developed the "maximum G" hypothesis include Hussain et al.[6], Palaniswamy and Knauss [7], Hwang *et al.*[8], and Hyashi and Nemat-Nasser [9]. The third category of mixed mode fracture theory proposes that a "strain energy density" factor, S, may be evaluated in the crack tip region, and it is this quantity which controls cracking behavior. The original "minimum S" hypothesis was proposed by Sih [10], and modified versions have been proposed by numerous workers including Jayatilaka et al.[11], and Theocaris and Andrianopoulos [12].

Here it is particularly pointed out that, on the basis of the test study of high or medium strength steels as well as of nodular cast iron performed by means of three-point and four-point bend specimens, Gao *et al.*[13] found that the extending direction of the cracks appears to agree well with the three original existing theories, but the crack extension resistance increases with the increase of the K_{II} / K_I ratio, thus deviating obviously from the theoretical prediction. According to the original three fracture theories [2,5,10], the crack extension resistance K_R can be expressed by

$$K_R = f(K_{Ic}) \tag{1}$$

Where K_{Ic} is the material fracture toughness. In formula (1), the function f can be determined easily through the fracture theories. For the maximum $\sigma_{\theta\theta}$ theory, for example, the crack extension resistance K_R is

$$K_R = K_{Ic} \tag{2}$$

DOI: 10.1201/9781003356721-6

While the crack extension force K_e is

$$K_e = \cos\frac{\theta_0}{2}\left(K_I \cos^2\frac{\theta_0}{2} - \frac{3}{2}K_{II}\sin\theta_0\right) \tag{3}$$

where θ_0 is the crack extension angle. According to the maximum $\sigma_{\theta\theta}$ theory, thus, the mixed crack extension equation can be written to be

$$K_e = K_R \tag{4}$$

Obviously, the crack extension resistance K_R by the formula (1) does not reveal the finding by Gao *et al.* [13] that "the crack extension resistance K_R increases with the increase of K_{II} / K_I ratio".

According to the above case, thus, this study specially focuses on a fracture equation of mixed mode cracks similar to (4). It is required that not only the crack extension force, K_e, varies with K_I and K_{II}, such as (3), but also the crack extension resistance, K_R, varies with the K_{II} / K_I ratio.

6.2 AN EMPIRICAL FRACTURE EQUATION OF MIXED MODE CRACKS

On the basis of the multiaxial fatigue limit equation by Liu and Yan [14], in this section, an empirical fracture equation of mixed mode cracks is proposed.

6.2.1 The Multiaxial Fatigue Limit Equation by Liu and Yan

Based on the invariant model of fatigue life prediction under multiaxial loading proposed by Liu and Yan [15], recently, a multiaxial fatigue limit prediction equation for metallic materials under loading ratio $R = -1$ is presented. Here, the multiaxial fatigue limit prediction equation is described briefly.

Under the loading ratio $R = -1$, the invariant model of fatigue life prediction under multiaxial loading can be expressed as

$$\log\sigma_{e,a} = A_\rho \log N + C_\rho \tag{5}$$

Where σ_e is a mechanical quantity which is a measure of stress states under multiaxial loading, and here, the von Mises equivalent stress is adopted; $\sigma_{e,a}$ is the amplitude of the mechanical quantity, and ρ is a multiaxial parameter, defined as follows:

$$\rho = \frac{\sigma_{11,a}}{\sigma_{e,a}} \tag{6}$$

where $\sigma_{11,a}$ is the amplitude of the first invariant of stress tensor. It is evident that for the axial and pure shear fatigue conditions, the values of multiaxial parameter ρ are equal to 1 and 0, respectively. A_ρ, C_ρ in the Eq. (5) are material parameters, which appear to be constants. In fact, they are varied with the multiaxial parameter ρ.

Under the axial and pure shear loadings, Eq.(5) can be simplified as

$$\log \sigma_a = A_1 \log N + C_1 \tag{7}$$

and

$$\log\left(\sqrt{3}\tau_a\right) = A_0 \log N + C_0 \tag{8}$$

In view of the complexity of fatigue life analysis in the multiaxial stress states, and at the same time, taking into account that the existing literature has accumulated a large number of fatigue experimental data under the axial and pure shear loadings, from the viewpoint of application, it is assumed that the material parameters in Eq. (5) can be obtained by interpolating the material parameters in Eqs. (7) and (8), i.e.:

$$A_\rho = A_1 \cdot \rho + A_0 \cdot (1-\rho) \tag{9}$$

$$C_\rho = C_1 \cdot \rho + C_0 \cdot (1-\rho) \tag{10}$$

If the applied stress σ_a, under the unaxial loading with loading ratio $R = -1$, is lower than a certain stress amplitude, σ_o, then failure should not occur up to a number of cycles (for example N_f) to failure theoretically equal to infinity; such a reference threshold is named the fatigue limit. Under the pure shear loading, a similar definition is a material shear fatigue limit, τ_o.

If a multiaxial fatigue limit is here denoted by σ_{eo}, the following equations can be obtained from Eqs. (5), (7) and (8):

$$\log \sigma_{eo} = A_\rho \log N_f + C_\rho \tag{11}$$

$$\log \sigma_o = A_1 \log N_f + C_1 \tag{12}$$

$$\log\left(\sqrt{3}\tau_o\right) = A_0 \log N_f + C_0 \tag{13}$$

By substituting (9) and (10) into (11) and by using (12) and (13), we can obtain

$$\begin{aligned}\log \sigma_{eo} &= (A_1 \cdot \rho + A_0 \cdot (1-\rho)) \log N_f + C_1 \cdot \rho + C_0 \cdot (1-\rho) \\ &= (A_1 \log N_f + C_1)\rho + (A_0 \log N_f + C_0)(1-\rho) \\ &= \rho \log \sigma_o + (1-\rho)\log\left(\sqrt{3}\tau_o\right) \\ &= \log(\sigma_o{}^\rho \left(\sqrt{3}\tau_o\right)^{(1-\rho)})\end{aligned}$$

Thus further we can obtain

$$\sigma_{eo} = \sigma_o{}^\rho \left(\sqrt{3}\tau_o\right)^{(1-\rho)} \tag{14}$$

which is an equation used to predict a multiaxial fatigue limit of metallic materials, σ_{eo}, by means of a tensile fatigue limit σ_o, a shear fatigue limit τ_o, and multiaxial parameter ρ.

The experimental verifications [14] showed that the multiaxial fatigue limit prediction equation (14) is not only simple in computation but also high in accuracy (see Figure 6.1). In Figure 6.1, $\sigma_{e,\exp}$ and $\sigma_{e,pre}$ are experimental and predicted results, respectively. $\sigma_{e,\exp}$ is calculated from experimental data σ_a and τ_a with the von-Mises stress definition, and the calculation of $\sigma_{e,pre}$ is carried out according to the right of Eq. (14) by using the tensile fatigue limit σ_o, and shear fatigue limit τ_o and multiaxial parameter ρ.

6.2.2 An Empirical Fracture Equation of Mixed Mode Cracks

On the basis of the multiaxial fatigue limit equation by Liu and Yan [14], as described in Section 2.1, here, an empirical fracture equation of mixed mode cracks is proposed.

It is assumed that an empirical fracture equation of mixed mode cracks is expressed to be:

$$K_e = K_R \tag{15}$$

where the extension crack force under Mode I and Mode II loading is taken to be:

$$K_e = K_e\left(K_I, K_{II}\right) = \sqrt{(K_I)^2 + 3(K_{II})^2} \tag{16}$$

The definition is the same as that of the von Mises equivalent stress, i.e., $\sigma_{e,a}$ in Eq.(5). While the extension crack resistance is assumed to be:

$$K_R = K_R\left(K_{Ic}, K_{IIc}, \rho\right) = \left(K_{Ic}\right)^{\rho}\left(\sqrt{3}K_{IIc}\right)^{1-\rho} \tag{17}$$

The definition is the same as that of σ_{eo} (see (14)). A multiaxial parameter ρ in (17) is defined as:

$$\rho = K_I / K_e \tag{18}$$

The definition is the same as that of ρ in (6).

For a crack of Mode I loading, $\rho = 1$ the empirical fracture equation (15) is simplified as

$$K_I = K_{Ic} \tag{19}$$

For a crack of Mode II loading, $\rho = 0$, the empirical fracture equation (15) is simplified as

$$K_{II} = K_{IIc} \tag{20}$$

For a crack under Mode I and Mode III loading, obviously, the serious equations similar to (15) to (20) are obtained easily and not listed here.

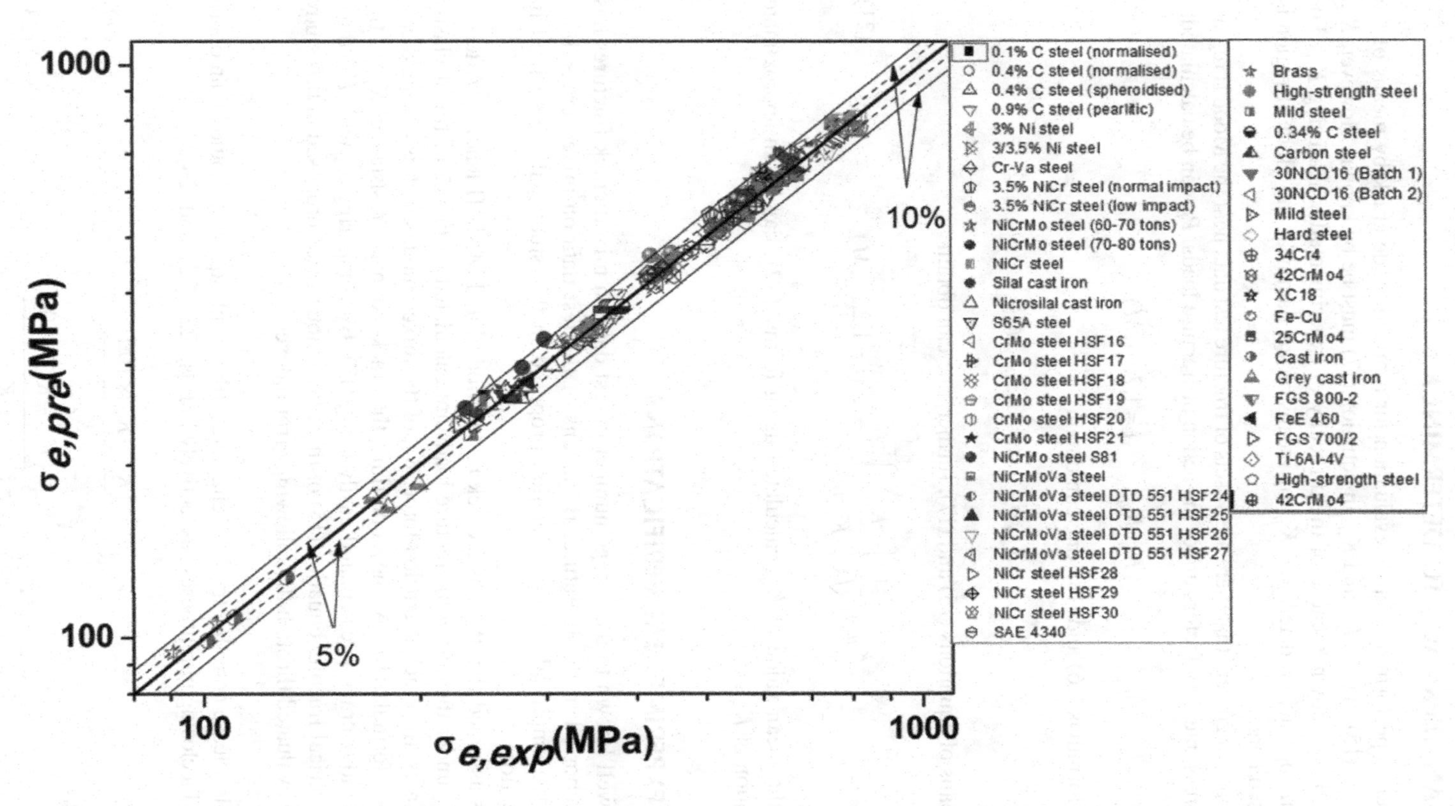

FIGURE 6.1 Comparison of multiaxial fatigue limits of metallic materials.

Source: (Data from liu and Yan [14])

6.3 AN APPROACH TO DETERMINE K_{IIc}

In order to perform a fracture evaluation for a mixed mode crack by means of the equations (15) to (18), K_{Ic} and K_{IIc} in equation (17) must be be known. However, K_{IIc} is usually unknown because of the difficulty of experimentally measuring K_{IIc}. For this, an approach to determine K_{IIc} by means of test data near the Mode II region is proposed here.

It is assumed here that there are M sets of fracture test data near the Mode II region; thus stress intensity factors under the action of fracture loads, P_{fi}, can be calculated:

$$K_{Ii}, K_{IIi}, \quad (i = 1,2,\cdots,M)$$

Using formulas (16) and (18), one can obtain

$$K_{ei}, \rho_i, \quad (i = 1,2,\cdots,M)$$

By means of equations (15) and (17), further, one can obtain

$$K_{IIc} = \frac{1}{\sqrt{3}}\left\{\frac{K_{ei}}{(K_{Ic})^{\rho_i}}\right\}^{\frac{1}{1-\rho_i}} \quad (i = 1,2,\cdots,M) \tag{21}$$

Then the mean value of K_{IIc} calculated using formula (21) can be an approximate estimation of K_{IIc}.

6.4 EXPERIMENTAL VERIFICATIONS

As is well known to us, a large number of test data on mixed crack fractures has been reported in the literature. By means of the test data on mixed crack fractures, the empirical fracture equation proposed in this study will be verified in this section.

It is assumed that there is a crack under Mode I and Mode II loading. After K_I and K_{II} under the action of fracture loads are calculated by means of linear elastic mechanics, the crack extension force, K_e, of the mixed mode crack is obtained easily using formula (16). At the same time, the crack extension resistance, K_R, of the mixed mode crack is calculated easily using (17). By comparing K_e with K_R, thus, the empirical fracture equation (15) of the mixed mode crack proposed in this study can be evaluated through the following error indexes:

[COMP: Please note that the junk characters (the blank squares) should be an opening and a closing parentheses respectively in Eqs. (22), (23), and (24).]

$$\text{R.E.} = \frac{(K_R - K_e)\times 100}{K_e} \tag{22}$$

$$M.E. = \frac{1}{N}\sum_{i=1}^{N}\frac{(K_{Ri} - K_{ei})\times 100}{K_{ei}} \tag{23}$$

$$M.A.E. = \frac{1}{N}\sum_{i=1}^{N}\left|\frac{(K_{Ri} - K_{ei})\times 100}{K_{ei}}\right| \tag{24}$$

$$E.R. = \{Min(R.E.), Max(R.E.)\} \tag{25}$$

which are a relative error, a mean error, an absolute mean error and error range, respectively. N in formulas (23) and (24) is the number of test cases.

Note: unit of K_I, K_{II}, K_{III}, K_{Ic}, K_{IIc}, K_{IIIc}, K_e and K_R in this study is MPam$^{0.5}$ except those labeled in the text.

6.4.1 Experimental Verifications by the Disk Test

In 1987, Awaji and Sato [16] used the Disk Test to study mixed-mode fractures of graphite, plaster and marble. Mixed-mode stress states ranging from pure Mode I to pure Mode II can be achieved in this specimen by selecting the angle of inclination of the central through-crack relative to the diametral line of compression loading (see Figure 6.2). Analytical solutions for Mode I and Mode II stress-intensity factors for through-cracks in the diametral-compression specimen had been reported by Libatskii and Kovchik [17], Atkinson *et al.*[18] and Awaji and Sato [16]. In the following, analytical solutions by Atkinson *et al.* [18] are given below:

The stress-intensity factors under Mode I and Mode II loading is calculated from the initial precrack length, a, the fracture load, P_f, and the crack-inclination angle, α, using the following relationships developed by Atkinson *et al.*[18]

$$K_I = \frac{P_f\sqrt{a}}{\sqrt{\pi}RB}N_I \tag{26}$$

$$K_{II} = \frac{P_f\sqrt{a}}{\sqrt{\pi}RB}N_{II} \tag{27}$$

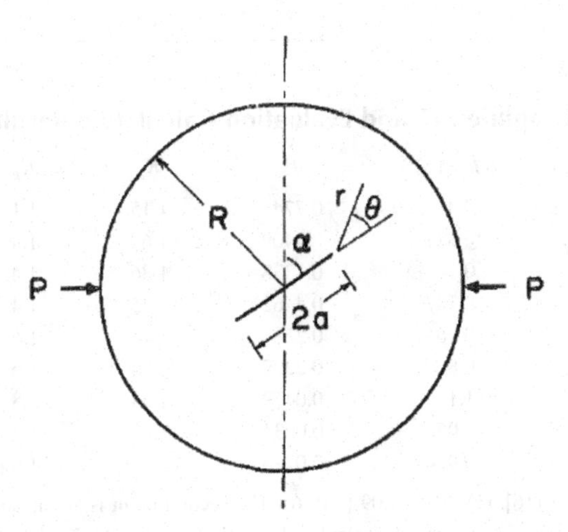

FIGURE 6.2 Inclined crack in a disc specimen and the associated system.

Source: (Data from Singh and Shetty [20])

where R and B are the disk radius and thickness, respectively, and N_I and N_{II} are nondimensional coefficients that are functions of the relative crack length (a/R) and the crack-inclination angle, α. Atkinson *et al.*[18] developed series solutions for N_I and N_{II} and gave numerical solutions for cracks in the size range, $a/R = 0.1$ to 0.6, using a five-term approximation.

Up to now, it has been proven from the previous researches [16–20] that the center-cracked disk test in diametral compression provides a simple and effective method to study mixed-mode fractures of brittle materials. In the following, some test results by the Disc Test in the literature will be shown in the form of examples to verify the empirical fracture equation proposed in this study.

Example 1: Graphite 747

For graphite 747 by Awaji and Sato [16], stress intensity factors K_I and K_{II} under the action of fracture loads are listed in Table 6.1. At the same time, the evaluation calculation results given by means of the empirical fracture equation proposed in this study are also listed in Table 6.1, in which $K_{Ic} = 0.943$ is a measured value [16] and $K_{IIc} = 1.09$ is also a measured value [16], which is in fact equal to that calculated from the seventh, eighth and ninth test data in this table and by means of formula (21). The calculated error indexes are: M.E. = –1%, M.A.E. = 4.2% and E.R. = [–6,6.2]%. The results, obviously, are very satisfactory.

Note here that the second point is an unusual test one. The reader observes test data in the first and second row, and then finds that $K_{II} = 0.871$ in the second row is an unusual test one.

Example 2: Soda-Lime Glass

For soda-lime glass by Shetty *et al.* [19], stress intensity factors K_I and K_{II} under the action of fracture loads are listed in Table 6.2. At the same time, the evaluation calculation results given by using the empirical fracture equation proposed

TABLE 6.1
Test Results of Graphite 747 and Evaluation Calculation Results

No.	K_I[16]	K_{II}[16]	ρ	K_e	K_R	R.E. (%)
1	0.887	0.419	0.774	1.15	1.1	–3.7
2	0.741	<u>0.871</u>	0.4409	1.68	1.4	–<u>17.3</u>
3	0.592	0.71	0.4338	1.36	1.4	2.4
4	0.645	0.747	0.4462	1.45	1.4	–4.2
5	0.391	0.964	0.228	1.71	1.6	–6
6	0.354	0.865	0.2299	1.54	1.6	4.5
7	0.093	1.12	0.0479	1.94	1.8	–6
8	0.115	1.05	0.0631	1.82	1.8	–0.8
9	0.121	0.974	0.0715	1.69	1.8	6.2

Note: (*a*) $K_{Ic} = 0.943$ [16]. (*b*) $K_{IIc} = 1.09$ [16]. (*c*) The second point is an unusual test one. (*d*) After deleting the unusual test point, the calculated error indexes are: M.E. = –1%, M.A.E. = 4.2% and E.R. = [–6,6.2]%.

TABLE 6.2
Test Results of Soda-Lime Glass and Evaluation Calculation Results

No.	K_I[19]	K_{II}[19]	ρ	K_e	K_R	R.E.(%)
1	0.73	0	1	0.73	0.73	0
2	0.715	0.13	0.9538	0.75	0.76	0.9
3	0.67	0.255	0.8349	0.8	0.83	3.1
4	0.63	0.371	0.7001	0.9	0.92	1.8
5	0.575	0.48	0.5688	1.01	1.01	0.2
6	0.5	0.56	0.4582	1.09	1.1	0.9
7	0.425	0.638	0.359	1.18	1.19	0.3
8	0.348	0.71	0.2723	1.28	1.27	−0.8
9	0.349	0.72	0.2695	1.3	1.27	−1.9
10	0.25	0.77	0.1842	1.36	1.36	−0.1
11	0.252	0.78	0.1834	1.37	1.36	−1.3
12	0.174	0.8	0.1246	1.4	1.42	1.6
13	0.1	0.823	0.07	1.43	1.48	3.4
14	0.11	0.87	0.0728	1.51	1.48	−2.4
15	0	0.9	0	1.56	1.56	0

Note: (*a*) K_{Ic} = 0.73 [19]. (*b*) K_{IIc} = 0.9 [19]. (*c*) M.E. = 0.4%, M.A.E. = 1.2% and E.R. = [−2.4,3.4]%.

in this study are also listed in Table 6.2, in which K_{Ic} = 0.73 is a measured one [19] and K_{IIc} = 0.9 is also a measured one [19]. The calculated error indexes are: M.E. = 0.4%, M.A.E. = 1.2% and E.R. = [−2.4,3.4]%. The results, obviously, are very satisfactory.

Example 3: Polycrystalline Ceramics (Alumia and Ce0,-TZP)

For alumia by Singh and Shelly [20], stress intensity factors K_I and K_{II} under the action of fracture loads are listed in Table 6.3. At the same time, the evaluation calculation results carried out by means of the empirical fracture equation proposed in this study are also given in Table 6.3. The evaluation calculations given here include two aspects. One is the evaluation calculations by means of K_{Ic} = 3.35 [20] and K_{IIc} = 6.7 [20], with the error indexes: M.E. = 7.1%, M.A.E. = 7.4% and E.R. = [−1.5,18]%. The other is the evaluation calculations by means of K_{Ic} = 3.35 and K_{IIc} = 6.35 calculated from the tenth, eleventh and twelfth test data and by means of formula (21), with the error indexes: M.E. = 3.4%, M.A.E. = 6.0% and E.R. = [−5.2,14]%. The two results, obviously, are satisfactory and the evaluation calculation results of the latter are a little better than those of the former.

For Ce0,-TZP by Singh and Shelly [20], similar results to those of Alumia are shown in Table 6.4. The evaluation calculations include two aspects. One is the evaluation calculations by means of K_{Ic} = 7.3 [20] and K_{IIc} = 11 [20], with the error indexes: M.E. = 5.8%, M.A.E. = 6.5% and E.R. = [−3.9,12.1]%. The other is the evaluation calculations by means of K_{Ic} = 7.3 and K_{IIc} = 10.7 which is calculated from tenth to thirteenth test data and by means of formula (21), with the error indexes: M.E. = 4.4%, M.A.E. = 5.8% and E.R. = [−3.9,11.4]%. The two results, obviously, are satisfactory and the evaluation calculation results of the latter are a little better than those of the former.

TABLE 6.3
Test Results of Alumia and Evaluation Calculation Results

No.	K_I[20]	K_{II}[20]	ρ	K_e	K_R (1)	R.E.(1) (%)	K_R (2)	R.E.(2) (%)
1	3.35	0	1	3.35	3.35	0	3.35	0
2	3.36	0.75	0.9327	3.6	3.64	1.1	3.63	0.7
3	3.1	1.55	0.7559	4.1	4.54	10.6	4.48	9.2
4	2.7	3.15	0.4435	6.09	6.69	9.9	6.49	6.6
5	2.21	3.47	0.3451	6.4	7.56	18	7.3	14
6	1.8	4.42	0.2289	7.86	8.73	11	8.38	6.5
7	1.3	4.7	0.1577	8.24	9.54	15.7	9.12	10.6
8	1.29	5.41	0.1364	9.46	9.8	3.6	9.35	−1.1
9	0.9	6.1	0.0849	10.6	10.44	−1.5	9.94	−6.2
10	0.2	5.7	0.0203	9.87	11.32	14.6	10.74	8.7
11	0.3	6.33	0.0274	10.97	11.22	2.3	10.65	−2.9
12	0	6.7	0	11.6	11.6	0	11	−5.2

Note 1: (*a*) K_{Ic} = 3.35 [20]. (*b*) K_{IIc} = 6.7 [20]. (*c*) The calculated error indexes are: M.E. = 7.1%, M.A.E. = 7.4% and E.R. = [−1.5,18]%.

Note 2: (*a*) K_{Ic} = 3.35 [20]. (b) K_{IIc} = 6.35, calculated from the tenth, eleventh and twelfth test data. (*c*) M.E. = 3.4%, M.A.E. = 6.0%, E.R. = [−5.2,14]%.

TABLE 6.4
Test Results of Ce0,-TZP and Evaluation Calculation Results

No.	K_I[20]	K_{II}[20]	ρ	K_e	K_R(1)	R.E.(1)(%)	K_R(2)	R.E.(2)(%)
1	7.6	0	1	7.6	7.3	−3.9	7.3	−3.9
2	6.98	0	1	6.98	7.3	4.6	7.3	4.6
3	6.8	1.7	0.9177	7.41	7.88	6.4	7.88	6.4
4	6.25	3	0.769	8.13	9.06	11.5	9.06	11.5
5	5.75	4	0.6386	9	10.23	13.6	10.23	13.6
6	5.41	5.91	0.4673	11.58	12.01	3.7	12.01	3.7
7	4.2	6.66	0.3421	12.28	13.5	10	13.5	10
8	4	7.6	0.2907	13.76	14.16	2.9	14.16	2.9
9	3.18	7.8	0.2291	13.88	15	8.1	15	8.1
10	1	9.8	0.0588	17	17.59	3.5	17.59	3.5
11	0.55	10.2	0.0311	17.68	18.05	2.1	18.05	2.1
12	0.45	10.8	0.024	18.71	18.17	−2.9	18.17	−2.9
13	0	11	0	19.05	18.58	−2.5	18.58	−2.5

Note 1: (*a*) K_{Ic} = 7.3 [20]. (b) K_{IIc} = 11 [20]. (*c*) the calculated error indexes are: M.E. = 5.8%, M.A.E. = 6.5% and E.R. = [−3.9,12.1]%.

Note 2: (*a*) K_{Ic} = 7.3 [20]. (*b*) K_{IIc} = 10.73, calculated from tenth to thirteenth test data. (*c*) M.E. = 4.4%, M.A.E. = 5.8%, E.R. = [−3.9,11.4]%.

6.4.2 Experimental Verifications by the AS4P Test

In 1978, Mixed Mode I/II testing was carried out on edge-cracked bend bar specimens using antisymmetric four-point loading developed by Gao *et al.*[13] to obtain Mode II and mixed Mode I/II, and using conventional symmetric four-point loading to obtain Mode I.

Figure 6.3 shows the loading arrangement for antisymmetric four-point (AS4P) and symmetric four-point (S4P) bend specimens, together with the corresponding shear force and bending moment diagrams. Given that the bending moment, M, is associated with Mode I, and that the shear force, Q, is associated with Mode II, it is

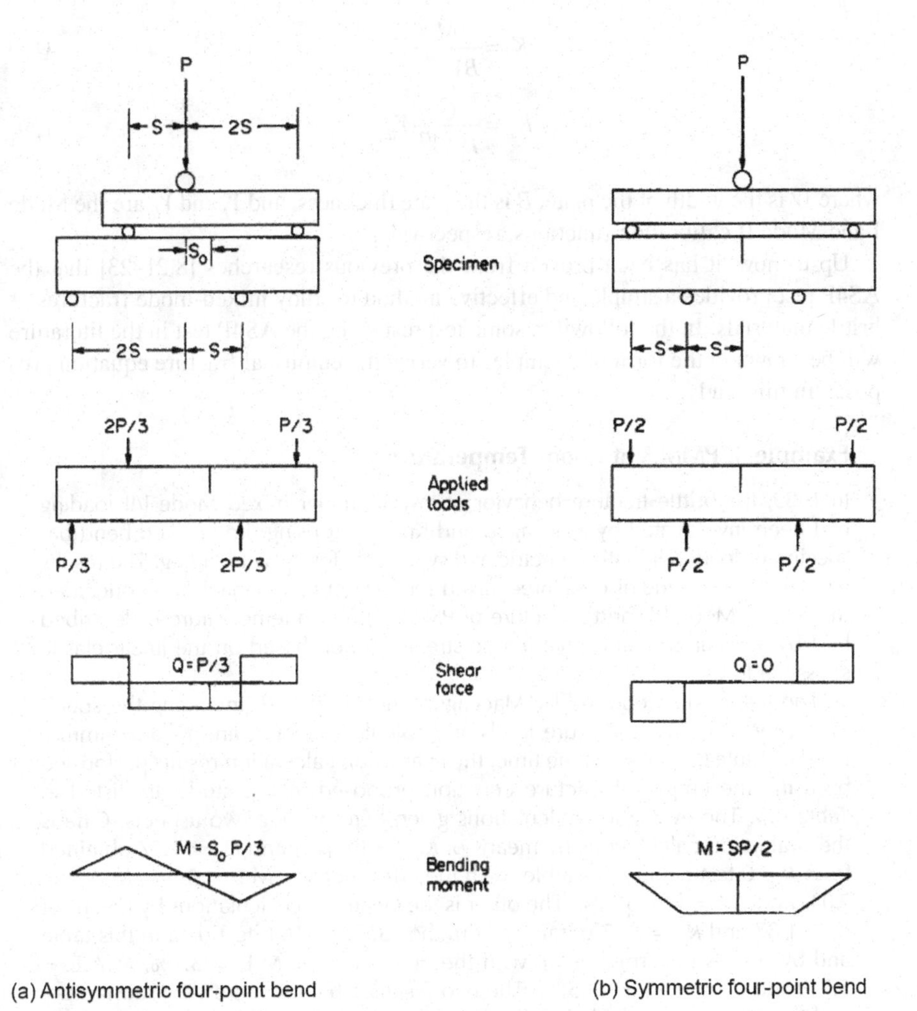

FIGURE 6.3 Loading configurations for mixed Mode I/II testing: (a) antisymmetric four-point bend; and (b) symmetric four-point bend.

Source: (Data from Maccagno and Knott [22])

evident that for the AS4P case (Figure 6.3*a*) the ratio of Mode I to Mode II varies with the positioning of the crack relative to the load points. Along the center axis of the loading arrangement $M = 0$, there is a substantial Q. Therefore, a crack positioned exactly on the central axis is loaded in pure Mode II. If the crack is positioned away from the center, Q remains constant and M increases, and therefore the Mode I to Mode II ratio increases. Pure Mode I is obtained from the S4P loading arrangement by positioning the crack in the central region where $Q = 0$, and therefore where there is no Mode II (Figure 6.3*b*).

According to Wang et a1.[21], K_I and K_{II}, for a straight edge crack in a plate can be determined from M and Q through the expressions:

$$K_I = \frac{M}{BW^{3/2}} Y_I \tag{28}$$

$$K_{II} = \frac{Q}{BW^{1/2}} Y_{II} \tag{29}$$

where W is the width of the plate, B is the plate thickness, and Y_I and Y_{II} are the Mode I and Mode II calibration functions, respectively.

Up to now, it has been proven from the previous researches [8,21–23] that the AS4P test provides a simple and effective method to study mixed-mode fractures of brittle materials. In the following, some test results by the AS4P test in the literature will be shown in the form of examples to verify the empirical fracture equation proposed in this study.

Example 4: PMMA at Room Temperature

In 1989, the brittle fracture behavior of PMMA under mixed Mode I/II loading had been investigated by Maccagno and Knott [22] using pre-cracked bend bar specimens loaded in antisymmetric and symmetric four-point loading. The results were used to test and discuss three mixed-mode fracture criteria. It was concluded that mixed Mode I/II brittle fracture of PMMA at room temperature is described best by a maximum tangential tensile stress criterion based on the linear elastic stress field.

The test results reported by Maccagno and Knott [22], including the specimen geometric sizes, fracture loads and calculations of K_I and K_{II} are summarized in Table 6.5. At the same time, the evaluation calculation results performed by using the empirical fracture equation proposed in this study are listed in Table 6.5. The evaluation calculations given here include two aspects. One is the evaluation calculations by means of $K_{Ic} = 1.87$ [22] and $K_{IIc} = 1.77$ obtained from $\rho = 0$ test data in this table, with the error indexes: M.E. = 5.9%, M.A.E. = 7.0% and E.R. = [−5, 5.7]%. The other is the evaluation calculations by means of $K_{Ic} = 1.87$ and $K_{IIc} = 1.67$ obtained from the 15th and 16th test data in this table and by means of formula (21), with the error indexes: M.E. = 3.7%, M.A.E. = 5.7% and E.R. = [−5.6, 12.5]%. The two results, obviously, are very satisfactory and the evaluation calculation results of the latter are a little better than those of the former.

Note here that the 13th and 14th test points are unusual.

TABLE 6.5
Test Results of PMMA at Room Temperature and Evaluation Calculation Results

No.	W (mm)	B (mm)	a (mm)	S_0(mm)	P_f(kN) [22]	K_I[22]	K_{II}[22]	R.E.(%)
1	20	4.9	7.2		0.352	1.91	0	−2.1
2	20	4.9	8.8		0.259	1.75	0	6.9
3	20	4.9	8		0.306	1.84	0	1.6
4	20	5	9.3		0.281	1.969	0	−5
5	20	5.1	10.8	12	0.506	1.782	0.484	−0.5
6	20	5.1	7.4	11.5	0.767	1.576	0.417	12.8
7	20	5.1	10.4	6.5	0.874	1.564	0.789	1.7
8	19.3	4.9	9.9	5.8	0.852	1.443	0.791	7.7
9	20	5	10.2	3.4	1.142	1.053	1.02	15.7
10	19.3	4.9	10.1	3.3	1.063	1.06	1.017	15.6
11	20	4.8	10.3	1.9	1.439	0.773	1.339	7.3
12	19.9	5.1	9.9	1.9	1.547	0.761	1.317	9
13	20	4.8	9.7	1	1.446	0.384	1.265	26.5
14	19.4	5	10	1	1.429	0.42	1.336	19.3
15	19.8	4.9	12.3	0.3	1.226	0.153	1.53	12.2
16	19.9	4.8	9.5	0	2.082	0	1.769	0.1

Note 1: (*a*) K_{Ic} = 1.87 [22]. (*b*) K_{IIc} = 1.77, obtained from $\rho = 0$ test data. (*c*) The 13th and 14th test points are unusual. (*d*) After deleting the unusual test points, the calculated error indexes are: M.E. = 5.9%, M.A.E. = 7.0% and E.R. = [−5, 15.7]%

Note 2: (*a*) K_{Ic} = 1.87 [22]. (*b*) K_{IIc} = 1.67, calculated from the 15th and 16th test data in this table. (*c*) The 13th and 14th test points are unusual. (*d*) After deleting the unusual test points, the calculated error indexes are: M.E. = 3.7%, M.A.E. = 5.7% and E.R. = [−5.6, 12.5]%.

Example 5: En3B Mild Steel, ICr-IMo-O.3V and a C-Mn Weld at −196°C

In 1991, cleavage fracture of En3B mild steel, ICr-IMo-O.3V and a C-Mn weld subjected to mixed Mode I/II loading was investigated by Maccagno and Knott [23] using edge-cracked bend bar specimens loaded in antisymmetric and symmetric four-point bend configurations. All specimens were tested at −196°C. These results were found to agree with predictions made according to a maximum tangential tensile stress ($\sigma_{\theta\theta}$) criterion based on the HRR elastic-plastic stress field.

For En3B mild steel by Maccagno and Knott [23], the test results including the specimen geometric sizes, fracture loads and calculations of K_I and K_{II} are summarized in Table 6.6. At the same time, the evaluation calculation results given by using the empirical fracture equation proposed in this study are listed in Table 6.6, in which K_{Ic} = 24.9 is calculated from the first, second and third test data, and K_{IIc} = 25.8 is calculated from the 14th and 15th test data and by means of formula (21). The calculated error indexes are: M.E. = −0.2%, M.A.E. = 5.5% and E.R. = [−9.4,11.3]%, which shows that the evaluation calculation results are very satisfactory.

For ICr-IMo-0.3V steel with small grain size, stress intensity factors K_I and K_{II} under the action of fracture loads are listed in Table 6.7. At the same time, the

TABLE 6.6
Test Tesults of En3B Mild Steel at Low Temperature and Evaluation Calculation Results

No.	W (mm)	B (mm)	a (mm)	S_0(mm)	P_f(kN) [23]	K_I[23]	K_{II}[23]	R.E.(%)
1	20	9.9	10		5.7	22.6	0	10.2
2	20	9.9	10		6.1	25	0	−0.4
3	20.1	9.9	10.2		6.6	27.1	0	−8.1
4	19.8	9.9	10.2	11.9	14.3	23.3	6.5	2
5	19.8	9.8	10.3	11.5	13	21.7	6.1	9.4
6	19.9	9.8	9.3	5.6	29.9	20.9	12.1	−0.1
7	20	9.9	10.2	5.6	28.1	21.7	12.7	−4.1
8	20	9.9	10	3.3	38.4	16.8	16.8	−0.8
9	20	9.9	10.6	3.6	31.9	16.8	15.3	4.1
10	20	10	10	1.9	51.4	12.8	22.3	−8.7
11	20	9.9	10	1.9	50.7	12.8	22.2	−8.4
12	20	9.9	11.9	1.7	36.5	11.2	20.7	−0.1
13	20.1	9.9	12.6	1.8	44.2	16.7	27.8	−27.6
14	19.7	9.8	11	1.2	38.8	7.5	20	11.3
15	20	9.9	11	1	50	7.8	25.3	−9.4

Note: (*a*) K_{Ic} = 24.9, calculated from the first, second and third test data. (*b*) K_{IIc} = 25.8, calculated from the 14th and 15th test data and by means of formula (21). (*c*) The 13th point is unusual.(*d*) After deleting the unusual test point, the calculated error indexes are: M.E. = −0.2%, M.A.E. = 5.5%, E.R. = [−9.4,11.3]%.

TABLE 6.7
Test Results of Cr-lMo-0.3V Steel with Small Grain Size and Evaluation Calculation Results

No.	K_I [23]	K_{II}[23]	ρ	K_e	K_R	R.E. (%)
1	23.6	0	1	23.6	24	1.7
2	24.4	0	1	24.4	24	−1.6
3	25.6	6.7	0.9108	28.11	25.99	−7.5
4	22	13.4	0.6879	31.98	31.73	−0.8
5	21.9	12.9	0.7	31.29	31.39	0.3
6	14.4	13.6	0.5216	27.61	36.82	33.4
7	14.7	13.5	0.5322	27.62	36.47	32.1
8	8.2	19.1	0.2406	34.08	47.34	38.9
9	3.3	32.1	0.0592	55.7	55.68	0

Note: (*a*) K_{Ic} = 24, calculated from the first and second test data. (*b*) K_{IIc} = 33.9, calculated from the ninth test point data and by means of formula (21). (*c*) The sixth, seventh and eighth test points are unusual.(*d*) After deleting the unusual test points, the calculated error indexes are: M.E. = −1.3%, M.A.E. = 2.0%, E.R. = [−7.5,1.7]%.

evaluation calculation results given by using the empirical fracture equation proposed in this study are listed also in Table 6.7, in which K_{Ic} = 24 is calculated from the first and second test point data, and K_{IIc} = 33.9 is calculated from the ninth test data and by means of formula (21). The calculated error indexes are: M.E. = –1.3%, M.A.E. = 2.0% and E.R. = [–7.5,1.7]%, which shows that the evaluation calculation results are very satisfactory.

Note here that the sixth, seventh and eighth test points are unusual.

For ICr-IMo-0.3V steel with large grain size, similar results to those for ICr-IMo-0.3V steel with small grain size can be obtained and are shown in Table 6.8. Here, K_{Ic} = 23.85 is calculated from the first and second test data, and K_{IIc} = 31.6.8 is calculated from the 11th and 12th test data and by means of formula (21). The calculated error indexes are: M.E. = 0.7%, M.A.E. = 4.5% and E.R. = [–7.9, 9.8]%, which shows that the evaluation calculation results are very satisfactory.

Note here that the fourth, seventh and eighth test points are unusual.

For C-Mn steel weld by Maccagno and Knott [23], stress intensity factors K_I and K_{II} under the action of fracture loads are listed in Table 6.9. At the same time, the evaluation calculation results obtained by means of the empirical fracture equation proposed in this study are listed also in Table 6.9, in which K_{Ic} = 21.6 is calculated from the first and second test data, and K_{IIc} = 47 is calculated from the ninth test data and by means of formula (21). The calculated error indexes are: M.E. = 0.2%, M.A.E. = 5.3% and E.R. = [–5.7,10]%, which shows that the evaluation results are very satisfactory.

Note here that the fourth test point is unusual.

TABLE 6.8
Test Results of ICr-IMo-0.3V Steel with Large Grain Size and Evaluation Calculation Results

No.	K_I[23]	K_{II}[23]	ρ	K_e	K_R	R.E. (%)
1	21.8	0	1	21.8	23.85	9.4
2	25.9	0	1	25.9	23.85	–7.9
3	23.5	6.3	0.907	25.91	25.77	–0.6
4	18.4	4.8	0.9113	20.19	25.67	27.2
5	19.9	10.6	0.735	27.08	29.72	9.8
6	19.1	17.3	0.5375	35.53	35.02	–1.4
7	11	19	0.317	34.7	42.06	21.2
8	10.7	18.4	0.3183	33.62	42.02	25
9	6.5	29.8	0.1249	52.02	49.34	–5.2
10	9.3	25.6	0.2053	45.31	46.15	1.9
11	2.9	30.9	0.0541	53.6	52.33	–2.4
12	2.7	29.5	0.0528	51.17	52.38	2.4

Note: (*a*) K_{Ic} = 23.85, calculated from the first and second test data. (*b*) K_{IIc} = 31.68, calculated from the 11th and 12th test data and by means of formula (21). (*d*) The fourth, seventh and eighth test points are unusual. (e) After deleting the unusual test points, the calculated error indexes are: M.E. = 0.7%, M.A.E. = 4.5%, E.R. = [–7.9, 9.8]%.

TABLE 6.9
Test Results of C-Mn Steel Weld at Low Temperature and Evaluation Calculation Results

No.	K_I [23]	K_{II} [23]	ρ	K_e	K_R	R.E. (%)
1	20.4	0	1	20.4	21.6	5.9
2	22.7	0	1	22.7	21.6	−4.8
3	20.8	5.4	0.912	22.81	24.27	6.4
4	28.9	7.9	0.9038	31.98	24.54	−23.2
5	23.5	13.4	0.7115	33.03	31.67	−4.1
6	23.9	13.6	0.7122	33.56	31.64	−5.7
7	19.6	17.3	0.5474	35.81	39.38	10
8	22.7	20.5	0.5386	42.14	39.84	−5.5
9	16.9	29.3	0.316	53.49	53.53	0.1

Note: (*a*) $K_{Ic} = 21.6$, calculated from the first and second test data. (*b*) $K_{IIc} = 47$, calculated from the ninth test point data and by means of formula (21). (*c*) The fourth test point is unusual. (*d*) After deleting the unusual test point, the calculated error indexes are: M.E. = 0.2%, M.A.E. = 5.3%, E.R. = [−5.7,10]%.

Example 6: High-Strength Steel (GC-4), Medium-Strength Steel (30CrMoV) and Nodular Cast Iron

In 1978, the extention of the I/II mode ($K_{II} / K_I = 0$ to14) cracks had been investigated by Gao et al. [13] by means of three-point and four-point bend specimens of high-strength steel (GC-4), medium-strength steel (30CrMoV) and nodular cast iron. All specimens were broken with brittle fractures under linear elastic plane strain conditions. The finding by Gao *et al.* [13] was that the crack extension resistance increases with the increase of the K_{II} / K_I ratio, thus deviating obviously from the theoretical prediction.

The brittle fracture of the steels studied by Gao *et al.*[13], here, will be analyzed by means of the empirical fracture equation proposed in this study.

For nodular cast iron, stress intensity factors K_I and K_{II} under the action of fracture loads are listed in Table 6.10. At the same time, the evaluation calculation results obtained by means of the empirical fracture equation proposed in this study are listed also in Table 6.10, in which $K_{Ic} = 97\ \text{kg} \cdot mm^{-1.5}$ is a measured one by the three-point bend test [13], and $K_{IIc} = 216.2\ \text{kg} \cdot mm^{-1.5}$ is calculated from the fifth test data and by means of formula (21). The calculated error indexes are: M.E. = −5.6%, MA.E. = 7.5% and E.R. = [−15.3,4.8]%, which shows that the evaluation calculation results are satisfactory.

For high-strength steel (GC-4), stress intensity factors K_I and K_{II} under the action of fracture loads are listed in Table 6.11. At the same time, the evaluation calculation results given by using the empirical fracture equation proposed in this study are listed also in Table 6.11. The evaluation calculations given here include two aspects. One is the evaluation calculations by means of $K_{Ic} = 160\ \text{kg} \cdot mm^{-1.5}$ [13] and $K_{IIc} = 324.5\ \text{kg} \cdot mm^{-1.5}$, obtained from the eighth test data and by means of formula (21), with the error indexes: M.E. = 7.3%, MA.E. = 8.8% and E.R. = [−5.3,15.4]%. The other is the evaluation calculations by means of $K_{Ic} = 160\ \text{kg} \cdot mm^{-1.5}$ and $K_{IIc} = 290.9\ \text{kg} \cdot mm^{-1.5}$, obtained from the seventh and eighth test data and by means of formula (21), with the error indexes: M.E. = 1.8%,

TABLE 6.10
Test Results of Nodular Cast Iron and Evaluation Calculation Results

No.	K_I[13]	K_{II}[13]	ρ	K_e	K_R	R.E. (%)
1	96	150	0.3466	276.98	234.46	−15.3
2	51.3	196	0.1494	343.34	306.02	−10.9
3	48	188	0.1458	329.14	307.51	−6.6
4	24	186	0.0743	323.05	338.71	4.8
5	13.8	205	0.0388	355.34	355.32	0

Note: (*a*) Unit of K_I, K_{IIc}, K_{Ic}, K_{IIc}, K_e, K_R in this table is $kg \cdot mm^{-1.5}$. (*b*) $K_{Ic} = 97$ [13]. (*c*) $K_{IIc} = 216.2$, calculated from the fifth test data and by means of formula (21). (*d*): M.E. = −5.6%, MA.E. = 7.5%, E.R. = [−15.3,4.8]%.

TABLE 6.11
Test Results of High-Strength Steel (GC-4) and Evaluation Calculation Results

No.	K_I[13]	K_{II}[13]	ρ	K_e	K_R(1)	R.E.(1)(%)	K_R(2)	R.E.(2)(%)
1	137	108	0.5909	231.86	267.52	15.4	255.82	10.3
2	138	52.2	0.8365	164.98	196.5	19.1	193.01	17
3	158	69.2	0.7967	198.32	206.56	4.2	202.02	1.9
4	146	82.2	0.7159	203.93	228.62	12.1	221.63	8.7
5	169	123	0.6215	271.93	257.43	−5.3	247	−9.2
6	132	160	0.43	306.96	327.43	6.7	307.66	0.2
7	96	183	0.2899	331.18	390.48	17.9	361.32	9.1
8	84	251	0.1897	442.79	442.84	0	405.31	−8.5

Note 1: (*a*) Unit of K_{Ic}, K_{II}, K_{Ic}, K_{IIc}, K_e, K_R in this table is $kg \cdot mm^{-1.5}$. (*b*) $K_{Ic} = 160$ is a measured one [13]. (*c*) $K_{IIc} = 324.5$, obtained from the eighth test data and by means of formula (21). (*c*) The second test point is unusual. (*d*) After deleting the unusual test point, the calculated error indexes are: M.E. = 7.3%, MA.E. = 8.8%, E.R. = [−5.3,15.4]%.

Note 2: (*a*) $K_{IIc} = 290.9$, calculated from the seventh and eighth test data and by means of formula (21). (*b*) The second test point is unusual. (*c*) After deleting the unusual test point, the calculated error indexes are: M.E. = 1.8%, MA.E. = 6.8%, E.R. = [−9.2,10.3]%.

MA.E. = 6.8% and E.R. = [−9.2,10.3]%. The two results, obviously, are satisfactory and the evaluation calculation results of the latter are a little better than those of the former.

Note here that the second test point is unusual.

For medium-strength steel (30CrMoV), stress intensity factors K_I and K_{II} under the action of fracture loads are listed in Table 6.12. At the same time, the evaluation calculation results obtained by using the empirical fracture equation proposed in this study are listed also in Table 6.12. The evaluation calculations performed in this example include two aspects. One is the evaluation calculations by means of $K_{Ic} = 145\ kg \cdot mm^{-1.5}$ [13] and $K_{IIc} = 404\ kg \cdot mm^{-1.5}$, obtained from the 11th test data and by means of formula (21), with the error indexes: M.E. = −7.6%, M.A.E. = 8.3%

TABLE 6.12
Test Results of Medium-Strength Steel (30CrMoV) and Evaluation Calculation Results

No.	K_I[13]	K_{II}[13]	ρ	K_e	$K_R(1)$	R.E.(1)(%)	$K_R(2)$	R.E.(2)(%)
1	159.6	21.3	0.9743	163.81	150.98	−7.8	151.42	−7.6
2	159	25.5	0.9635	165.02	153.57	−6.9	154.2	−6.6
3	166	37.5	0.9313	178.25	161.57	−9.4	162.82	−8.7
4	156.4	26.4	0.9598	162.95	154.47	−5.2	155.16	−4.8
5	175.6	24.6	0.9718	180.7	151.58	−16.1	152.06	−15.8
6	162.1	25.4	0.9651	167.96	153.19	−8.8	153.79	−8.4
7	158.1	28.5	0.9546	165.63	155.75	−6	156.55	−5.5
8	149.4	25.1	0.9602	155.6	154.38	−0.8	155.07	−0.3
9	169	29.9	0.9561	176.76	155.37	−12.1	156.14	−11.7
10	176.4	183.6	0.4851	363.65	326.1	−10.3	345.51	−5
11	137.4	227.6	0.3291	417.47	416.82	−0.2	449.43	7.7

Note 1: (*a*) Unit of K_I, K_{II}, K_{Ic}, K_{IIc}, K_e, K_R in this table is $kg \cdot mm^{-1.5}$. (*b*) $K_{Ic} = 145$ is a measured one [13]. (*c*) $K_{IIc} = 404$, obtained from the 11th test data and by means of formula (21). (*c*) M.E. = −7.6%, M.A.E. = 8.3%, E.R. = [−16.1,−0.2]%.

Note 2: (*a*) $K_{IIc} = 452$, calculated from the tenth and 11th test data and by means of formula (21). (*b*) The calculated error indexes are: M.E. = −6.0%, M.A.E. = 7.5%, E.R. = [−15.8,7.7]%.

and E.R. = [−16.1,−0.2]%. The other is the evaluation calculations by means of $K_{Ic} = 145\ kg \cdot mm^{-1.5}$ and $K_{IIc} = 452\ kg \cdot mm^{-1.5}$, calculated from the tenth and 11th test data and by means of formula (21), with the error indexes: M.E. = −6.0%, M.A.E. = 7.5% and E.R. = [−15.8,7.7]%. The two results, obviously, are satisfactory and the evaluation calculation results of the latter are a little better than those of the former.

6.4.3 Experimental Verifications by Circumferentially Notched Cylindrical Rods

As is well known to us, testing of circumferentially notched cylindrical rods loaded in combined tension torsion is one of the most important ones in multiaxial fracture and fatigue of notches [24–27]. Here, the mixed-mode fracture behavior of (cyclic) fatigue precracked ceramic specimens given by Suresh and Tschegg [24] will be evaluated here by means of the empirical fracture equation proposed in this study.

Example 7: Combined Mode I-Mode III Fracture of Fatigue-Precracked Alumina

For polycrystalline Alumina, stress intensity factors K_I and K_{III} under the action of fracture loads are listed in Table 6.13. At the same time, the evaluation calculation results obtained by using the empirical fracture equation proposed in this study are shown also in Table 6.13. The evaluation calculations performed in this example include two aspects. One is the evaluation calculations by means of $K_{Ic} = 3.35$ [24] and $K_{IIIc} = 7.63$, obtained from $\rho = 0$, with the error indexes: M.E. = 6.8%,

TABLE 6.13
Test Results of Polycrystalline Alumina and Evaluation Calculation Results

No.	K_I [24]	K_{III}[24]	ρ	K_e	$K_R(1)$	R.E.(1)(%)	$K_R(2)$	R.E.(2)(%)
1	3.35	0	1	3.35	3.35	0	3.35	0
2	2.1	4.45	0.262875	7.99	9.21	15.3	8.8	10.2
3	1.3	5.65	0.131685	9.87	11.03	11.7	10.45	5.9
4	0	7.63	0	13.22	13.22	0	12.42	−6

Note 1: (*a*) K_{Ic} = 3.35 is a measured one [24]; (*b*) K_{IIIc} = 7.63, obtained from ρ = 0; (*c*) M.E. = 6.8%, MA.E. = 6.8%, E.R. = [0,15.3]%.

Note 2: (*a*) K_{IIIc} = 7.17, calculated from the third and fourth test data and by means of formula (21); (*b*) M.E. = 2.4%, MA.E. = 5.5%, E.R. = [−6, 10.2] %.

MA.E. = 6.8% and E.R. = [0,15.3] (%. The other is the evaluation calculations by means of K_{Ic} = 3.35 and K_{IIIc} = 7.17, calculated from the third and fourth test data and by means of formula (21), with the error indexes: M.E. = 2.4%, MA.E. = 5.5% and E.R. = [−6,10.2] %. The two results, obviously, are satisfactory and the evaluation calculation results of the latter are a little better than those of the former.

6.5 FINAL COMMENTS

From the experimental verifications made in Section 4, it is concluded that the empirical fracture equation (15) of mixed-mode cracks proposed in this study is not only simple in computation but also high in accuracy. What is the reason for it? It is due to the following respects:

(a) Von Mises definition of *K* factors (16) is a proper property which is used to characterize the extension (fracture failure) force of mixed mode cracks.
(b) The right of (17) is a proper property which is used to characterize the extension resistance of mixed-mode cracks.
(c) The origin of the equation (15) is the equation (14), which has been proven to have high accuracy in predicting multiaxial fatigue limits.

Here, one has naturally at heart that whether or not the type of equation (14) can be used to perform the static strength analysis of plain specimens under multiaxial loading. For it, two examples of tension/torsion test are given in the Appendix *A* to illustrate that the type of equation (14) can be used to perform the multiaxial static strength evaluation of plain specimens.

It is worth mentioning that equation (21) can be used to properly estimate K_{IIc} by means of test data near Mode II region.

Finally, the author is concerned with the following question presented by a reviewer:

> *"All test data investigated in this study is obtained from brittle materials which have relatively small fracture process zone. Could the proposed approach be able to predict the fracture behavior of materials which have large fracture process zone such as rock or concrete? Please explain."*

According to this question, Appendix B is added to this study. In Appendix B, an attempt is made to test whether or not the empirical fracture equation of mixed-mode cracks proposed in this study is used to perform the fracture analysis of cracked specimens made of plastic materials. Two examples are given. From the study in Appendix B, the author considers that the empirical fracture equation of mixed-mode cracks proposed in this study appears to be used to perform the fracture analysis of cracked specimens made of plastic materials. However, further experimental verifications are needed.

In addition, in the paper "A unified lifetime estimation equation for a low/medium/high cycle fatigue of metallic materials under uniaxial and multiaxial loading" published on the author's blog, a description on the practicability of establishing the unified prediction equation for a low/medium/high cycle fatigue of metallic materials (see Appendix C) perhaps is helpful for the reader to understand it is practical for the fracture analysis of cracked specimens made of plastic materials to be performed by using the empirical fracture equation of mixed-mode cracks proposed in this study.

REFERENCES

1 Irwin, G.R. Fracture. In: Fulgge, S., editor. *Encyclopedia of physics*, Vol. 6. Ikrlin: Springer; 1958. p. 551–590.
2 Erdogan, F., Sih, G.C. On the crack extension in plates under plane loading and transverse shear. *J. Basic Engng* 85, 519–527 (1963).
3 Williams, J.G., Ewing, P.D. Fracture under complex stress-the angle crack problem. *Int. J. Fracture Mech.* 8, 441–446 (1972).
4 Maiti, S.K., Smith, R.A. Comparison of the criteria for mixed mode brittle fracture based on the preinstability stress-strain field Part I: Slit and elliptical cracks under uniaxial tensile loading. *Inf. J. Fracfure* 23, 281–295 (1983).
5 Griffith, A.A. The phenomenon of rupture and flow in solids. *Phil. Trans. R. Sot.* 221, 163–198 (1920).
6 Hussain, M.A., Pu, S.L., Underwood, J. Strain energy release rate for a crack under combined mode I and mode II. *ASTM STP* 560, 2–28 (1974).
7 Palaniswamy, K., Knauss, W.G. On the problem of crack extension in brittle solids under general loading. In: Nemat-Nasser, S., editor. *Mechanics today-vol. 4*. Oxford: Pergamon Press; 1978. p. 87–148.
8 Hwang, K.C., Yu, S.W., Hua, D.H. On the maximum energy release rate fracture criterion for combined Mode I-II-III cracks. *Acta Mechanica Solida Sinica* 3, 313–321 (1983).
9 Hyashi, K., Nemat-Nasser, S. On branched, interface cracks. *J. Appl. Mech.* 48, 520–524 1981).
10 Sih, G.C. Strain-energy-density factor applied to mixed mode crack problems. *Int. J. Fracture* 10, 305–321 (1974).
11 Jayatilaka, A. de S., Jenkins, I.J., Prasad, S.V. Determination of crack growth in mixed mode loading system. In: *Proc. 4th Inr. Conf Fracture*, Waterloo, Vol. 3. New York: Pergamon Press; 1977. p. 15–23.
12 Theocaris, P.S., Andrianopoulos, N.P. The mises elastic-plastic boundary as the core region in fracture criteria. *Engng. Fracture Mech.* 16, 425–432 (1982). doi: 10.1016/0013-7944(82)90120-5
13 Gao, H., Wangg, Z.Q., Yang, C.S., Zhou, A.H. An investigation on the brittle fracture of K_I-K_{II} composite mode cracks. *Acta Metallurgica Sinica* 15(3), 380–391 (1979).

14 Liu, B.W.;, Yan, X.Q. A multi-axial fatigue limit prediction equation for metallic materials. *ASME Journal of Pressure Vessel Technology*, 142, 034501-6 (2020).
15 Liu, B.W., Yan, X.Q. A new model of multiaxial fatigue life prediction with influence of different mean stresses. *Int. J. Damage Mech.* 28(9), 1323–1343 (2019).
16 Awaji, H., Sato, S. Combined mode fracture toughness measurement by the disk test. *ASME J. Eng. Muter. Technol.* 100, 175–182 (1978).
17 Libatskii, L.L., Kovchik, S.E. Fracture of disks containing cracks. *Soviet Muter. Sci.* 3, 334–339 (1967).
18 Atkinson, C., Smelser, R.E., Sanchez, J. Combined mode fracture via the cracked brazitian disk test. *Int. J.Fracture* 18, 279–291 (1982).
19 Shetty, D.K., Rosenfield, A.R., Duckworth, W.H. Mixed-mode fracture in biaxial stress state: Application of the diametral-compression (Brazilian disk) test. *Engng. Fract. Mech.* 26, 825–840 (1987).
20 Singh, D., Shetty, D.K. Fracture toughness of polycrystalline ceramics in combined mode I and mode II loading. *J. Am. Ceram. Soc.* 72, 78–84 (1989).
21 Wang, K.J., Hsu, C.L., Gao, H. Calculation of stress intensity factors for combined mode specimens. In: *Proc. 4th Int. Conf: Fracture*, Waterloo, Vol. 4. New York: Pergamon Press; 1977. p. 123–133.
22 Maccagno, T.M., Knott, J. F. The fracture behaviour of PMMA in mixed modes I and II. *Engng. Fract. Mech.* 34, 65–86 (1989).
23 Maccagno, T.M., Knott, J.F. The low temperature brittle fracture behaviour of steel in mixed modes I and II. *Engng. Fract. Mech.* 38, 111–128 (1991).
24 Suresh, S., Tschegg, E.K. Combined mode I-mode I11 fracture of fatigue-precracked alumina. *J. Am. Ceram.* SOC., 70(10), 726–733 (1987).
25 Atzori, B., Meneghetti, G., Susmel, L. Fatigue behaviour of AA356-T6 cast aluminium alloy weakened by cracks and notches. *Engng. Fract. Mech.* 71, 759–768 (2004).
26 Berto, F. Some recent results on the fatigue strength of notched specimens made of 40CrMoV13.9 steel at room and high temperature. *Physical Mesomechanics* 18, 105–126 (2015).
27 Gao, Z., Qiu, B., Wang, X., Jiang, Y. An investigation of fatigue of a notched member. *International Journal of Fatigue* 32, 1960–1969 (2010).
28 Kitae, T. Kim, Suh, Jeong. Fracture of alumina tubes under combined tension/torsion. *J Am ceram Soc.* 75(4), 896–902 (1992).
29 Petrovic, J.J., Stout, M.G. Fracture of AI_2O_3 under combined tension/torsion: I, experiments. *Journal of the American Ceramic Society-Petrouic and Stout*, 64(11), 656–660 (1981).
30 Maccagno, T.M., Knott, J.F. The mixed mode I/II fracture behaviour of lightly tempered HY 130 steel at room temperature. *Engineering Fracture Mechanics* 41(6), 805–820 (1992).
31 Bhattacharjee, D., Knott, J.F. Ductile fracture in HY100 steel under mixed mode 1/ mode II loading. *Acta Metall, Mater.* 42(5), 1747–1754 (1994).

APPENDIX A: EXPERIMENTAL INVESTIGATIONS: THE EMPIRICAL FAILURE CONDITION IS WELL SUITED FOR THE FAILURE ANALYSIS FOR PLAIN MATERIALS

An attempt is made to test whether or not the type of equation (14) is used to perform the static strength analysis of plain specimens under multiaxial loading. Two examples are given below.

Example 1A: Fracture of Alumina Tubes Under Combined Tension/Torsion

The test data of fracture of Alumina tubes under combined tension/torsion by Kim and Suh [28] are listed in Table 1A. At the same time, the evaluation calculation

TABLE 1A

The Test Data of Fractured Alumina Tubes Under Combined Tension/Torsion and Evaluation Calculation Results

No.	σ [28]	τ [28]	ρ	σ_e	σ_R (1)	R.E.(1)(%)	σ_R (2)	R.E.(2)(%)
1 (1)	100.1	87.5	0.5511	181.63	205.04	12.9	194.62	7.2
2	97.8	87.6	0.5418	180.52	206.37	14.3	195.67	8.4
3	61.6	120.4	0.2833	217.45	246.69	13.4	226.98	4.4
4	31.5	154	0.1173	268.59	276.65	3	249.68	−7
5	30.3	137.7	0.126	240.42	274.98	14.4	248.43	3.3
6	30.3	136	0.1276	237.5	274.69	15.7	248.21	4.5
1 (2)	140.5	28.2	0.9446	148.75	156.27	5.1	155.27	4.4
2	130.1	59.2	0.7854	165.65	174.42	5.3	170.13	2.7
3	127.2	57.9	0.7853	161.98	174.43	7.7	170.13	5
4	119.4	55.1	0.7811	152.85	174.93	14.4	170.54	11.6
5	118.4	54.2	0.7836	151.1	174.64	15.6	170.3	12.7
6	98.3	85.6	0.5526	177.89	204.83	15.1	194.46	9.3
7	96.6	97.1	0.4981	193.95	212.69	9.7	200.64	3.5
8	94.9	84.6	0.5436	174.58	206.11	18.1	195.46	12
9	93.6	94.4	0.4968	188.4	212.88	13	200.79	6.6
10	67.6	134	0.2796	241.74	247.31	2.3	227.45	−5.9
11	63.2	126	0.2782	227.21	247.57	9	227.65	0.2
12	63.1	126.1	0.2776	227.34	247.67	8.9	227.73	0.2
13	38.4	167.6	0.1311	292.82	274.01	−6.4	247.7	−15.4
14	35.1	151.7	0.1324	265.09	273.77	3.3	247.52	−6.6
15	34.5	150.5	0.1312	262.95	274	4.2	247.69	−5.8
15	34.1	148.1	0.1318	258.77	273.89	5.8	247.61	−4.3
17	31.3	137	0.1308	239.35	274.08	14.5	247.75	3.5
1 (3)	121.1	60.1	0.7583	159.69	177.71	11.3	172.79	8.2
2	75.2	112.8	0.3592	209.35	234.09	11.8	217.3	3.8
3	45.1	135.3	0.189	238.65	263.29	10.3	239.61	0.4
4	0	165.4	0	286.48	299.98	4.7	267.07	−6.8

Note 1: (*a*) Unit of σ, τ, $\sigma_e\sigma_R$, σ_t and τ_{tor} in this table is MPa. (*b*) $\sigma_t = 150.4$ is a measured one [28]. (*c*) $\tau_{tor} = 173.2$ is a measured one [28]. (*d*) M.E. = 12.3%, MA.E. = 12.3% and E.R. = [3.0,15.7]% for loading path 1, M.E. = 8.6%, MA.E. = 9.3% and E.R. = [−6.4,18.1]% for loading path 2, M.E. = 9.5 %, MA.E. = 9.5% and E.R. = [4.7,11.8]% for loading path 3.

Note 2: (*a*) $\tau_{tor} = 154.2$, calculated from the fourth, fifth and sixth test data on loading path 1 and by means of formula (21). (*b*) M.E. = 3.5 %, MA.E. = 5.8% and E.R. = [−7,8.4]% for loading path 1, M.E. = 2.0 %, MA.E. = 6.4% and E.R. = [−15.4,12]% for loading path 2, M.E. = 1.4 %, MA.E. = 4.8% and E.R. = [−6.8, 8.2]% for loading path 3.

results obtained by using the strength equation similar to equation (2–10) are listed in the Table 1A. The evaluation calculations performed here include two aspects. One is the evaluation calculations by means of σ_t = 150.4 MPa, which is a measured one [28] and τ_{tor} = 173.2 MPa, which is a measured one [28]. The calculated error indexes are: M.E. = 12.3%, MA.E. = 12.3% and E.R. = [3,15.7]% for loading path 1, M.E. = 8.6 %, MA.E. = 9.3% and E.R. = [−6.4,18.1] % for loading path 2, M.E. = 9.5 %, MA.E. = 9.5% and E.R. = [4.7,11.8]% for loading path 3. The other is the evaluation calculations by means of σ_t = 150.4 MPa, and τ_{tor} = 154.2 MPa, calculated from the fourth, fifth and sixth test data on loading path 1 and by means of formula (21). The calculated error indexes are: M.E. = 3.5%, MA.E. = 5.8% and E.R. = [−7,8.4]% for loading path 1, M.E. = 2.0%, MA.E. = 6.4% and E.R. = [−15.4,12]% for loading path 2, M.E. = 1.4%, MA.E. = 4.8% and E.R. = [−6.8, 8.2]% for loading path 3. The two results, obviously, are satisfactory and the evaluation calculation results of the latter are a little better than those of the former.

Example 2A: Fracture of Al_2O_3 under Combined Tension/Torsion

For the study of fracture of Al_2O_3 under combined tension/torsion by Petrovic and Stout [29], similar results to those in Table 1A are given in Table 2A. A description of the evaluation calculation results is shown in the note of this table. The same conclusion as that to Example 1A can be obtained.

TABLE 2A
Test Results of Fracture of Al_2O_3 Under Combined Tension/Torsion and Evaluation Calculation Results

No.	σ [29]	τ [29]	ρ	σ_e	σ_R (1)	R.E.(1)(%)	σ_R (2)	R.E.(2)(%)
1	185	92.29	0.7567	244.49	269.87	10.4	260.37	6.5
2	145	144.1	0.5023	288.65	332.23	15.1	308.75	7
3	137	134	0.5083	269.51	330.61	**22.7**	307.51	**14.1**
4	136	134.7	0.5036	270.05	331.88	**22.9**	308.49	**14.2**
5	58.09	232.3	0.1429	406.53	445.68	9.6	392.83	−3.4
6	55.84	224.4	0.1422	392.66	445.93	13.6	393.01	0.1
7	54.18	216.9	0.1427	379.57	445.74	17.4	392.87	3.5

Note 1: (*a*) Unit of σ, τ, $\sigma_e \sigma_R$, σ_t and τ_{tor} in this table is MPa. (*b*) σ_t = 221.2 is a measured one [29]. (*c*) τ_{tor} = 289.2 is a measured one [29]. (*c*) The third and fourth test points are unusual (failure at grip). (*d*) After deleting the unusual test points, the calculated error indexes are: M.E. = 13.2%, MA.E. = 13.2% and E.R. = [9.6,17.4]%.

Note 2: (*a*) τ_{tor} = 249.6, calculated from the fifth, sixth and seventh test data and by means of formula (21). (*b*) After deleting the unusual test points, the calculated error indexes are: M.E. = 2.7%, MA.E. = 4.1% and E.R. = [−3.4,7]%.

APPENDIX B: EXPERIMENTAL INVESTIGATIONS: THE FAILURE CONDITION IS WELL SUITED FOR THE FAILURE ANALYSIS FOR CRACKED SPECIMENS MADE OF PLASTIC MATERIALS

An attempt is made to test whether or not the empirical fracture equation of mixed-mode cracks proposed in this study is used to perform the fracture analysis of cracked specimens made of plastic materials. Two examples are given below.

Example 1B: The Mixed-Mode I/II Fracture Behavior of Lightly Tempered HY 130 Steel at Room Temperature

The abstract in a paper (T. M. MACCAGNO and J. F. KNOTT, THE MIXED MODE I/II FRACTURE BEHAVIOUR OF LIGHTLY TEMPERED HY 130 STEEL AT ROOM TEMPERATURE, Engineering Fracture Mechanics 41(6):805–820 (1992)) stated: "The mixed Mode I/II fracture behavior at room temperature of HY 130 steel tempered at 350°C has been investigated using edge-cracked bend bar specimens loaded in anti-symmetric and symmetric four point bend configurations. In all cases fracture occurred by a localized shear decohesion mechanism that could not be characterized by the stress intensity factors, K_I and K_{II}, but for which the crack tip displacements, δ_I, and δ_{II}, appear to provide a first level of characterization. The results suggest that fracture is described by a maximum shear criterion, and this is consistent with the present understanding of fibrous fracture micromechanisms in the material".

Here, the mixed Mode I/II fracture behavior at room temperature of HY 130 steel tempered at 350°C reported by MACCAGNO and KNOTT [30] is analyzed by means of the empirical fracture equation of mixed-mode cracks proposed in this study. Test data, including the geometric sizes of antisymmetric and symmetric four-point bend configurations, fracture loads, and Y_I and Y_{II} (the Mode I and Mode II calibration functions) are listed in Table 1B. Comparison of the extension crack force with the extension crack resistance at fracture is shown in Table 2B, with the calculation error indexes: M.E. = 0.0%, M.A.E. = 1.4%, E.R. = [−3,2.5]%, which shows that the results calculated by using the empirical fracture equation of mixed-mode cracks are in excellent agreement with ones measured experimentally.

TABLE 6.1B

Fracture Test Data Reported in Ref.[30] and Values of Calibration Functions Y_I and Y_{II}

No.	W	B	a	S_0	P_{max}	Y_1	Y_2
1	20	10	9.6	12.3	98.5	10.51	1.725
2	20	10	9.7	5.7	164.4	10.66	1.75
3	20	10	10	3.3	190.1	11.12	1.838
4	20	10	10.3	1.9	200.5	11.711	1.924
5	20	10	9.9	0.9	200.6	10.97	1.924
6	20	10	10	0	205	11.12	1.838

TABLE 6.2B
Comparison of the Extension Crack Force with the Extension Crack Resistance at Fracture

No.	K_I	K_{II}	ρ	K_e	K_R	R.E. (%)
1	150.06	40.05	0.9077	165.32	167.92	1.6
2	117.72	67.81	0.7079	166.3	165.59	−0.4
3	82.21	82.36	0.4993	164.64	163.2	−0.9
4	52.58	90.93	0.3167	166.03	161.13	−3
5	23.34	90.97	0.1465	159.28	159.23	0
6	0	88.81	0	153.82	157.61	2.5

Note: (a) K_I and K_{II} in this table are calculated by means of test data and calibration functions shown in Table 1B; (b) $K_{Ic} = 169$, $K_{IIc} = 91$ are obtained by using formula (21); (d) M.E. = 0.%, M.A.E. = 1.4%, E.R. = [−3,2.5]%

Example 2B: Ductile Fracture in HY100 Steel Under Mixed Mode 1/Mode II Loading

In 1994, DUCTILE FRACTURE IN HY100 STEEL UNDER MIXED MODE 1/MODE II LOADING was investigated by BHATTACHARJEE and KNOTT [31] by using antisymmetric four-point bend configurations. The abstract of the paper [31] stated "A number of criteria have been proposed which predict the direction of cracking under mixed Mode I/Mode II loading. All have been evaluated for brittle materials, in which a crack subjected to tension and shear propagates normal to the maximum tensile stress (i.e. fracture is of the Mode I type). In a ductile material, however, a notch subjected to mixed Mode I/Mode II loading may initiate a crack in the direction of maximum shear. This paper shows that the profile of the notch tip changes with increasing mixed mode load in such a way that one side of the tip blunts while the other sharpens. Various specimens, subjected to the same mixed mode ratio, were unloaded from different points on the load-displacement curves to study the change in notch-tip profile. Studies under the Scanning Electron Microscope (SEM) have shown that cracks initiate at the sharpened end, along a microscopic shear band. Using a dislocation pile-up model for deeohesion of the carbide-matrix interface, a micromechanical model has been proposed for crack initiation in the shear band. It is shown that a theoretical prediction of the shear strain required for decohesion gives a result that is, of magnitude, similar to that of the shear strain at crack initiation measured in the experiments".

Here, DUCTILE FRACTURE IN HY100 STEEL UNDER MIXED MODE 1/MODE II LOADING reported by BHATTACHARJEE and KNOTT [31] is analyzed by means of the empirical fracture equation of mixed-mode cracks proposed in this study. Fracture test data and the evaluation calculation results given here are shown in Table 3B. By comparing the extension crack force with the extension crack resistance at fracture (see Table 3B), it is found that the predicted results are in excellent agreement with ones measured experimentally, with the calculation error indexes: M.E. = 0.1%, M.A.E. = 3.4%, E.R. = [−7.1,4.5]%.

TABLE 6.3B
Fracture Test Data Reported in Ref.[31] and the Evaluation Calculation Results

No.	S_0(mm)[31]	P_f(KN) [31]	K_I	K_{II}	ρ	K_e	K_R	R.E. (%)
1	10	137.5	180.19	59.57	0.8678	207.64	195.66	−5.8
2	7	153.5	140.81	66.5	0.774	181.92	187.49	3.1
3	5	173	113.36	74.95	0.6578	172.34	177.82	3.2
4	5	178.8	117.16	77.46	0.6578	178.12	177.82	−0.2
5	3	186.5	73.32	80.8	0.4641	157.99	162.81	3.1
6	3	184	72.34	79.71	0.4641	155.87	162.81	4.5
7	2	192.3	50.4	83.31	0.3298	152.84	153.15	0.2
8	1	201.5	26.41	87.29	0.172	153.49	142.54	−7.1

Note: (*a*) Asymmetric four-point bend specimen configuration: W = 20 mm, B = 10 mm, L = 110 mm, a/W ratio within 0.45 and 0.55; (b) K_I and K_{II} in this table are calculated by means of the calibration functions $Y_I = 11.12$, $Y_2 = 1.838$ when $a/W = 0.5$; (c) $K_{Ic} = 207.8$, $K_{IIc} = 76.1$ are obtained by using formula (21); (d) M.E. = 0.1%, M.A.E. = 3.4%, E.R. = [−7.1,4.5]%

APPENDIX C: THE PRACTICABILITY OF ESTABLISHING THE UNIFIED PREDICTION EQUATION FOR A LOW/MEDIUM/HIGH CYCLE FATIGUE OF METALLIC MATERIALS

(See Appendix A in Chapter 1)

7 An Empirical Failure Equation to Assess Mixed-Mode Fracture of Notched Components

Nomenclature

k_e	Stress intensity parameter (SIP)
k_R	Material intensity parameter (MIP)
K_I, K_{II}, K_{III}	Stress intensity factors for Mode I, Mode II and Mode III cracks
K_{Ic}, K_{IIc}, K_{IIIc}	Fracture toughness for Mode I, Mode II and Mode III cracks
K_{Iv}, K_{IIv}, K_{IIIv}	Notch stress intensity factors for Mode I, Mode II and Mode III V-notches
K_{Ivc}, K_{IIvc}, K_{IIIvc}	Notch fracture toughness for Mode I, Mode II and Mode III V-notches
K_{IU}, K_{IIU}, K_{IIIU}	Notch stress intensity factors for Mode I, Mode II and Mode III U-notches
K_{IUc}, K_{IIUc}, K_{IIIUc}	Notch fracture toughness for Mode I, Mode II and Mode III U-notches
λ_1, λ_2, λ_3	Mode I, II, III Williams' eigenvalues for stress distribution at V-notches
ξ	Multiaxial parameter
σ_e	The von Mises equivalent stress
$\sigma_{e,a}$	Amplitude of the von Mises equivalent stress
$\sigma_{kk,a}$	Amplitude of the first invariant of stress tensor
σ_o, τ_o	Material tension and torsion fatigue limits
σ_t, τ_t	Material tension and torsion critical stresses
σ_a, τ_a	Amplitude of bending (or tension) stress and torsion stress
σ_{eo}	Stress intensity parameter for multiaxial fatigue limit analysis
σ_R	Material intensity parameter for multiaxial fatigue limit analysis

7.1 INTRODUCTION

V-notches that are common in mechanical components, such as bolts, nuts and screws, decrease dramatically the load-bearing capacity of components due to the concentration of stress in the vicinity of their tips. A reliable prediction of the mechanical failure like crack formation and growth in the vicinity of V-notches has been a topic of great interest to researchers.

DOI: 10.1201/9781003356721-7

Since the beginning of the last century, Neuber [1,2] and Peterson [3] made significant advances in the local approaches to deal with brittle failure of notched components, in particular, the CD [Critical Distance] approach [4,5] (often called TCD [Theory of Critical Distance]) and the volume-based SED approach [6–8] [Strain Energy Density]. From Refs [4–8], it is seen that the mixed-mode brittle failure of notched components is assumed to be the Mode I dominance, which is wholly the same as that made for mixed–mode cracks [9,10].

Concerning mixed-mode fracture failure of notched components, here, Torabi and his co-authors' work [11–16] needs to be emphasized. But to the author's knowledge, their failure criteria based on the maximum tangential stress (MTS) and the mean-stress (MS) are also assumed to be the Mode I dominance.

For the materials (for example, high-or medium-strength steels as well as nodular cast iron reported by Gao *et al.*[17]) whose crack extension resistance varies seriously with the increase of the K_{II} / K_I ratio, the failure load calculated by using the Mode I dominance criteria is obviously different from that measured experimentally near the region of pure II cracks.

Recently, the author was concerned with material failure and the equation for multiaxial stresses state, including cracked and notched members. This originated from the fact that the multiaxial fatigue limit equation proposed by Liu and Yan [18] employing the well-known Wöhler equation and the multiaxial fatigue life equation as also proposed by Liu and Yan [19] were both simple in computation form and highly accurate, falling within ±10% interval: By using 53 sets of experimental data from the literature (see Figure 7.1), experimental verifications given in Ref 18 showed that the multiaxial fatigue limits predicted by the empirical failure equation are in excellent agreement with those measured experimentally. This accuracy was so exciting that the empirical failure equation was extended to the static failure model for the multiaxial stresses state, including cracked materials [20] and notched materials, which is reported in this study.

By the way, a new concept of the material intensity parameter [MIP] is introduced in this study. It is a material property parameter that depends on both Mode-I fracture toughness, Mode-II (or Mode-III) fracture toughness and the multiaxial parameter to characterize the variation of the material failure resistance with the multiaxial stresses state. Thus the empirical failure equation to assess mixed-mode fracture of notched components proposed in this study is in fact a notch failure condition, which states that the stress intensity parameter [SIP] is equal to the MIP.

7.2 A BRIEF DESCRIPTION OF THE MULTIAXIAL FATIGUE LIFE EQUATION

The multiaxial fatigue life equation proposed by Liu and Yan [18] was

$$\sigma_{eo} = \sigma_o^{\xi} \left(\sqrt{3}\tau_o\right)^{(1-\xi)} \tag{1}$$

where the **stress intensity parameter** at the multiaxial stresses state is taken to be the well-known von Mises effective stress, σ_{eo}, which is defined at tension-torsion or bending-torsion loading to be

$$\sigma_{eo} = \sqrt{(\sigma_a)^2 + 3(\tau_a)^2} \tag{2}$$

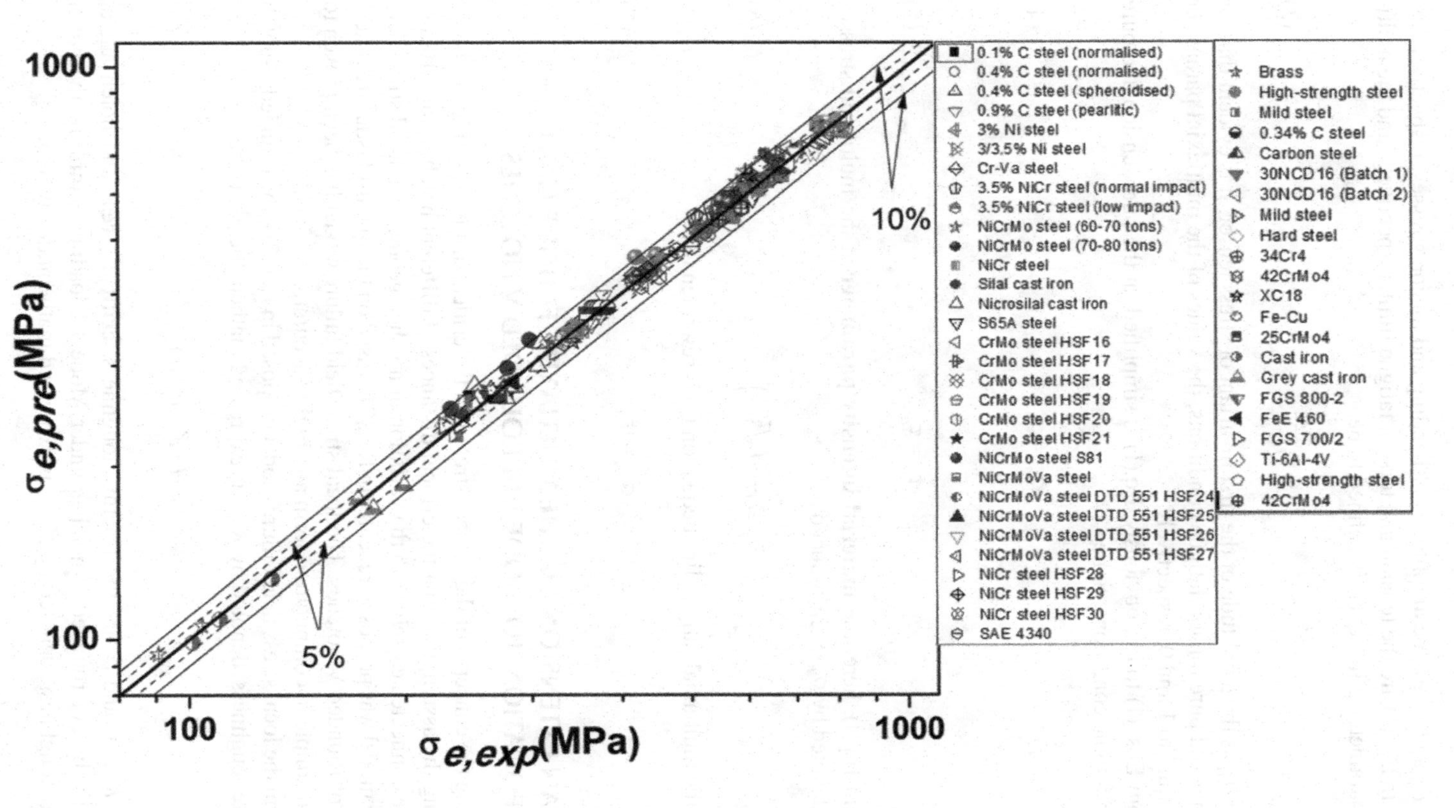

FIGURE 7.1 Comparison of multiaxial fatigue limits of metallic materials.

Source: (Data from liu and Yan [18])

where σ_a and τ_a are the amplitudes of bending and torsion stresses, respectively. σ_o and τ_o in Eq. (1) are the tension and torsion fatigue limit, respectively, and the multiaxial parameter ξ in Eq. (1) is defined to be

$$\xi = \frac{\sigma_{kk}}{\sigma_{eo}} \tag{3}$$

Where σ_{kk} is the amplitude of the first invariant of stress tensor. It is evident that for the axial and pure torsion fatigue conditions, the values of the multiaxial parameter ξ are equal to 1 and 0, respectively.

From Eqs. (1) to (3), obviously, Eq.(1) is simplified, at the pure tension and pure torsion fatigue conditions, to be

$$\sigma_a = \sigma_o \tag{4}$$

and

$$\tau_a = \tau_o \tag{5}$$

From Eq.(1), here, the **material intensity parameter** at the multiaxial stresses state, denoted by σ_R, is introduced as

$$\sigma_R = \sigma_o^{\xi} \left(\sqrt{3}\tau_o\right)^{(1-\xi)} \tag{6}$$

Thus, the multiaxial fatigue limit equation (1) is rewritten to be

$$\sigma_{eo} = \sigma_R \tag{7}$$

7.3 AN EXTENSION OF THE MULTIAXIAL FATIGUE LIMIT EQUATION TO MODE I/III ROUNDED V-NOTCHES

By using the concepts of the critical stress and the critical distance by Taylor [4] and by using the assumption that the critical distances of different notches are different and are connected closely with the variation of notched geometry, a local stress field failure model was proposed recently in Ref.[21] based on the linear-elastic stress field ahead of rounded V-notches. The local stress field failure model has been proven to be both simple in computation form and highly accurate.

From the local stress field failure model proposed in Ref.[21], the rounded V-notch fracture toughness, denoted by K_{IVc}, is expressed mathematically to be

$$K_{\text{IVc}} = H_{\text{I}}(K_{\text{Ic}}, \sigma_t, \alpha) \tag{8}$$

where K_{Ic}, σ_t and α are the fracture toughness, **critical stress** and notch angle, respectively. For rounded V-notches under Mode III loading, similarly, the notch fracture toughness, denoted by K_{IIIVc}, is expressed mathematically to be

$$K_{\text{IIIVc}} = H_{\text{III}}(K_{\text{IIIc}}, \tau_t, \alpha) \tag{9}$$

Because the notch stress intensity factors, K_{Iv} and K_{IIIv}, have different units, the **stress intensity parameter** for Mode I/III V-notches is taken to be a nondimensional von Mises definition:

$$k_e = \sqrt{(\bar{k}_{\mathrm{Iv}})^2 + 3(\bar{k}_{\mathrm{IIIv}})^2} \tag{10}$$

where

$$\bar{k}_{\mathrm{Iv}} = \frac{K_{\mathrm{Iv}}}{K_{\mathrm{Ivc}}}, \quad \bar{k}_{\mathrm{IIIv}} = \frac{K_{\mathrm{IIIv}}}{K_{\mathrm{IIIvc}}} \tag{11}$$

And the corresponding **material intensity parameter** becomes

$$k_R = k_R(\bar{k}_{\mathrm{IVc}}, \bar{k}_{\mathrm{IIIVc}}, \xi) = (\sqrt{3})^{1-\xi} \tag{12}$$

which is obtained by replacing σ_o and τ_o in Eq. (6) by $\bar{k}_{\mathrm{IVc}} = 1$ and $\bar{k}_{\mathrm{IIIVc}} = 1$, respectively. Then a fracture failure condition for rounded V-notches can be written as

$$k_e = k_R \tag{13}$$

Here, the multiaxial parameter ξ is defined as

$$\xi = \bar{k}_{\mathrm{Iv}} / k_e \tag{14}$$

For rounded V-notches under Mode I loading, $\xi = 1$, the fracture condition (13) becomes

$$K_{\mathrm{Iv}} = K_{\mathrm{Ivc}} \tag{15}$$

While for rounded V-notches under Mode III loading, $\xi = 0$, the fracture condition (13) becomes

$$K_{\mathrm{IIIv}} = K_{\mathrm{IIIvc}} \tag{16}$$

Obviously, the fracture conditions (15) and (16) are the same as K criteria for pure Mode I cracks and pure Mode III cracks, respectively.

From Eqs. (8) and (15), the fracture condition for rounded V-notches under Mode I loading can be expressed mathematically to be

$$G_{\mathrm{I}}(K_{\mathrm{IV}}, K_{\mathrm{Ic}}, \sigma_t, \alpha) = 0 \tag{17}$$

From Eqs. (9) and (16), similarly, the fracture condition for rounded V-notches under Mode III loading can be expressed mathematically to be

$$G_{\mathrm{III}}(K_{\mathrm{IIIV}}, K_{\mathrm{IIIc}}, \tau_t, \alpha) = 0 \tag{18}$$

From Eqs. (10) to (14), the failure condition for rounded V-notches under Mode I/III loading can be expressed mathematically to be

$$S(K_{\mathrm{IV}}, K_{\mathrm{IIIV}}, K_{\mathrm{IVc}}, K_{\mathrm{IIIc}}) = 0 \tag{19}$$

From Eqs. (8), (9) and (19), further, the failure condition for rounded V-notches under Mode I/III loading can be also expressed mathematically to be

$$G(K_{\mathrm{IV}}, K_{\mathrm{IIIV}}, K_{\mathrm{Ic}}, K_{\mathrm{IIIc}}, \sigma_t, \tau_t, \alpha) = 0 \tag{20}$$

From Eq. (17), it is seen that the failure condition for rounded V-notches under Mode I loading is controlled by the fracture toughness K_{Ic} and the critical stress σ_t. From Eq. (18), similarly, the failure condition for rounded V-notches under Mode III loading is controlled by the fracture toughness K_{IIIc} and the critical stress τ_t. While from Eq. (20), the failure condition for rounded V-notches under Mode I/III loading is controlled by the fracture toughness, K_{Ic} and K_{IIIc}, and the critical stresses, σ_t and τ_t.

For rounded V-notches under Mode I/II loading, obviously, the series equations similar to Eqs. (8) to (14) are obtained easily and are not listed here.

7.4 EXPERIMENTAL VERIFICATIONS

By using test data of mixed-mode fracture failure of notched components from the literature, the empirical failure equation to assess mixed-mode fracture of notched components proposed in this study is verified.

It is assumed that there is a rounded V-notch under Mode I/II loading. After the notch stress intensity factors, K_{Iv} and K_{IIv}, under the action of fracture loads are calculated, the **stress intensity parameter**, k_e, is determined easily by formula (10) once K_{Ivc} and K_{IIvc} are known. At the same time, the **material intensity parameter**, k_R, is calculated easily by formula (12). By comparing k_e with k_R, thus, the empirical failure equation to assess mixed-mode fracture of notched components proposed in this study can be evaluated through the following error indexes:

$$\text{R.E.} = \frac{(k_R - k_e)\times 100}{k_e} \tag{21}$$

$$M.E. = \frac{1}{N}\sum_{i=1}^{N}\frac{(k_{Ri} - k_{ei})\times 100}{k_{ei}} \tag{22}$$

$$M.A.E. = \frac{1}{N}\sum_{i=1}^{N}\left|\frac{(k_{Ri} - k_{ei})\times 100}{k_{ei}}\right| \tag{23}$$

$$\text{E.R.} = \{\text{Min(R.E.)}, \text{Max(R.E.)}\} \tag{24}$$

which are a relative error, a mean error, an absolute mean error and error range, respectively. N in formulas (22) to (23) is the number of test cases.

For the sake of clear illustrations and discussions, failure analysis for mixed-mode notched components is given in the forms of examples.

Example 1: Mixed-Mode Fracture for Rounded-Tip V-Notched Components

In 2010, the rounded-tip V-notched Brazilian disc (RV-BD) specimens made of PMMA (see Figure 7.2), were used by Ayatollahi and Torabi [16] for the fracture failure investigation of mixed-mode notches. In Figure 7.2, β is the angle between the loading direction and the notch bisector line and the parameters D, $d/2$ and P are the disc diameter, the notch length and the applied compressive load, respectively. When the direction of applied load P is along the notch bisector line (i.e., $\beta = 0$), the upper and the lower corners of the rhombic hole are subjected to pure Mode I deformation. If the angle β is not zero, the notch is subjected to mixed Mode I/II loading. When the angle β gradually increases, the loading conditions change from pure Mode I towards pure Mode II. For a specific angle called β_{II}, pure Mode II deformation is achieved. The Mode II loading angle β_{II} is always less than 90° and depends on the notch length and its opening angle and also on the notch tip radius.

The analyses by Ayatollahi and Torabi [16] indicated that for a constant notch opening angle 2α, the angle β_{II} is not significantly affected by the notch tip radius. The finite element calculations for the RV-BD specimens of $d/D = 0.5$ showed that

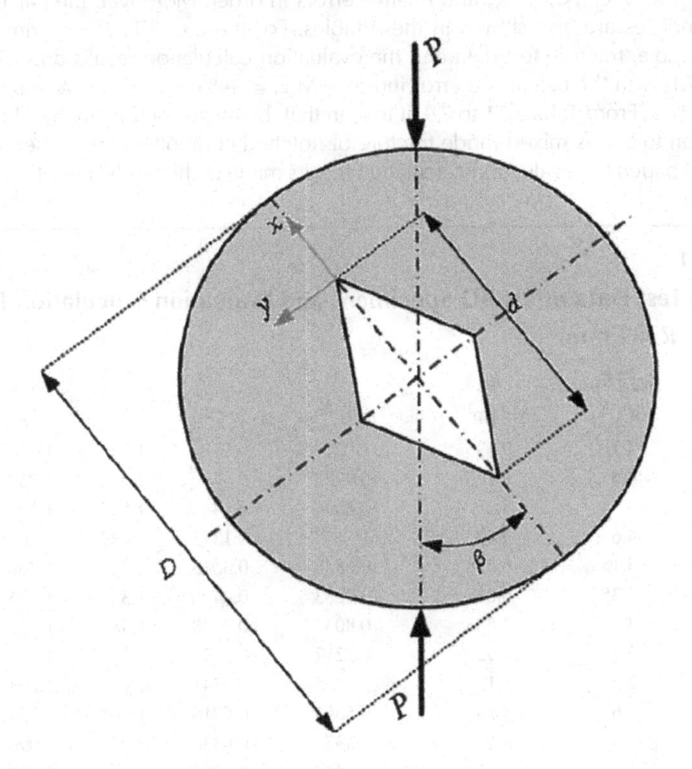

FIGURE 7.2 Schematic of geometry of the RV-BD specimen.

Source: (Data from Ayatollahiand Torab 16).

β_{II} depended on the notch opening angle and was about 25°, 30° and 35° for the notch opening angles 2α = 30°, 60° and 90°, respectively.

In Ref.16, for all the RV-BD specimens, the disc diameter (*D*), the notch length (*d*/2) and the thickness were 80 mm, 20 mm and 10 mm.

To study the effects of the notch opening angle and the notch tip radius on the fracture behavior of the RV-BD specimens, three values of notch opening angle 2α = 30°, 60°, 90° and three values of notch radius R = 1, 2 and 4 mm were considered for preparing the specimens. A total number of 162 mixed-mode fracture tests were performed for various notch geometry parameters and different loading angles β from 0 (pure Mode I) to β_{II} (pure Mode II). For 2α = 30°, experiments were performed according to the loading angles (β) equal to 0°, 5°, 10°, 15°, 20°, 25°. Similarly, for 2α = 60° and 90°, fracture tests were conducted for various angles (β) of 0°, 5°, 10°, 15°, 25°, 30° and 0°, 5°, 10°, 15°, 25°, 35°, respectively. For each geometry shape and loading angle at least two fracture tests were performed.

Fracture test data, including K_{Iv}/K_{Ivc}, K_{IIv}/K_{Ivc} and K_{Ivc}, of all cases reported by Ayatollahi and Torabi [16] are shown in Tables 7.1 to 7.9. At the same time, the evaluation calculation results made in this study using the empirical failure equation to assess mixed-mode fracture of notched components proposed in the present chapter are also given in these tables: including the computations of K_{IIvc}, K_{IIv}, K_{IIv}/K_{IIvc}, ξ, k_e, k_R and relative errors in order. Moreover, the calculated error indexes are also shown in these tables. For the 2α =30°, R = 1 mm case, for example, fracture test data and the evaluation calculation results are listed in Table 7.1, with the calculated error indexes: M.E. = −4%, M.A.E. = 6% and E.R. = [−18,8.1]%. From Tables 7.1 to 7.9, it is seen that, by means of the empirical failure equation to assess mixed-mode fracture of notched components proposed in the present paper, the evaluation calculation results made in this study are satisfactory.

TABLE 7.1
Fracture Test Data of RV-BD Specimens and Evaluation Calculation Results (2α =30°, R = 1 mm)

K_{Iv}/K_{Ivc}	K_{IIv}/K_{Ivc} ($m^{\lambda_1-\lambda_2}$)	K_{IIv} ($MPam^{1-\lambda_2}$)	K_{IIv}/K_{IIvc}	ξ	k_e	k_R	R.E. (%)
0	4.7	9.635	1.0217	0	1.77	1.732	2.1
0	4.5	9.225	0.9783	0	1.694	1.732	−2.2
0.22	4	8.2	0.8696	0.1445	1.522	1.6	−5.1
0.24	4.6	9.43	1	0.1373	1.749	1.606	8.1
0.375	3.15	6.458	0.6848	0.3015	1.244	1.468	−18
0.46	3.35	6.868	0.7283	0.3426	1.343	1.435	−6.9
0.5	3.7	7.585	0.8043	0.3378	1.48	1.439	2.8
0.614	2.4	4.92	0.5217	0.562	1.093	1.272	−16.4
0.66	2.5	5.125	0.5435	0.5741	1.15	1.264	−9.9
0.7	2.6	5.33	0.5652	0.5816	1.203	1.258	−4.6
0.88	1.55	3.178	0.337	0.8334	1.056	1.096	−3.8
0.88	1.57	3.219	0.3413	0.8301	1.06	1.098	−3.6
0.884	1.57	3.219	0.3413	0.8313	1.063	1.097	−3.2
0.98	0	0	0	1	0.98	1	-2
1.02	0	0	0	1	1.02	1	2

Note: (*a*) 2α =30°, R = 1 mm, λ_1 = 0.5015, λ_2 = 0.5982, K_{Ivc} = 2.05 $MPam^{1-\lambda_1}$; (*b*) K_{IIvc} = 9.43 $MPam^{1-\lambda_2}$, calculated from ξ = 0; (*c*) M.E. = −4.0%, M.A.E. = 6.0%, E.R. = [−18,8.1]%.

TABLE 7.2
Fracture Test Data of RV-BD Specimens and Evaluation Calculation Results ($2\alpha = 30^{\circ}$, R = 2 mm)

$K_{\mathrm{Iv}} / K_{\mathrm{Ivc}}$	$K_{\mathrm{IIv}} / K_{\mathrm{Ivc}}$ ($\mathrm{m}^{\lambda_1-\lambda_2}$)	K_{IIv} ($\mathrm{MPam}^{1-\lambda_2}$)	$K_{\mathrm{IIv}} / K_{\mathrm{IIvc}}$	ξ	k_e	k_R	R.E. (%)
0	5.1	12.75	1.03	0	1.785	1.732	2.9
0	4.8	12	0.97	0	1.68	1.732	−3.1
0.28	4.4	11	0.889	0.1789	1.565	1.57	−0.3
0.29	4.5	11.25	0.909	0.1811	1.601	1.568	2.1
0.428	3.08	7.7	0.622	0.3691	1.16	1.414	−22
0.44	3.15	7.875	0.636	0.3707	1.187	1.413	−19.1
0.52	3.65	9.125	0.737	0.3771	1.379	1.408	−2.1
0.656	2.33	5.825	0.471	0.6269	1.046	1.227	−17.3
0.68	2.5	6.25	0.505	0.6137	1.108	1.236	−11.6
0.734	2.7	6.75	0.545	0.6135	1.196	1.237	−3.4
0.84	1.55	3.875	0.313	0.8401	1	1.092	−9.2
0.88	1.61	4.025	0.325	0.8422	1.045	1.091	−4.4
0.97	1.8	4.5	0.364	0.8387	1.157	1.093	5.5
0.97	0	0	0	1	0.97	1	−3.1
0.98	0	0	0	1	0.98	1	−2
1.08	0	0	0	1	1.08	1	7.4

Note: (*a*) $2\alpha = 30^{\circ}$, $R = 2$ mm, $\lambda_1 = 0.5015$, $\lambda_2 = 0.5982$, $K_{\mathrm{Ivc}} = 2.5\ \mathrm{MPam}^{1-\lambda_1}$; (*b*) $K_{\mathrm{IIvc}} = 12.375\ \mathrm{MPam}^{1-\lambda_2}$, calculated from $\xi = 0$; (*c*) M.E. = −5.0%, M.A.E. = 7.2%, E.R. = [−22,7.4]%.

TABLE 7.3
Fracture Test Data of RV-BD Specimens and Evaluation Calculation Results ($2\alpha = 30^{\circ}$, R = 4 mm)

$K_{\mathrm{Iv}} / K_{\mathrm{Ivc}}$	$K_{\mathrm{IIv}} / K_{\mathrm{Ivc}}$ ($\mathrm{m}^{\lambda_1-\lambda_2}$)	K_{IIv} ($\mathrm{MPam}^{1-\lambda_2}$)	$K_{\mathrm{IIv}} / K_{\mathrm{IIvc}}$	ξ	k_e	k_R	R.E. (%)
0	4.95	15.84	0.937	0	1.623	1.732	−6.7
0	5.4	17.28	1.022	0	1.771	1.732	2.2
0	5.5	17.6	1.041	0	1.804	1.732	4
0.28	4.95	15.84	0.937	0.17	1.647	1.578	4.2
0.282	5	16	0.947	0.1695	1.664	1.578	5.2
0.54	3.6	11.52	0.682	0.4159	1.298	1.378	−6.2
0.56	3.8	12.16	0.72	0.4099	1.366	1.383	−1.2
0.636	4.2	13.44	0.795	0.4192	1.517	1.376	9.3
0.69	2.05	6.56	0.388	0.7162	0.963	1.169	−21.3
0.72	2.1	6.72	0.398	0.7226	0.996	1.165	−16.9
0.79	2.2	7.04	0.417	0.7384	1.07	1.155	−7.9
0.84	1.5	4.8	0.284	0.8629	0.973	1.078	−10.8
0.88	1.6	5.12	0.303	0.8589	1.025	1.081	−5.5
0.97	1.8	5.76	0.341	0.8542	1.136	1.083	4.6
0.97	0	0	0	1	0.97	1	−3.1
0.98	0	0	0	1	0.98	1	−2
1.08	0	0	0	1	1.08	1	7.4

Note: (*a*) $2\alpha = 30^{\circ}$, $R = 4$ mm, $\lambda_1 = 0.5015$, $\lambda_2 = 0.5982$, $K_{\mathrm{Ivc}} = 3.2\ \mathrm{MPam}^{1-\lambda_1}$; (*b*) $K_{\mathrm{IIvc}} = 16.9\ \mathrm{MPam}^{1-\lambda_2}$, calculated from $\xi = 0$; (*c*) M.E. = −2.6%, M.A.E. = 7.0%, E.R. = [−21.3,9.3]%.

TABLE 7.4

Fracture Test Data of RV-BD Specimens and Evaluation Calculation Results ($2\alpha = 60^{\circ}$, R = 1 mm)

K_{Iv} / K_{Ivc}	K_{IIv} / K_{Ivc} ($m^{\lambda_1-\lambda_2}$)	K_{IIv} ($MPam^{1-\lambda_2}$)	K_{IIv} / K_{IIvc}	ξ	k_e	k_R	R.E. (%)
0	11	24.2	1.048	0	1.815	1.732	4.5
0	11	24.2	1.048	0	1.815	1.732	4.5
0	9.5	20.9	0.905	0	1.567	1.732	−10.5
0.236	9	19.8	0.857	0.157	1.503	1.589	−5.7
0.25	9.9	21.78	0.943	0.1513	1.652	1.594	3.5
0.6	6.4	14.08	0.61	0.4941	1.214	1.32	−8.7
0.636	6.8	14.96	0.648	0.4932	1.289	1.321	−2.4
0.7	7.38	16.236	0.703	0.4985	1.404	1.317	6.2
0.738	4.6	10.12	0.438	0.6972	1.059	1.181	−11.6
0.74	4.68	10.296	0.446	0.692	1.069	1.184	−10.7
0.785	6	13.2	0.571	0.6214	1.263	1.231	2.5
0.9	2.5	5.5	0.238	0.9091	0.99	1.051	−6.2
0.975	2.6	5.72	0.248	0.9154	1.065	1.048	1.6
0.98	2.62	5.764	0.25	0.915	1.071	1.048	2.2
0.99	0	0	0	1	0.99	1	−1
1.01	0	0	0	1	1.01	1	1
1.07	0	0	0	1	1.07	1	6.5

Note: (*a*) $2\alpha = 60^{\circ}$, $R = 1$ mm, $\lambda_1 = 0.5122$, $\lambda_2 = 0.7309$, $K_{Ivc} = 2.2$ $MPam^{1-\lambda_1}$; (*b*) $K_{IIvc} = 23.1$ $MPam^{1-\lambda_2}$, calculated from $\xi = 0$; (*c*) M.E. = −1.4%, M.A.E. = 5.3%, E.R. = [−11.6,6.5]%.

TABLE 7.5

Fracture Test Data of RV-BD Specimens and Evaluation Calculation Results ($2\alpha = 60^{\circ}$, R = 2 mm)

K_{Iv} / K_{Ivc}	K_{IIv} / K_{Ivc} ($m^{\lambda_1-\lambda_2}$)	K_{IIv} ($MPam^{1-\lambda_2}$)	K_{IIv} / K_{IIvc}	ξ	k_e	k_R	R.E. (%)
0	12	35.16	1.04	0	1.8	1.73	3.9
0	11.7	34.28	1.01	0	1.76	1.73	1.4
0	11	32.23	0.95	0	1.65	1.73	−4.9
0.21	9.4	27.54	0.81	0.1472	1.43	1.6	−12
0.22	9.6	28.13	0.83	0.1509	1.46	1.59	−9.3
0.56	7.2	21.1	0.62	0.46	1.22	1.35	−10.5
0.634	7.9	23.15	0.68	0.4714	1.35	1.34	0.6
0.66	8.4	24.61	0.73	0.4637	1.42	1.34	5.7
0.69	4.3	12.6	0.37	0.7302	0.95	1.16	−22.7
0.76	4.7	13.77	0.41	0.7328	1.04	1.16	−11.7
0.785	4.9	14.36	0.42	0.7296	1.08	1.16	−7.8
0.9	3.12	9.14	0.27	0.887	1.01	1.06	−4.9
0.908	3.16	9.26	0.27	0.8863	1.02	1.06	−3.9
0.91	3.185	9.33	0.28	0.8852	1.03	1.07	−3.6
0.98	0	0	0	1	0.98	1	−2
0.99	0	0	0	1	0.99	1	−1
1.03	0	0	0	1	1.03	1	2.9

Note: (*a*) $2\alpha = 60^{\circ}$, $R = 2$ mm, $\lambda_1 = 0.5122$, $\lambda_2 = 0.7309$, $K_{Ivc} = 2.93$ $MPam^{1-\lambda_1}$; (*b*) $K_{IIvc} = 33.8$ $MPam^{1-\lambda_2}$, calculated from $\xi = 0$; (*c*) M.E. = −4.7%, M.A.E. = 6.4%, E.R. = [−22.7,5.7]%.

TABLE 7.6
Fracture Test Data of RV-BD Specimens and Evaluation Calculation Results ($2\alpha = 60^\circ$, R = 4 mm)

K_{Iv} / K_{Ivc}	K_{IIv} / K_{Ivc} ($m^{\lambda_1-\lambda_2}$)	K_{IIv} ($MPam^{1-\lambda_2}$)	K_{IIv} / K_{IIvc}	ξ	k_e	k_R	R.E. (%)
0	13.2	46.464	1.05	0	1.82	1.73	5.1
0	12.4	43.648	0.99	0	1.71	1.73	−1
0	12	42.24	0.96	0	1.66	1.73	−4.4
0.236	12	42.24	0.96	0.1408	1.68	1.6	4.3
0.25	13	45.76	1.04	0.1378	1.81	1.61	11.5
0.58	7.75	27.28	0.62	0.4761	1.22	1.33	−9.4
0.64	8.5	29.92	0.68	0.4783	1.34	1.33	0.5
0.64	8.8	30.976	0.7	0.4656	1.37	1.34	2.4
0.66	5.2	18.304	0.42	0.6763	0.98	1.19	−22.4
0.69	5.25	18.48	0.42	0.689	1	1.19	−18.5
0.86	3	10.56	0.24	0.9007	0.95	1.06	−10.6
0.94	3.4	11.968	0.27	0.8944	1.05	1.06	−0.8
1	3.8	13.376	0.3	0.8853	1.13	1.07	5.7
0.99	0	0	0	1	0.99	1	−1
1.01	0	0	0	1	1.01	1	1

Note: (*a*) $2\alpha = 60^\circ$, R = 4 mm, $\lambda_1 = 0.5122$, $\lambda_2 = 0.7309$, $K_{Ivc} = 3.52 MPam^{1-\lambda_1}$; (*b*) $K_{IIvc} = 44.1\ MPam^{1-\lambda_2}$, calculated from $\xi = 0$; (*c*) M.E. = −2.5%, M.A.E. = 6.6%, E.R. = [−22.4,5.7]%.

TABLE 7.7
Fracture Test Data of RV-BD Specimens and Evaluation Calculation Results $2\alpha = 90^\circ$, R = 1 mm)

K_{Iv} / K_{Ivc}	K_{IIv} / K_{Ivc} ($m^{\lambda_1-\lambda_2}$)	K_{IIv} ($MPam^{1-\lambda_2}$)	K_{IIv} / K_{IIvc}	ξ	k_e	k_R	R.E. (%)
0	21.5	56.975	1	0	1.73	1.73	0
0.58	13	34.45	0.6	0.4846	1.2	1.33	−10.9
0.68	16	42.4	0.74	0.4668	1.46	1.34	8
0.68	10.5	27.825	0.49	0.6267	1.09	1.23	−13.1
0.75	11	29.15	0.51	0.6462	1.16	1.21	−4.6
0.835	4.98	13.197	0.23	0.9014	0.93	1.06	−14
0.845	3.05	8.0825	0.14	0.9603	0.88	1.02	−16.1
0.91	3.1	8.215	0.14	0.9644	0.94	1.02	−8.1
0.98	5.08	13.462	0.24	0.9228	1.06	1.04	1.8
0.99	0	0	0	1	0.99	1	−1
1.01	0	0	0	1	1.01	1	1
1	0	0	0	1	1	1	0

Note: (*a*) $2\alpha = 90^\circ$, R = 1 mm, $\lambda_1 = 0.5445$, $\lambda_2 = 0.9085$, $K_{Ivc} = 2.65\ MPam^{1-\lambda_1}$; (*b*) $K_{IIvc} = 57\ MPam^{1-\lambda_2}$,, calculated from $\xi = 0$; (*c*) M.E. = −4.8%, M.A.E. = 6.6%, E.R. = [−16.1,8]%.

TABLE 7.8
Fracture Test Data of RV-BD Specimens and Evaluation Calculation Results ($2\alpha = 90^\circ$, R = 2 mm)

K_{Iv} / K_{Ivc}	K_{IIv} / K_{Ivc} ($m^{\lambda_1-\lambda_2}$)	K_{IIv} ($MPam^{1-\lambda_2}$)	K_{IIv} / K_{IIvc}	ξ	k_e	k_R	R.E. (%)
0	23	82.8	1.08	0	1.87	1.73	7.2
0	22	79.2	1.03	0	1.79	1.73	3
0	19	68.4	0.89	0	1.54	1.73	−12.3
0.636	14	50.4	0.66	0.4883	1.3	1.32	−1.7
0.66	15	54	0.7	0.4765	1.39	1.33	3.8
0.75	11	39.6	0.52	0.6431	1.17	1.22	−4.3
0.82	11.7	42.12	0.55	0.6534	1.25	1.21	3.6
0.84	5	18	0.23	0.9004	0.93	1.06	−13.2
0.85	3	10.8	0.14	0.9613	0.88	1.02	−15.5
0.902	3.1	11.16	0.15	0.9632	0.94	1.02	−9
0.96	6	21.6	0.28	0.8918	1.08	1.06	1.4
0.99	0	0	0	1	0.99	1	−1
1.01	0	0	0	1	1.01	1	1
1	0	0	0	1	1	1	0
			0				

Note: (*a*) $2\alpha = 90^\circ$, $R = 2$ mm, $\lambda_1 = 0.5445$, $\lambda_2 = 0.9085$, $K_{Ivc} = 3.6$ $MPam^{1-\lambda_1}$; (*b*) $K_{IIvc} = 76.8$ $MPam^{1-\lambda_2}$, calculated from $\xi = 0$; (*c*) M.E. = −2.6%, M.A.E. = 5.5%, E.R. = [−15.5,7.2]%.

TABLE 7.9
Fracture Test Data of RV-BD Specimens and Evaluation Calculation Results ($2\alpha = 90^\circ$, R = 4 mm)

K_{Iv} / K_{Ivc}	K_{IIv} / K_{Ivc} ($m^{\lambda_1-\lambda_2}$)	K_{IIv} ($MPam^{1-\lambda_2}$)	K_{IIv} / K_{IIvc}	ξ	k_e	k_R	R.E. (%)
0	22.8	110.6	1	0	1.73	1.73	0
0.64	14.2	68.87	0.62	0.51	1.25	1.31	−4.3
0.836	12	58.2	0.53	0.676	1.24	1.19	3.4
0.9	6	29.1	0.26	0.892	1.01	1.06	−5.2
0.94	3	14.55	0.13	0.972	0.97	1.02	−5
0.96	6.6	32.01	0.29	0.886	1.08	1.06	1.7
0.97	3.2	15.52	0.14	0.97	1	1.02	−1.7
0.99	0	0	0	1	0.99	1	−1
1.01	0	0	0	1	1.01	1	1
1	0	0	0	1	1	1	0

Note: (*a*) $2\alpha = 90^\circ$, $R = 4$ mm, $\lambda_1 = 0.5445$, $\lambda_2 = 0.9085$, $K_{Ivc} = 4.85$ $MPam^{1-\lambda_1}$; (*b*) $K_{IIvc} = 110.58$ $MPam^{1-\lambda_2}$, calculated from $\xi = 0$; (*c*) M.E. = −1.1%, M.A.E. = 2.3%, E.R. = [−5.2,3.4]%.

EXAMPLE 2: MIXED-MODE FRACTURE OF U-NOTCHES

Here, the test data of mixed-mode fracture of U-notches reported by Berto et al. [22] are taken to check the accuracy of the empirical failure equation to assess mixed-mode fracture of notched components proposed in the present chapter.

The material chosen for the experimental programme by Berto et al. [22] was polymethyl-methacrylate (PMMA) tested at −60°C.

The geometries of the specimens are shown in Figure 7.3. In all the specimens, the thickness, B, was 14 mm, the size, W, was 28 mm and the notch height, a, was 14 mm.

The influence of the notch root radius, R, was examined by introducing seven different root radii. In the notch series (characterized by $b = 9$, 18, 27 and 36 mm), R was 0, 0.2, 0.3, 0.5, 1.0, 2.0 and 4.0 mm. In the two series with $b = -3$ and 3 mm, the geometry with $R = 0.2$ mm was not considered.

For mixed-mode U-notches, the failure evaluation calculation can be carried out employing the fracture equation of mixed cracks reported in Ref. [13] because the notch stress intensity factors, K_{IU} and K_{IIU}, have the same unit, $MPam^{0.5}$. According to the empirical failure equation of mixed cracks reported in Ref. [13], K_{IUc} and K_{IIUc} must be known prior in order to perform the failure assessment of mixed-mode U-notches. Taking into account that test data of K_{IUc} and K_{IIUc} are not given in the Ref.22, the values of K_{IUc} and K_{IIUc} can be estimated by means of numerically fitting techniques and formulas (10) to (14).

Fracture test data, including K_{IU} and K_{IIU}, of series specimens and evaluation calculation process data, including estimated results of K_{IUc} and K_{IIUc} and computation results of ξ, K_e, K_R and relative errors, are shown in Tables 7.10 to 7.16, which correspond to $R = 4$ mm, 2 mm, 1 mm, 0.5 mm, 0.3 mm, 0.2 mm, and 0 mm, respectively. From Tables 7.10 to 7.16, it is seen that the failure evaluation calculation results are very satisfactory. For $R = 4$ mm case, for example, the calculated error indexes are: M.E. = −0.2%, M.A.E. = 3% and E.R. = [−5,5.7]% (see Table 7.10), which show that the calculated values of the SIPs, K_e, obtained employing the fracture test data are in excellent agreement with the predicted values of the MIPs, K_R, obtained employing the fracture toughness, K_{IUc}, K_{IIUc} and the multiaxial parameter ξ.

It is necessary here to illustrate that, for the R = 0 mm case, $K_{Ic} = 1.7\ MPam^{0.5}$ is measured experimentally and $K_{IIc} = 1.155\ MPam^{0.5}$ is calculated employing an approach of determining K_{IIc} proposed in Ref. [13].

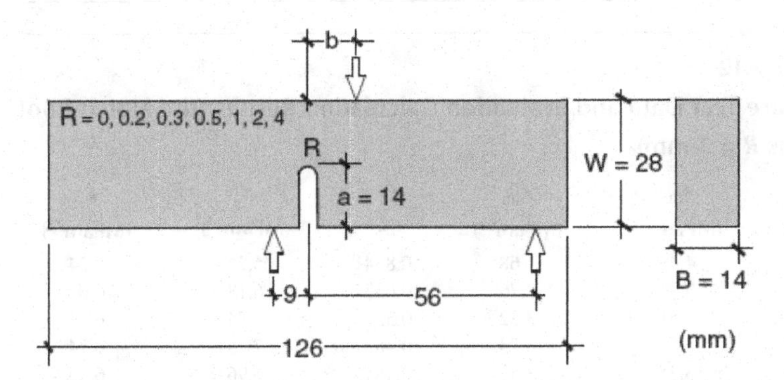

FIGURE 7.3 Geometry and load condition. Data in mm ($b = -3$, 3,9, 18, 27 and 36 mm).

Source: (Data from Berto et al.[22])

TABLE 7.10
Fracture Test Data and Evaluation Calculation Results (the Notch Root Radius R = 4 mm)

b (mm)	K_{IU} ($MPam^{0.5}$)	K_{IIU} ($MPam^{0.5}$)	ξ	K_e ($MPam^{0.5}$)	K_R ($MPam^{0.5}$)	R.E. (%)
−3	7.9	2.81	0.8514	9.28	9.64	−3.8
3	7.38	6.83	0.5293	13.94	13.14	5.7
9	6.2	7.91	0.4123	15.04	14.71	2.2
18	5.88	8.03	0.3894	15.1	15.04	0.4
27	5.57	7.62	0.3888	14.33	15.05	−5
36	5.82	7.91	0.391	14.89	15.02	−0.9

Note: (a) K_{IU} and K_{IIU} are from Ref.22; (b) K_{IUc} = 8.35 $MPam^{0.5}$ and K_{IIUc} = 12.64 $MPam^{0.5}$, which are calculated employing numerically fitting techniques and formulas (10) to (14); (c) M.E. = −0.2%, M.A.E. = 3.0%, E.R. = [−5,5.7]%.

TABLE 7.11
Fracture Test Data and Evaluation Calculation Results (the Notch Root Radius R = 2 mm)

b (mm)	K_{IU} ($MPam^{0.5}$)	K_{IIU} ($MPam^{0.5}$)	ξ	K_e ($MPam^{0.5}$)	K_R ($MPam^{0.5}$)	R.E. (%)
−3	6.14	2.31	0.8378	7.33	7.57	−3.3
3	5.48	4.34	0.5891	9.3	8.9	4.4
9	4.87	5.19	0.4763	10.22	9.57	6.4
18	4.33	5.06	0.4429	9.78	9.78	−0.1
27	4.05	4.71	0.4447	9.11	9.77	−7.3
36	4.3	4.96	0.4476	9.61	9.75	−1.5

Note: (a) K_{IU} and K_{IIU} are from Ref.22; (b) K_{IUc} = 6.81 $MPam^{0.5}$ and K_{IIUc} = 7.53 $MPam^{0.5}$, which are calculated employing numerically fitting techniques and formulas (10) to (14); (c) M.E. = −0.2%, M.A.E. = 3.8%, E.R. = [−7.3,6.4]%.

TABLE 7.12
Fracture Test Data and Evaluation Calculation Results (the Notch Root Radius R = 1 mm)

b(mm)	K_{IU} ($MPam^{0.5}$)	K_{IIU} ($MPam^{0.5}$)	ξ	K_e ($MPam^{0.5}$)	K_R ($MPam^{0.5}$)	R.E. (%)
−3	4.4	1.68	0.8341	5.28	5.24	0.6
3	3.91	2.76	0.6331	6.18	6.06	1.9
9	3.51	3.32	0.521	6.74	6.56	2.6
18	3.57	3.73	0.4837	7.38	6.74	8.7
27	3.35	3.52	0.4816	6.96	6.75	2.9
36	3.19	3.29	0.4885	6.53	6.72	−2.9

Note: (a) K_{IU} and K_{IIU} are from Ref.22; (b) K_{IUc} = 4.66 $MPam^{0.5}$ and K_{IIUc} = 5.51 $MPam^{0.5}$, which are calculated employing numerically fitting techniques and formulas (10) to (14); (c) M.E. = 2.3%, M.A.E. = 3.3%, E.R. = [−2.9,8.7]%.

TABLE 7.13
Fracture Test Data and Evaluation Calculation Results (the Notch Root Radius R = 0.5 mm)

b (mm)	K_{IU} (MPam$^{0.5}$)	K_{IIU} (MPam$^{0.5}$)	ξ	K_e (MPam$^{0.5}$)	K_R (MPam$^{0.5}$)	R.E. (%)
−3	3.32	1.29	0.8296	4	4.17	−4.3
3	3.33	2.2	0.658	5.06	4.54	10.4
9	2.75	2.4	0.5517	4.98	4.78	4.2
18	2.56	2.47	0.5135	4.99	4.87	2.4
27	2.59	2.53	0.5088	5.09	4.88	4.2
36	2.43	2.37	0.5094	4.77	4.88	−2.2

Note: (a) K_{IU} and K_{IIU} are from Ref.22; (b) K_{IUc} = 3.84 MPam$^{0.5}$ and K_{IIUc} = 3.61 MPam$^{0.5}$, which are calculated employing numerically fitting techniques and formulas (10) to (14); (c) M.E. = 2.4%, M.A.E. = 4.6%, E.R. = [−4.3,10.4]%.

TABLE 7.14
Fracture Test Data and Evaluation Calculation Results (the Notch Root Radius R = 0.3 mm)

b (mm)	K_{IU} (MPam$^{0.5}$)	K_{IIU} (MPam$^{0.5}$)	ξ	K_e (MPam$^{0.5}$)	K_R (MPam$^{0.5}$)	R.E. (%)
−3	2.79	1.13	0.8187	3.41	3.51	−3
3	2.79	1.84	0.6587	4.24	3.89	8.1
9	2.4	2.02	0.5657	4.24	4.13	2.6
18	2.31	2.18	0.5219	4.43	4.25	4
27	2.37	2.21	0.5264	4.5	4.24	5.8
36	2.37	2.21	0.5264	4.5	4.24	5.8

Note: (a) K_{IU} and K_{IIU} are from Ref.22; (b) K_{IUc} = 3.12 MPam$^{0.5}$ and K_{IIUc} = 3.44 MPam$^{0.5}$, which are calculated employing numerically fitting techniques and formulas (10) to (14); (c) M.E. = 3.9%, M.A.E. = 4.9%, E.R. = [−3,8.1]%.

TABLE 7.15
Fracture Test Data and Evaluation Calculation Results (the Notch Root Radius R = 0.2 mm)

b (mm)	K_{IU} (MPam$^{0.5}$)	K_{IIU} (MPam$^{0.5}$)	ξ	K_e (MPam$^{0.5}$)	K_R (MPam$^{0.5}$)	R.E. (%)
9	1.61	1.33	0.5729	2.81	2.83	−0.6
18	1.52	1.39	0.5339	2.85	2.98	−4.5
27	1.58	1.45	0.5325	2.97	2.98	−0.5
36	1.68	1.52	0.5379	3.12	2.96	5.2

Note: (a) K_{IU} and K_{IIU} are from Ref.22; (b) K_{IUc} = 1.62 MPam$^{0.5}$ and K_{IIUc} = 3.45 MPam$^{0.5}$, which are calculated employing numerically fitting techniques and formulas (10) to (14); (c) M.E. = −0.1%, M.A.E. = 2.7%, E.R. = [−4.5,5.2]%.

TABLE 7.16
Fracture Test Data and Evaluation Calculation Results (the Notch Root Radius R = 0 mm)

b (mm)	K_{IU} (MPam$^{0.5}$)	K_{IIU} (MPam$^{0.5}$)	ξ	K_e (MPam$^{0.5}$)	K_R (MPam$^{0.5}$)	R.E. (%)
−3	1.74	0.4	0.9291	1.87	1.72	8.2
3	1.65	0.58	0.8541	1.93	1.74	9.9
9	1.34	0.61	0.7853	1.71	1.76	−3.2
18	1.35	0.68	0.7535	1.79	1.77	1.2
27	1.36	0.68	0.7559	1.8	1.77	1.7
36	1.29	0.65	0.7534	1.71	1.77	−3.3

Note: (a) K_{IU} and K_{IIU} are from Ref.22; (b) K_{Ic} = 1.7 MPam$^{0.5}$; (c) K_{IIc} = 1.155 MPam$^{0.5}$, calculated by means of an approach of determining K_{IIc} employing fracture test data near the Mode II region proposed in Ref.20; (c) M.E. = 2.4%, M.A.E. = 4.6%, E.R. = [−3.3,9.9]%.

Example 3: Mixed-Mode Fracture of U-Notched Graphite Plates

In 2012, fracture of U-notched plates under mixed-mode loading was studied by Berto *et al.* [23]. The material used was a commercial isostatic graphite particularly used in mechanical applications for its high performances. The samples used were plates with a central blunt U-notch, as shown in Figure 7.4. Different notch radii, ρ, and notch angles, β, were used in order to investigate their influence on the failure. All the specimens are subjected to tensile load. By varying the notch angles, β, different mode-mixity can be produced. If the notch angle is equal to zero (see Figure 7.4), the notches are subjected to pure Mode I, while varying β the loading condition changes from pure Mode I towards mixed Mode I + II. For all the tested graphite specimens, the width, the distance between the tip of the notches and the thickness were 50 mm, 10 mm and 10 mm, respectively. Five values of notch radius ρ = 0.25, 0.5, 1, 2, 4 mm were considered for manufacturing the test specimens so that the effects of the notch tip radius on mixed-mode fracture of the graphite specimens are studied. With the aim to obtain different mode-mixity, four values of the angle β were considered to be $\beta = 0^\circ$, 30°, 45° and 60°.

It is necessary here to illustrate that because the values of K_{IUc} for all cases are known, K_{IIUc} can be estimated by means of an approach of determining K_{IIc} proposed in Ref.20.

Fracture test data, including P_f, K_{IU} / σ_{nom} and K_{IIU} / σ_{nom}, and evaluation calculation process data, including computation results of K_{IU}, K_{IIU} and relative errors, are shown in Tables 7.17 to 7.21, which correspond to ρ = 0.25 mm, 0.5 mm, 1 mm, 2 mm and 4 mm, respectively. From Tables 7.17 to 7.21, it is seen that the evaluation calculation results are satisfactory. For the ρ = 0.25 mm case, for example, the calculated error indexes are: M.E. = 2.6%, M.A.E. = 2.7% and E.R. = [−0.3,5.7]% (see Table 7.17), which shows that the calculated values of the SIPs, K_e, obtained employing the fracture test data, are in excellent agreement with the predicted values of the MIPs, K_R, obtained employing the fracture toughness, K_{IUc}, K_{IIUc} and the multiaxial parameter ξ.

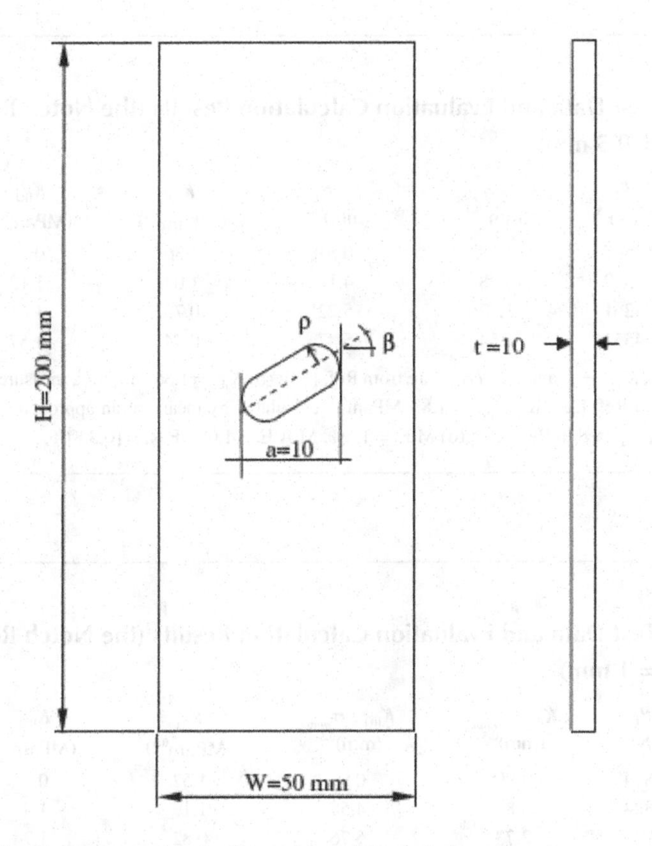

FIGURE 7.4 Geometry and main dimensions of the tested specimens.

Source: (Data from Berto *et al.*[23]).

TABLE 7.17
Fracture Test Data and Evaluation Calculation Results (the Notch Root Radius ρ = 0.25 mm)

β(°)	P_f (N)	K_{IU}/σ_{nom} (mm)$^{0.5}$	K_{IIU}/σ_{nom} (mm)$^{0.5}$	K_{IU} (MPam$^{0.5}$)	K_{IIU} (MPam$^{0.5}$)	R.E. (%)
0	4426	4.63	0	1.3	0	–0.3
30	4034	3.71	3.81	0.95	0.97	4.9
45	3927	2.71	4.89	0.67	1.21	5.7
60	3942	1.6	5.18	0.4	1.29	0

Note: (*a*) P_f, K_{IU}/σ_{nom} and K_{IIU}/σ_{nom} are from Ref. [23]; (b) K_{IUc} = 1.3 MPam$^{0.5}$, measured experimentally in Ref. [23]; (c) K_{IIUc} = 1.48 MPam$^{0.5}$, calculated by means of an approach of determining K_{IIc} proposed in Ref. [20]; (d) M.E. = 2.6%, M.A.E. = 2.7%, E.R. = [–0.3,5.7]%.

TABLE 7.18

Fracture Test Data and Evaluation Calculation Results (the Notch Root Radius ρ = 0.5 mm)

β(°)	P_f (N)	K_{IU}/σ_{nom} (mm)$^{0.5}$	K_{IIU}/σ_{nom} (mm)$^{0.5}$	K_{IU} (MPam$^{0.5}$)	K_{IIU} (MPam$^{0.5}$)	R.E. (%)
0	4505	4.78	0	1.36	0	0
30	4349	3.78	4.1	1.04	1.13	3.8
45	4261	2.72	5.22	0.73	1.41	2.5
60	4551	1.54	5.47	0.44	1.57	0

Note: (a) P_f, K_{IU}/σ_{nom} and K_{IIU}/σ_{nom} are from Ref. [23]; (b) K_{IUc} = 1.36 MPam$^{0.5}$, measured experimentally in Ref. [23]; (c) K_{IIUc} = 1.83 MPam$^{0.5}$, calculated by means of an approach of determining K_{IIc} proposed in Ref. [20]; (d) M.E. = 1.6%, M.A.E. = 1.6%, E.R. = [0,3.8]%.

TABLE 7.19

Fracture Test Data and Evaluation Calculation Results (the Notch Root Radius ρ = 1 mm)

β(°)	P_f (N)	K_{IU}/σ_{nom} (mm)$^{0.5}$	K_{IIU}/σ_{nom} (mm)$^{0.5}$	K_{IU} (MPam$^{0.5}$)	K_{IIU} (MPam$^{0.5}$)	R.E. (%)
0	4814	4.99	0	1.52	0	0
30	4824	3.89	4.58	1.19	1.4	9.7
45	4777	2.73	5.76	0.82	1.74	9.6
60	4779	1.44	5.94	0.44	1.8	0

Note: (a) P_f, K_{IU}/σ_{nom} and K_{IIU}/σ_{nom} are from Ref. [23]; (b) K_{IUc} = 1.52 MPam$^{0.5}$, measured experimentally in Ref. [23]; (c) K_{IIUc} = 2.04 MPam$^{0.5}$, calculated by means of an approach of determining K_{IIc} proposed in Ref. [20]; (d) M.E. = 4.8%, M.A.E. = 4.8%, E.R. = [0,9.7]%.

TABLE 7.20

Fracture Test Data and Evaluation Calculation Results (the Notch Root Radius ρ = 2 mm)

β(°)	P_f (N)	K_{IU}/σ_{nom} (mm)$^{0.5}$	K_{IIU}/σ_{nom} (mm)$^{0.5}$	K_{IU} (MPam$^{0.5}$)	K_{IIU} (MPam$^{0.5}$)	R.E. (%)
0	5516	5.31	0	1.85	0	0
30	5916	4.05	5.41	1.52	2.02	17.4
45	5603	2.73	6.71	0.97	2.38	13.5
60	5455	1.27	6.77	0.44	2.34	0

Note: (a) P_f, K_{IU}/σ_{nom} and K_{IIU}/σ_{nom} are from Ref. [23]; (b) K_{IUc} = 1.85 MPam$^{0.5}$, measured experimentally in Ref. [23]; (c) K_{IIUc} = 2.58 MPam$^{0.5}$, calculated by means of an approach of determining K_{IIc} proposed in Ref. [20]; (d) M.E. = 7.8%, M.A.E. = 7.8%, E.R. = [0,17.4]%.

TABLE 7.21
Fracture Test Data and Evaluation Calculation Results (the Notch Root Radius ρ = 4 mm)

β(°)	P_f (N)	K_{IU}/σ_{nom} (mm)$^{0.5}$	K_{IIU}/σ_{nom} (mm)$^{0.5}$	K_{IU} (MPam$^{0.5}$)	K_{IIU} (MPam$^{0.5}$)	R.E. (%)
0	6789	5.81	0	2.49	0	0.2
30	6886	4.3	6.85	1.87	2.98	15.1
45	6862	2.71	8.35	1.18	3.62	15.1
60	6749	0.97	8.17	0.41	3.49	0

Note: (a) P_f, K_{IU}/σ_{nom} and K_{IIU}/σ_{nom} are from Ref. [23]; (b) K_{IUc} = 2.49 MPam$^{0.5}$, measured experimentally in Ref. [23]; (c) K_{IIUc} = 3.73 MPam$^{0.5}$, calculated by means of an approach of determining K_{IIc} proposed in Ref. [20]; (d) M.E. = 7.6%, M.A.E. = 7.6%, E.R. = [0,15.1]%.

Example 4: Fracture Initiation at Sharp Notches Under Mild Mixed-Mode Loading

In 1997, Dunn et al. [24] carried out a companion experimental study to extract critical values of the Mode I and II stress intensities for a series of notched polymethyl methacrylate (PMMA) tensile and flexure specimens with notch angles of 90°, and further analyzed and tested a series of T-shaped structures containing 90° corners. Figure 7.5 shows geometry of the test specimens and notch details considered in Ref.24. By means of notched tensile specimens shown in Figure 7.5(a) and notched antisymmetric flexure specimens shown in Figure 7.5(b), the experimentally measured K_{Ivc} and K_{IIvc} are 0.769 $MPa \cdot m^{0.455}$ and 28.2 $MPa \cdot m^{0.0915}$, respectively.

For the uniaxially-loaded T-structure shown in Figure 7.5(c), the 90_ corner is subjected to mixed-mode loading. Dimensional considerations lead to the following forms for K_{Iv} and K_{IIv}:

$$K_{Iv} = \sigma^0 \cdot W_1^{1-\lambda_I} \cdot f_I(W_1/W_2) \tag{25}$$

$$K_{IIv} = \sigma^0 \cdot W_1^{1-\lambda_{II}} \cdot f_{II}(W_1/W_2) \tag{26}$$

where

$$\sigma^0 = \frac{P}{W_1 \cdot t} \tag{27}$$

and the order of the singularities λ_I and λ_{II} are 0.5445 and 0.9085, determined from the asymptotic analysis. By means of plane strain finite element calculations, the nondimensional factors $f_I(W_1/W_2)$ and $f_{II}(W_1/W_2)$ had been obtained in Ref. [24] (see Figure 7.6).

A series of tests with the TT-structures of Figure 7.6(c) were carried out in Ref. [24]. The tensile specimens are 100 mm long, W_2 = 50.8 mm and W_1 = 5.08 mm, 10.16 mm, 15.24 mm and 20.32 mm (W_1/W_2 = 0.1,0.2,0.3 and 0.4). Measured failure strengths for the tensile-loaded T-structures are shown in Table 7.22. Using the measured failure strengths, the notch stress intensity factors, K_{vI} and K_{vII}, are

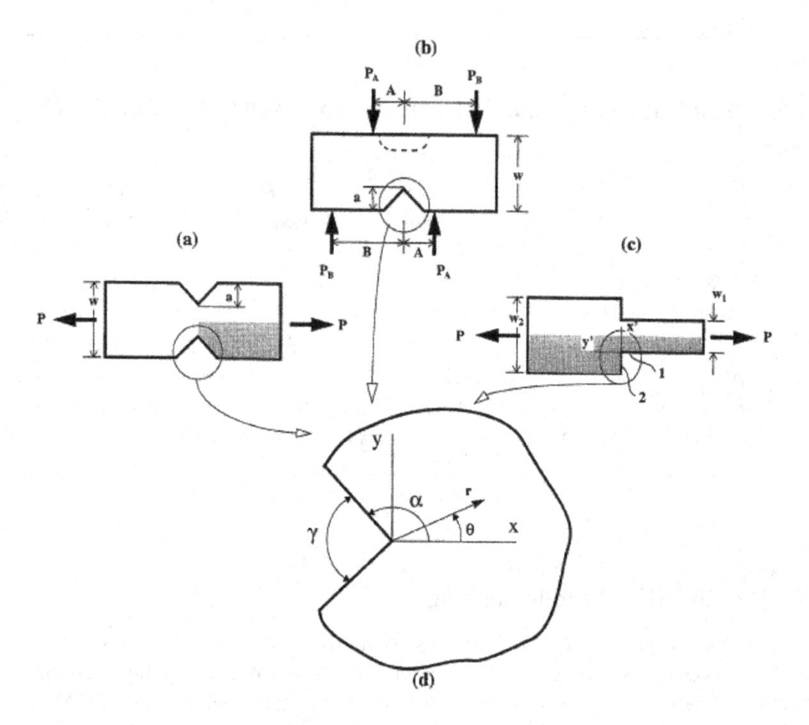

FIGURE 7.5 Geometry of the test specimens and notch details considered in Ref.24. Note the notch-tip coordinate system used for the asymptotic analysis. The thickness of all specimens is denoted by *t*.

Source: (Data from Dunn et al.[24])

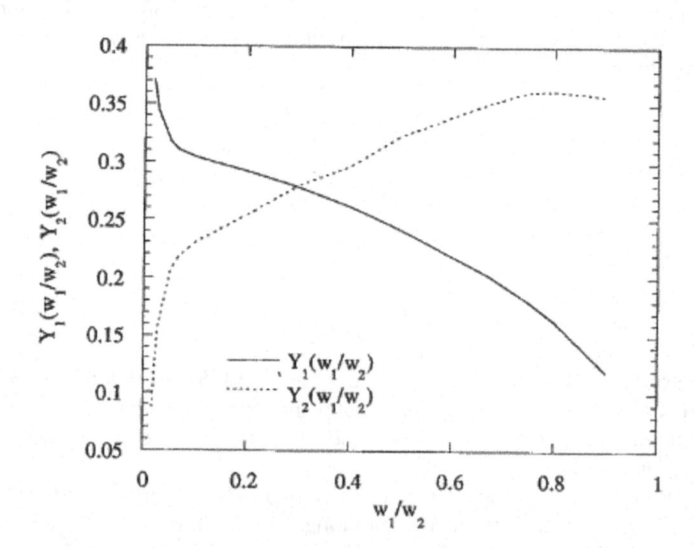

FIGURE 7.6 Dimensionless functions $f_I(W_1 / W_2)$ and Y $f_{II}(W_1 / W_2)$ for the TT-structure.

Source: (Data from Dunn *et al.*[24])

TABLE 7.22
Measured Failure Strengths and the Values of $f_I(W_1/W_2)$ and $f_{II}(W_1/W_2)$ for the Tensile-Loaded T-Structures

No.	W_1/W_2	$f_I(W_1/W_2)$	$f_{II}(W_1/W_2)$	W_1 (mm)	σ_f (MPa)
1	0.1	0.4	0.23	5.08	28.7
2	0.2	0.29	0.25	10.16	21.7
3	0.3	0.28	0.28	15.24	19.6
4	0.4	0.26	0.293	20.32	17.0

Note: (a) $f_I(W_1/W_2)$ and $f_{II}(W_1/W_2)$ are taken from Figure 7.6; (b) σ_f, the measured failure strengths, are from Ref.24 [4].

TABLE 7.23
Evaluation Calculation Process Data of the Tensile-Loaded T-Structures

No.	K_{Iv} $MPa \cdot m^{0.455}$	K_{IIv} $MPa \cdot m^{0.0915}$	$\bar{k}_{Iv}$	$\bar{k}_{IIv}$	ξ	k_e	k_R	R.E. (%)
1	0.75	2.95	0.975	0.105	0.983	0.992	1.009	−1.7
2	0.778	3.565	1.012	0.126	0.977	1.035	1.013	2.2
3	0.816	3.742	1.061	0.133	0.977	1.086	1.013	6.8
4	0.749	3.487	0.974	0.124	0.977	0.998	1.013	−1.5

Note: (*a*) K_{Iv} and K_{IIv} are calculated from formulas (25) and (26); (*b*) $K_{Ivc} = 0.:769\ MPa \cdot m^{0.455}$ and $K_{IIvc} = 28.2\ MPa \cdot m^{0.0915}$ are from Ref. [24]; (c) M.E. = 1.4%, M.A.E. = 3.0%, E.R. = [−1.7,6.8] %.

calculated from formulas (25) and (26) (see Table 7.23). At the same time, evaluation calculation process data, including $\bar{k}_{vI}$, $\bar{k}_{vII}$, ξ, k_e, k_R and relative errors are also listed in Table 7.23, with the calculated error indexes: M.E. = 1.4%, M.A.E. = 3.0% and E.R. = [−1.7,6.8]%, from which it is seen that the calculated values of the SIPs, K_e, obtained employing the fracture test data, are in excellent agreement with the predicted values of the MIPs, K_R, obtained employing the fracture toughness, K_{Ivc}, K_{IIvc} and the multiaxial parameter ξ.

7.5 FINAL COMMENTS

From this study and Refs. 18 and 20, it is found that the empirical failure equation at the multiaxial stresses state is not only simple in computation form but also highly accurate. What is the reason for it? It is due to the following respects:

1. The von Mises definition of the **stress intensity parameter** (for example, the definition of the SIP by formula (2) for multiaxial fatigue limit analysis) is a proper quantity to characterize the failure force.
2. A definition of the **material intensity parameter** (for example, the definition of the MIP by formula (6) for multiaxial fatigue limit analysis) is a proper material quantity to characterize the material failure resistance.

(3) The empirical failure equation at the multiaxial stresses state (for example, Eq.(7) for multiaxial fatigue limit analysis) originated from the multiaxial fatigue limit equation (1), which was both simple in computation form and highly accurate, falling within ±10% interval [18].
(4) The empirical failure equation at the multiaxial stresses state, for example, for fracture analysis for Mode I/II cracks, is obviously different from the failure equation obtained from the maximum $\sigma_{\theta\theta}$ theory originally proposed by Erdogan and Sih [9] and the strain energy density factor theory by Sih [10]. Employing the two well-known fracture criteria [9,10], the failure equation can be expressed mathematically to be

$$f(K_I, K_{II}, K_{Ic}) = 0 \tag{28}$$

that is to say that fracture failure of Mode I/II cracks is controlled by the fracture toughness K_{Ic} of Mode I cracks. For the materials (for example, high-or medium-strength steels as well as nodular cast iron reported by Gao *et al.*[17] whose crack extension resistance varies seriously with the increase of the K_{II} / K_I ratio), the failure load calculated by using the failure equation (28) is obviously different from that measured experimentally near the region of pure II cracks.

For rounded V-notches, according to the two well-known local approaches (TCD, Theory of Critical Distance [4,5], and the volume-based SED approach [6–8]), the failure equation can be expressed as

$$h(K_{IV\rho}, K_{IIV\rho}, K_{Ic}, \sigma_t) = 0 \tag{29}$$

Obviously, the failure equation (29) is different from Eq.(20).

REFERENCES

1 Neuber, H. Zur Theorie der technischen Formzahl. *Forschg Ing-Wes* 7, 271–281 (1936).
2 Neuber, H. *Theory of notch stresses: Principles for exact calculation of strength with reference to structural form and material.* 2nd ed. Berlin: Springer Verlag; 1958.
3 Peterson, R.E. Notch sensitivity. In: Sines, G., Waisman, J.L., editors. *Metal fatigue.* New York, USA: McGraw Hill; 1959. p. 293–306.
4 Taylor, D. *The theory of critical distances: A new perspective in fracture mechanics.* Oxford, UK: Elsevier; 2007.
5 Susmel, L., Taylor, D. The theory of critical distances to estimate the static strength of notched samples of Al6082 loaded in combined tension and torsion. *Part II: Multiaxial Static Assessment, Engineering Fracture Mechanics* 77: 470–478 (2010).
6 Berto, F., Lazzarin, P. Recent developments in brittle and quasi-brittle failure assessment of engineering materials by means of local approaches. *Materials Science and Engineering R* 75, 1–48 (2014).
7 Lazzarin, P., Zambardi, R. A finite-volume-energy based approach to predict the static and fatigue behaviour of components with sharp V-shaped notches. *International Journal of Fracture* 112, 275–298 (2001).
8 Lazzarin, P., Berto, F. Some expressions for the strain energy in a finite volume surrounding the root of Blunt V-notches. *International Journal of Fracture* 13, 161–185 (2005).

9 Erdogan, F., Sih, G. C. On the crack extension in plates under plane loading and transverse shear. *J. Basic Engng* 85, 519–527 (1963).
10 Sih, G.C. Strain-energy-density factor applied to mixed mode crack problems. *Int. J. Fracture* 10, 305–321 (1974). doi:10.1007/BF00035493
11 Saboori, B., Ayatollahi, M.R., Torabi, A.R., Berto, F. Mixed mode I/III brittle fracture in round-tip V-notches. *Theoretical and Applied Fracture Mechanics* 83, 135–151 (2016). doi:10.1016/j.tafmec.2015.12.002.
12 Torabi, A.R., Rahimi, A.S., Ayatollahi, M.R. Mixed mode I/II fracture prediction of blunt V-notched nanocomposite specimens with nonlinear behavior by means of the equivalent material concept. *Composites Part B: Engineering* 154, 363–373 (2018). doi:10.1016/j.compositesb.2018.09.025
13 Ayatollahi, M.R., Torabi, A.R. A criterion for brittle fracture in U-notched components under mixed mode loading. *Engineering Fracture Mechanics* 76(12), 1883–1896 (2009). doi:10.1016/j.engfracmech.2009.04.008
14 Torabi, A.R., Habibi, R. Investigation of ductile rupture in U-notched Al 6061-T6 plates under mixed mode loading. *Fatigue & Fracture of Engineering Materials & Structures* 39(5), 551–565 (2016).doi:10.1111/ffe.12376
15 Torabi, A.R., Kamyab, M. The fictitious material concept. *Engineering Fracture Mechanics* 209, 17–31 (2019).doi:10.1016/j.engfracmech.2019.01.022
16 Ayatollahi, M.R., Torabi, A.R. Investigation of mixed mode brittle fracture in rounded-tip V-notched components. *Engineering Fracture Mechanics* 77, 3087–3104 (2010).
17 Gao, H., Wang, Z.Q., Yang, C.S., Zhou A.H. An investigation on the brittle fracture of K_I-K_{II} composite mode cracks. *ACTA METALLURGICA SINICA* 15(3), 380–391 (1979).
18 Liu, B.W., Yan, X.Q. A multiaxial fatigue limit prediction equation for metallic materials. *ASME Journal of Pressure Vessel Technology* 142, 034501 (2020).
19 Liu, B.W., Yan, X.Q. A new model of multiaxial fatigue life prediction with influence of different mean stresses. *Int. J. Damage Mech.* 28(9), 1323–1343 (2019).
20 Yan, X.Q. An empirical fracture equation of mixed mode cracks. *Theoretical and Applied Fracture Mechanics* 116, 103146 (2021).
21 Yan, X.Q. A local approach for fracture analysis of V-notch specimens under mode-I loading. *Engineering Fracture Mechanics* 274, 108753 (2022).
22 Berto, F., Lazzarin, P., Gómez, F.J., Elices, M. Fracture assessment of U-notches under mixed mode loading: two procedures based on the 'equivalent local mode I' concept. *Int J Fract* 148, 415–433 (2007).
23 Berto, F., Lazzarin, P., Marangon, C. Brittle fracture of U-notched graphite plates under mixed mode loading. *Materials and Design* 41, 421–432 (2012).
24 Dunn, M.A., Suwito, W., Cunningham, S., May, C.W. Fracture initiation at sharp notches under mode I, mode II, and mild mixed mode loading. *International Journal of Fracture* 84, 367–381 (1997).

8 A New Type of *S-N* Equation and Its Application to Multiaxial Fatigue Life Prediction

8.1 INTRODUCTION

Many critical mechanical components experience multiaxial cyclic loading during their service life, such as railroad wheels, crankshafts, axles, turbine blades, etc. Different from the uniaxial fatigue problem, the multiaxial fatigue problem is more complex due to the complex stress states. In recent decades, a significant amount of research has been devoted to acquire a better understanding of the failure mechanisms under multiaxial loading [1–32], including theoretical model study (e.g., a stress invariant method [27,28], and a critical plane method [13,29,30]), fatigue tests of metallic materials under multiaxial loading (e.g., fatigue test for engineering steels by Gough [4], and fatigue of wrought high-tensile alloy steel by Frith [7]).

As is well known to us, the *S-N* equation is one of the most important equations in fatigue model investigation. A majority of fatigue models (e.g., [1,2,27–30]), including multiaxial fatigue models and mean effect models, are established on the basis of the *S-N* equation. Obviously, the accuracy of the *S-N* equation is very important. Taking into account that the *S-N* equation is, in fact, an empirical one in which the material constants are determined by fitting fatigue experimental data, in this study, the *S-N* equation can be improved, by further processing these fatigue experimental data, to present a new type of *S-N* equation that is more accurate than the *S-N* equation. The new type of *S-N* equation is called a similar *S-N* equation in this study. By using a large number of experimental data of metallic materials reported in the literature, an accuracy of the similar *S-N* equation has been proven.

8.2 A BRIEF DESCRIPTION OF A MULTIAXIAL FATIGUE MODEL

In order to illustrate the similar *S-N* curve concept presented in this study, in this section, a multiaxial fatigue limit model proposed recently by Liu and Yan [2] is described briefly, which is based on the invariant model of fatigue life prediction under multiaxial loading by Liu and Yan [1].

Under the multiaxial fatigue loading ratio $R = -1$, the invariant model of fatigue life prediction can be expressed as

$$\log \sigma_{ea} = A_p \log N + C_p \tag{1}$$

DOI: 10.1201/9781003356721-8

where σ_e is a mechanical parameter which is a measure of stress states under multiaxial loading, and here, the von Mises equivalent stress is adopted; σ_{ea} is the amplitude of the mechanical parameter, and ρ is a multiaxial parameter, defined as follows:

$$\rho = \frac{\sigma_{11,a}}{\sigma_{e,a}} \tag{2}$$

where $\sigma_{11,a}$ is the amplitude of the first invariant of stress tensor. It is evident that for the axial and shear fatigue conditions, the values of multiaxial parameter ρ are equal to 1 and 0, respectively. A_ρ, C_ρ in Eq. (1) are material parameters, which appear to be constants. In fact, they are varied with the multiaxial parameter ρ.

Under the axial and shear fatigue loading, Eq.1 can be simplified as

$$\log \sigma_a = A_1 \log N + C_1 \tag{3}$$

and

$$\log(\sqrt{3}\tau_a) = A_0 \log N + C_0 \tag{4}$$

In view of the complexity of multiaxial fatigue life analysis, and at the same time, taking into account that the existing literature has accumulated a large number of fatigue experimental data under the axial and shear loadings, from the point of application, it is assumed that the material parameters in Eq. (1) can be obtained by interpolating the material parameters in Eqs. (3) and (4), i.e.:

$$A_\rho = A_1 \cdot \rho + A_0 \cdot (1 - \rho) \tag{5}$$
$$C_\rho = C_1 \cdot \rho + C_0 \cdot (1 - \rho) \tag{6}$$

In this way, the multiaxial fatigue life of metal materials under the loading ratio $R = -1$ can be predicted by combining the mathematical invariant Eq.(1) with the fatigue experimental data under the axial and shear fatigue loading. The validation of Eq.(1) was verified by Liu and Yan [1] by using a large number of multiaxial fatigue experimental data reported in the literature.

If the applied stress σ_a, under the axial fatigue loading with loading ratio $R = -1$, is lower than a certain stress amplitude, σ_o, then failure should not occur up to a number of cycles (for example N_f) to failure theoretically equal to infinity; such a reference threshold is named the fatigue limit. Under the shear fatigue loading, a similar definition is a material shear fatigue limit, τ_o.

If a multiaxial fatigue limit is here denoted by σ_{eo}, the following equations can be obtained from Eqs. (1), (3) and (4):

$$\log \sigma_{eo} = A_\rho \log N_f + C_\rho \tag{7}$$
$$\log \sigma_o = A_1 \log N_f + C_1 \tag{8}$$
$$\log(\sqrt{3}\tau_o) = A_0 \log N_f + C_0 \tag{9}$$

By substituting Eqs. (5) and (6) into Eq. (7) and by using Eqs. (8) and (9), we can obtain

$$\begin{aligned}\log\sigma_{eo} &= (A_1\cdot\rho + A_0\cdot(1-\rho))\log N_f + C_1\cdot\rho + C_0\cdot(1-\rho)\\ &= (A_1\log N_f + C_1)\rho + (A_0\log N_f + C_0)(1-\rho)\\ &= \rho\log\sigma_o + (1-\rho)\log\left(\sqrt{3}\tau_o\right)\\ &= \log\left(\sigma_o^{\ \rho}\left(\sqrt{3}\tau_o\right)^{(1-\rho)}\right)\end{aligned}$$

Thus further we can obtain

$$\sigma_{eo} = \sigma_o^{\ \rho}\left(\sqrt{3}\tau_o\right)^{(1-\rho)} \tag{10}$$

which is an equation used to predict the multiaxial fatigue limit of metallic materials, σ_{eo}, by using a tensile fatigue limit σ_o, a shear fatigue limit, τ_o, and multiaxial parameter ρ.

Based on a large number of fatigue limit experimental data of various metallic materials reported in the literature, the multiaxial fatigue limit prediction equation (10) was verified by Liu and Yan [2] to be simple in computation and high in accuracy.

8.3 A NEW TYPE OF *S-N* EQUATION

As is well known to us, the *S-N* equation such as Eq. (3) is one of the most important equations in fatigue model investigation. A majority of fatigue models (e.g., [1,2,27–30]), including multiaxial fatigue models and mean effect models, are established on the basis of the *S-N* equation. Obviously, the accuracy of the *S-N* equation is very important. Taking into account that the *S-N* equation is, in fact, an empirical one in which the material constants (e.g., A_1 and C_1 in Eq. (3)) are determined by numerical fitting fatigue experimental data, in this study, the *S-N* equation can be improved, by further processing these fatigue experimental data, to present a new type of *S-N* equation that is more accurate than the *S-N* equation when used to perform the fatigue life assessment.

Assume that there are M sets of fatigue experimental data under the axial fatigue loading ratio $R = -1$, in which the axial loading amplitude and the corresponding fatigue life are

$$\{\sigma_a\}_i, \{N\}_i, \qquad (i = 1,2,\ldots,M) \tag{11}$$

After the material constants, A_1 and C_1, in Eq. (3) are determining by numerical fitting M sets of fatigue experimental data, the following M sets of data can be obtained

$$\{\sigma_a\}_i, \{\bar{\sigma}_a\}_i, \{N\}_i, \qquad (i = 1,2,\ldots,M) \tag{12}$$

where

$$\log\{\bar{\sigma}_a\}_i = A_1 \log\{N\}_i + C_1, \qquad (i = 1,2,\ldots,M) \tag{13}$$

Here, $\{\bar{\sigma}_a\}_i$ has the same unit as $\{\sigma_a\}_i$, and is called a fictitious stress. Assume that the following relation between $\{\bar{\sigma}_a\}_i$ and $\{\sigma_a\}_i$ exists

$$\frac{\bar{\sigma}_a}{\sigma_a} = \alpha_1 + \beta_1 \cdot \frac{\sigma_a}{\sigma_0} \tag{14}$$

where α_1 and β_1 are numerical fitting constants between $\left\{\frac{\bar{\sigma}_a}{\sigma_a}\right\}_i$ and $\left\{\frac{\sigma_a}{\sigma_0}\right\}(i = 1,2,\ldots,M)$.

In this study, the following equation

$$\log\sigma_a\left(\alpha_1 + \beta_1 \cdot \frac{\sigma_a}{\sigma_0}\right) = A_1 \log N + C_1 \tag{15}$$

will be used to replace Eq. (3) in performing fatigue life assessment. Eq. (15) is called a similar *S-N* equation.

For the shear fatigue, an equation similar to Eq. (15) can be described as follows:

Assume that there are L sets of fatigue experimental data under the shear fatigue loading ($R = -1$), in which the shear loading amplitude and the corresponding fatigue life are

$$\{\tau_a\}_i, \{N\}_i, \qquad (i = 1,2,\ldots,L) \tag{16}$$

After determining the material constants, A_0 and C_0, in Eq. (4) by using L sets of fatigue experimental data, the following L sets of data can be obtained

$$\{\tau_a\}_i, \{\bar{\tau}_a\}_i, \{N\}_i, \qquad (i = 1,2,\ldots,L) \tag{17}$$

where

$$\log\left\{\sqrt{3}\bar{\tau}_a\right\}_i = A_0 \log\{N\}_i + C_0, \qquad (i = 1,2,\ldots,L) \tag{18}$$

Here, $\{\bar{\tau}_a\}_i$ has the same unit as $\{\tau_a\}_i$, and is also called here a fictitious stress. Assume that the following relation between $\{\bar{\tau}_a\}_i$ and $\{\tau_a\}_i$ exists

$$\frac{\bar{\tau}_a}{\tau_a} = \alpha_0 + \beta_0 \cdot \frac{\tau_a}{\tau_0} \tag{19}$$

where α_0 and β_0 are numerical fitting constants between $\left(\frac{\bar{\tau}_a}{\tau_a}\right)_i$ and $\left\{\frac{\tau_a}{\tau_0}\right\}_i$, $(i = 1,2,\ldots,L)$.

In this study, the following equation

$$\log\sqrt{3}\tau_a\left(\alpha_0+\beta_0\cdot\frac{\tau_a}{\tau_0}\right)=A_0\log N+C_0 \tag{20}$$

will be used to replace Eq. (4) in performing fatigue life assessment. Eq. (20) is a similar *S-N* equation under the shear fatigue loading.

After establishing the similar *S-N* equations (15) and (20) under the axial and shear fatigue loading, the multiaxial fatigue life model can be written as

$$\log\sigma_{ea}\left(\alpha_\rho+\beta_\rho\frac{\sigma_{ea}}{\sigma_0^\rho(\sqrt{3}\tau_0)^{1-\rho}}\right)=A_\rho\log N+C_\rho \tag{21}$$

where α_ρ and β_ρ are computed by means of the following interpolating functions:

$$\alpha_\rho=\alpha_1\cdot\rho+\alpha_0\cdot(1-\rho) \tag{22}$$
$$\beta_\rho=\beta_1\cdot\rho+\beta_0\cdot(1-\rho) \tag{23}$$

From Eqs. (21), (22) and (23), it can be seen that, for the axial and shear fatigue loading, Eq. (21) is simplified as Eq. (15) and Eq. (20), respectively.

8.4 EXPERIMENTAL VERIFICATIONS AND DISCUSSIONS

Based on the experimental fatigue data of metallic materials from the literature, in this section, fatigue life analysis will be carried out by using the similar *S-N* equations (15), (20) and (21) presented in this study. At the same time, the corresponding fatigue life analysis will also be performed by means of the *S-N* equations (3), (4) and (1). The comparison of two results will be given to illustrate the accuracy of the similar *S-N* equations.

In order to quantitatively evaluate the accuracy of the fatigue prediction, the following error indexes are defined:

$$ER1=\frac{(N_{cal}-N_{\exp})\times 100}{N_{\exp}} \tag{24}$$

$$ER2=\frac{(N'_{cal}-N_{\exp})\times 100}{N_{\exp}} \tag{25}$$

$$MER1=\frac{1}{n}\sum_{i=1}^{n}(ER1)_i \tag{26}$$

$$MER2=\frac{1}{n}\sum_{i=1}^{n}(ER2)_i \tag{27}$$

where n is the number of experimental cases, $N_{\exp}$ is experimental fatigue life; N'_{cal} and N_{cal} are computed fatigue lives, by the similar *S-N* equation and *S-N* equation, respectively.

As is well known to us, there are unavoidably some *unusual experimental data* in experimental data of a physical problem, e.g., fatigue experiment data reported in the literature, while the *unusual experimental data* usually cover the law of the studied problem. Thus the *unusual* experimental data should be eliminated in the handling of the experimental data. Concerning the fatigue life experimental data of metallic materials handled here, obviously, such experimental data with **untimely fatigue failure** are the *unusual experimental data*, which will be eliminated in the processing of experimental fatigue data in this study. The *unusual experimental data* are abbreviated to *unusual data* in this study.

In the following, you will see that, in the processing of experimental fatigue data, eliminating the *unusual data* will promote obviously the accuracy of the fatigue life prediction equation obtained. This illustrates that eliminating the *unusual data* is necessary to obtain the accurate fatigue life prediction equation. Please note that data handled here are experimental data; from mathematical statistics, the *unusual data* in the experimental data should be eliminated to obtain the law of the studied problem. Of course, this is only one of many forms of handling the *unusual data.* It is important to analyze the reason why the *unusual data* occur, which is not concerned with this study.

In this section, the fatigue life prediction of five metallic materials, 1045 steel, AlCu4Mg1, SM45C, Z12CNDV12-2, and SAE1045 under uniaxial and multiaxial loading are carried out. For the sake of clear discussions, they are described in the forms of examples, respectively. The material constants determined by using the experimental fatigue data of these materials are listed in Tables 8.1 to 8.2.

TABLE 8.1
The Material Constants in the Fatigue Model Under the Axial Fatigue Loading

Materials	A_1	C_1	α_1	β_1	σ_0
1045 steel	−0.02122	2.58948	1.22159	−0.20992	280
AlCu4Mg1	−0.10085	2.87014	1.23286	−0.17756	153.6
SM45C	−0.09431	3.01046	1.03048	−0.02462	278.3
Z12CNDV12-2	−0.08854	3.17482	1.00364	−0.00217	440
SAE 1045	−0.09922	3.04875	1.09882	−0.08044	284

TABLE 8.2
The Material Constants in the Fatigue Model Under the Shear Fatigue Loading

Materials	A_0	C_0	α_0	β_0	τ_0
1045 steel	−0.02529	2.63105	1.19728	−0.18939	167
AlCu4Mg1	−0.12717	3.03093	1.13073	−0.10133	85.2
SM45C	−0.04841	2.87184	1.10336	−0.09171	220.2
Z12CNDV12-2	−0.05814	3.07799	1.00963	−0.00438	193
SAE 1045	−0.0537	2.84618	1.00008	−0.0001	193

Note here σ_0 in (14) and τ_0 in (19) are the axial and shear fatigue limits, respectively, which can be evaluated by means of Eqs. (8) and (9). For example, for 1045 steel, let N_f = 5000000 (cycles), and the following two values can be obtained

$$\sigma_0 = 280 \text{ (MPa)}, \ \tau_0 = 167 \text{ (MPa)}$$

Example 1: 1045 Steel

For 1045 steel reported by Verreman and Guo [33], predicted results of fatigue life are given in Table 8.3 under the axial, shear and multiaxial fatigue loading.

Under the axial fatigue loading, the error ranges of fatigue life, *ER*1 and *ER*2, are, respectively, [−46,155] (%) and [−35,137] (%), which illustrates that the predictions by the similar *S-N* equation (15) are better than those by the *S-N* equation (3).

Under the shear fatigue loading, the error ranges of fatigue life, *ER*1 and *ER*2, are, respectively, [−57,165] (%) and [−65,151] (%), from which the advantage of the similar *S-N* equation (20) over the *S-N* equation (4) cannot be seen. But the mean errors of fatigue life, *MER*1 and *MER*2 are, respectively, 26% and 18%, which shows that the predicted results by the similar *S-N* equation (20) are better than those by the *S-N* equation (4).

Under the multiaxial fatigue loading, the error ranges of fatigue life, *ER*1 and *ER*2, are, respectively, [123,623] (%) and [44,367] (%), from which it can be seen that the predicted results by the similar *S-N* equation (21) are much better than those by the *S-N* equation (1).

TABLE 8.3
Predicted Results of Fatigue Life of 1045 Steel

σ_a	τ_a	N_{exp}	*ER*2 (%)	*ER*1 (%)
0	166.5	4917324	−27	18
0	180	497000	−35	−46
0	180	440920	−27	−39
0	175	317943	137	155
0	180	257679	23	3
275	0	5417503	29	119
285	0	5148389	−65	−57
300	0	379178	−32	−47
290	0	366512	151	165
300	0	238236	7	−17
300	0	214908	19	−8
220	110	573978	44	123
220	110	177263	367	623

Example 2: AlCu4Mg1

For AlCu4Mg1 reported in [34], predicted results of fatigue life under the axial fatigue loading are listed in Table 8.4. The error ranges of fatigue life, *ER*1 and *ER*2, are, respectively, [−68,221] (%) and [−67,145] (%), which illustrates that the predicted results by the similar *S-N* equation (15) are better than those by the *S-N* equation (3).

Under the shear fatigue loading, predicted results of fatigue life are given in Table 8.5. The error ranges of fatigue life, *ER*1 and *ER*2, are, respectively, [−72,326] (%) and [−69,231] (%), which shows that the predicted results by the similar *S-N* equation (20) are better than those by the *S-N* equation (4).

Under the multiaxial fatigue loading, predicted results of fatigue life are shown in Table 8.6. The error ranges of fatigue life, *ER*1 and *ER*2, are, respectively, [−69,104] (%) and [−67,33] (%), from which it can be seen that the predicted results by the similar *S-N* equation (21) are much better than those by the *S-N* equation (1).

Example 3: SM45C

For SM45C reported by Lee [35], predicted results of fatigue life are given in Table 8.7 under the axial, shear and multiaxial fatigue loading.

TABLE 8.4
Predicted Results of Fatigue Life of AlCu4Mg1 Under the Axial Fatigue Loading

σ_a	N_{exp}	*ER*2 (%)	*ER*1 (%)
242.5	63602	65	2
226.4	97373	76	32
227.8	157229	4	−23
242.8	163634	−36	−61
242.8	202469	−48	−68
226.7	222239	−23	−43
204.9	253878	41	36
192.3	331307	76	96
177.6	444016	145	222
203.9	438146	−14	−17
192.4	462104	25	40
204	520910	−28	−31
177.7	679782	59	109
178.7	776557	33	73
168.6	935617	76	156
191.6	1490769	−59	−55
169.6	1971504	−20	14
168.9	4939029	−67	−52
155.2	6276071	−48	−13

TABLE 8.5
Predicted Results of Fatigue Life of AlCu4Mg1 Under the Shear Fatigue Loading

τ_a	N_{exp}	*ER*2 (%)	*ER*1 (%)
137.4	108364	68	29
138.2	127248	37	5
137.4	137891	32	1
138.2	143541	22	−7
138.2	145476	20	−8
128.9	180225	53	28
128.9	262183	6	−12
128.9	829096	−67	−72
102.4	1494234	−12	−6
93.5	1778279	39	62
82.7	1778279	232	326
102.4	2841112	−54	−50
102.4	3291864	−60	−57
83.2	4078168	39	77
93	5548634	−54	−46
93	8515939	−70	−65
83.2	8864912	−36	−18
75.9	8402695	30	77
75.9	8071918	35	84
75.9	6873996	58	116

TABLE 8.6
Predicted Results of Fatigue Life of AlCu4Mg1 Under the Multiaxial Fatigue Loading

σ_a	τ_a	N_{exp}	*ER*2 (%)	*ER*1 (%)
170.5	85.2	358127	−50	−57
169.6	84.8	529217	−65	−70
168.7	84.3	558510	−65	−70
161.6	80.8	751157	−65	−66
160.8	80.4	814374	−66	−68
160.8	80.4	847946	−68	−69
146.9	73.4	792909	−32	−22
146.1	73	1220009	−54	−47
145.3	72.6	1802849	−68	−62
132.7	66.4	1270596	−8	24
132.7	66.4	1396186	−17	13
133.4	66.7	3048729	−63	−51
121.9	60.9	1709038	33	105

TABLE 8.7
Experimental Data and Predicted Results of Fatigue Life of SM45C

σ_a	τ_a	N_{exp}	*ER*2 (%)	*ER*1 (%)
413.4	0	14921	8	1
389.1	0	25972	15	10
373.5	0	52288	−13	−15
366.8	0	73480	−25	−27
354.1	0	91209	−14	−15
337.3	0	101076	28	29
324.1	0	163940	19	22
314.9	0	210131	25	29
313	0	323068	−13	−11
295.2	0	446301	15	20
294.2	0	714643	−26	−22
0	280.4	10370	−13	−35
0	257.7	19918	112	95
0	270.2	23486	−25	−38
0	256.8	30169	49	38
0	249.5	108937	−30	−30
0	231.8	140230	115	147
0	247.7	163940	−46	−46
0	233.1	332893	−19	−7
0	227.9	402734	3	22
0	220.4	1127312	−31	−13
390	151	8500	−39	−48
349	148	24000	−45	−52
325	153	32000	−39	−45
372	93	38000	−41	−45
309	134	100000	−53	−56

Under the axial fatigue loading, the error ranges of fatigue life, *ER*1 and , are, respectively, [−27,29] (%) and [−25,28] (%), which illustrates that the predicted results by the similar *S-N* equation (15) almost are the same as those by the *S-N* equation (3). Here, *R*-square of the *S-N* fitting curve is 0.966 (see the comparison of experimental data and the *S-N* curve shown in Figure 8.1).

Under the shear fatigue loading, the error ranges of fatigue life, *ER*1 and *ER*2, are, respectively, [−46, 147] (%) and [−46,115] (%), from which it can be seen that the predicted results by the similar *S-N* equation (20) are a little better than those by the *S-N* equation (4). Here, *R*-square of the *S-N* fitting curve is 0.886 (see the comparison of experimental data and the *S-N* curve shown in Figure 8.2).

Under the multiaxial fatigue loading, the error ranges, *ER*1 and *ER*2, are, respectively, [−56,−45] (%) and [−53,−39] (%); and the mean errors, *MER*1 and *MER*2, are, respectively, −49% and −43%, from which it can be seen that the predicted results by the similar *S-N* equation (21) are a little better than those by the *S-N* equation (1) (also see Figure 8.3).

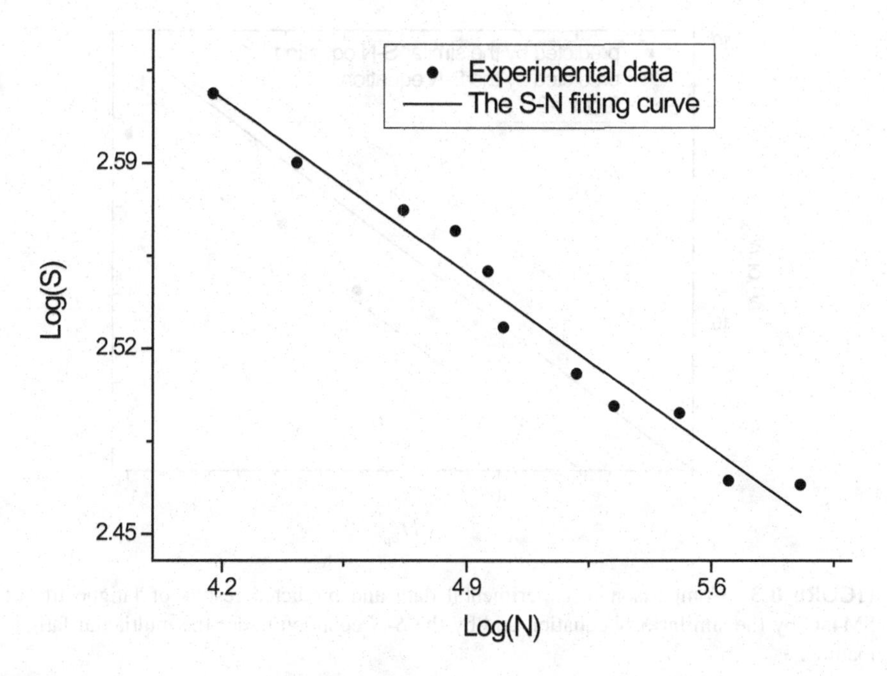

FIGURE 8.1 Comparison of experimental data and the *S-N* curve of SM45C under the axial fatigue loading.

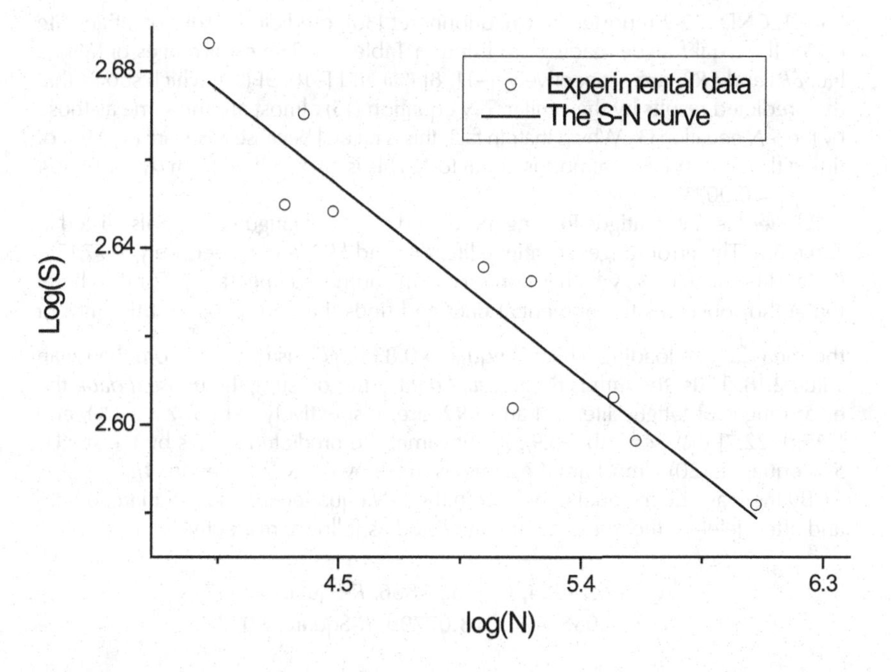

FIGURE 8.2 Comparison of experimental data and the *S-N* curve of SM45C under the shear fatigue loading.

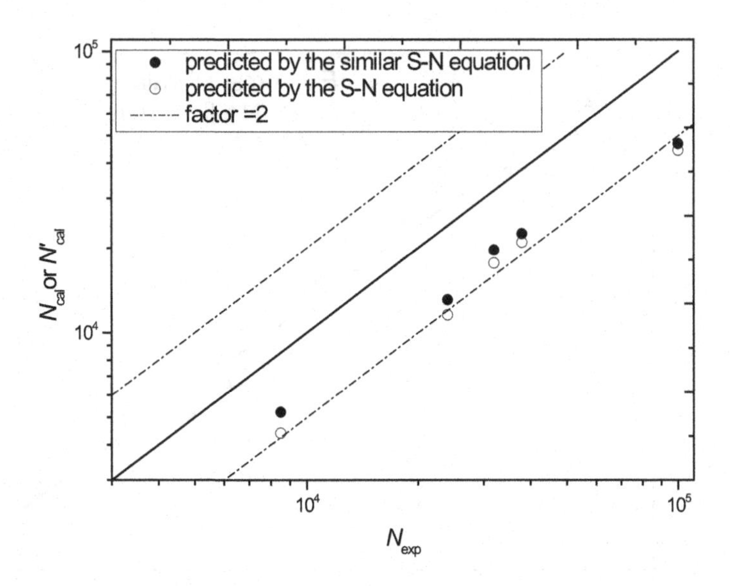

FIGURE 8.3 Comparison of experimental data and predicted results of fatigue life of SM45C by the similar *S-N* equation and by the *S-N* equation under the multiaxial fatigue loading.

Example 4: Z12CNDV12-2

For Z12CNDV12-2 reported by Chaudonneret [36], predicted results of fatigue life under the axial fatigue loading are listed in Table 8.8. The error ranges of fatigue life, *ER*1 and *ER*2, are, respectively, [−11, 8] (%) and [−10, 9] (%), which shows that the predicted results by the similar *S-N* equation (15) almost are the same as those by the *S-N* equation (3). Why is that? In fact, this is natural because *R*-Square (= 0.99) of fitting the *S-N* equation almost is equal to 1. This is also verified from $\alpha_1 = 1.00364$ and $\beta_1 = 0.00217$.

Under the shear fatigue loading, predicted results of fatigue life are also listed in Table 8.8. The error ranges of fatigue life, *ER*1 and *ER*2, are, respectively, [−77,155] (%) and [−78,192] (%), which is contrary to the author's expectancy. For this, here, the author observes the numerical data and finds that the ratio $\frac{\tau_a}{\tau_0}$ at the fifth of the shear fatigue loading which is equal to 0.83 largely is different from the mean value 2.18. Thus, the fifth is the *unusual data*. After deleting the *unusual data*, the error ranges of fatigue life, *ER*1 and *ER*2, are, respectively [−15.3, 23.3] (%), and [−15.0, 22.7] (%) (see Table 8.9). At this time, the predicted results by the similar *S-N* equation (20) almost are the same as those by the *S-N* equation (4).

By the way, the material constants in the *S-N* equation (4) and *R*-Square, before and after deleting the *unusual data*, are listed as follows, respectively,

$$A_0 = -0.14254,\ C_0 = 3.37886,\ R\text{-Square} = 0.87$$
$$A_0 = 0.05814,\ C_0 = 3.07799,\ R\text{-Square} = 0.99$$

Under the multiaxial fatigue loading, predicted results after deleting the *unusual data* are given in Table 8.10. The error ranges of fatigue life, *ER*1 and *ER*2, are,

TABLE 8.8
Experimental Data and Predicted Results of Fatigue Life of Z12CNDV12-2

σ_a	τ_a	N_{exp}	$ER2$ (%)	$ER1$ (%)	$\frac{\sigma_a}{\sigma_0}$ or $\frac{\tau_a}{\tau_0}$
870	0	477	−4	−5	1.977273
775	0	1556	8	8	1.761364
641	0	15951	−11	−10	1.456818
627	0	16929	8	9	1.425
0	470	754	192	155	2.437317
0	432	3800	−2	−8	2.240257
0	410	6415	−20	−22	2.12617
0	373	42070	−78	−77	1.934296
0	160	1570000	54	135	0.829725

TABLE 8.9
Experimental Data and Predicted Results of Fatigue Life of Z12CNDV12-2 After Deleting the *Unusual Data* (Shear Fatigue Loading)

τ_a	N_{exp}	$ER2$ (%)	$ER1$ (%)
470	754	2	0
432	3800	−15	−15
410	6415	23	23
373	42070	−6	−4

TABLE 8.10
Experimental Data and Predicted Results of Fatigue Life of Z12CNDV12-2 After Deleting the *Unusual Data* (Multiaxial Fatigue Loading)

σ_a	τ_a	N_{exp}	$ER2$ (%)	$ER1$ (%)
463	263	21673	−29.3	−28.5
556	312	1732	−2.7	−2.7

respectively, [−28.5, −2.7] (%) and [−29.3, −2.7] (%), from which it can be seen that the predicted results by the similar *S-N* equation (21) almost are the same as those by the *S-N* equation (1). This result is natural because, under the axial fatigue loading, *R*-Square of fitting the *S-N* equation (3) almost is equal to 1, and under the shear fatigue loading, *R*-Square of fitting the *S-N* equation (4) almost is also equal to 1 after deleting the *unusual data*.

If the fifth of the shear fatigue data is not deleted, fatigue lives predicted by means of the *S-N* equation (1) are given in Table 8.11. The error range under the shear fatigue loading is [−77, 155] (%), which is much more than that ([−15.3, 23.3] (%)) after deleting the *unusual data*. The error range under the multiaxial fatigue

TABLE 8.11
Experimental Data and Predicted Results of Fatigue Life of Z12CNDV12-2 by the *S-N* Equation at Not Deleting the *Unusual Data*

σ_a	τ_a	N_{exp}	*ER*1(%)
0	470	754	155
0	432	3800	−8
0	410	6415	−22
0	373	42070	−77
0	160	1570000	135
463	263	21673	−49
556	312	1732	16

loading is [−49,16] (%), which is also much more than that ([−28.5, −2.7] (%)) after deleting the *unusual data*.

Example 5: SAE 1045

For SAE 1045 reported in [37], predicted results of fatigue life under the axial fatigue loading are listed in Table 8.11. The error ranges of fatigue life, *ER*1 and *ER*2, are, respectively, [−38,82] (%) and [−38,63] (%), which shows that the predicted results by the similar *S-N* equation (15) are a little better than those by the *S-N* equation (3).

Under the shear fatigue loading, predicted results of fatigue life are shown in Table 8.12. The error ranges of fatigue life, *ER*1 and *ER*2, are, respectively, [−73,295] (%) and [−77,76] (%). Here, that the minimum of *ER*2 is less than that of *ER*1 is contrary to the author's expectancy. For this, here, the author observes the numerical data and finds that the ratio $\frac{\tau_a}{\tau_0}$ at the fourth of the shear fatigue loading which is equal to $\underline{1.137697}$ largely is different from other values (see Table 8.12). Thus, the fourth is the *unusual data*. After deleting the *unusual data*, the error ranges of fatigue life, *ER*1 and *ER*2, are, respectively, [−2.6,2.4] (%), [−2.4,2.5] (%) (see Table 8.13). At this time, the predicted results by the similar *S-N* equation (20) almost are the same as those by the *S-N* equation (4).

By the way, the *material constants* in the *S-N* equation (4) *and R*-Square before and after deleting the *unusual data* are listed as follows, respectively,

$$A2 = -0.09394,\ C2 = 3.01734,\ R\text{-Square} = 0.56;$$
$$A2 = -0.0537,\ C2 = 2.84618,\ R\text{-Square} = 0.99.$$

Under the multiaxial fatigue loading, predicted results after deleting the *unusual data* are given in Table 8.14. The error ranges of fatigue life, *ER*1 and *ER*2, are, respectively, [−51, 2] (%) and [−36,−6] (%), which shows that the predicted results by the similar *S-N* equation (21) are a little better than those by the *S-N* equation (1).

TABLE 8.11
Experimental Data and Predicted Results of Fatigue Life of SAE 1045 Under the Axial Fatigue Loading

σ_a	N_{exp}	ER2 (%)	ER1 (%)
413.8	22800	19	−1
350.1	121000	1	1
350.1	196700	−38	−38
310.4	225000	63	82
310.4	459800	−20	−11

TABLE 8.12
Experimental Data and Predicted Results of Fatigue Life of SAE 1045 Under the Shear Fatigue Loading

τ_a	N_{exp}	ER2 (%)	ER1 (%)	$\frac{\tau_a}{\tau_0}$
238.7	19500	44	−5	0.995225
238.7	18550	52	0	0.999905
202.9	391900	−77	−73	0.883231
159.2	350000	76	295	1.137697

TABLE 8.13
Experimental Data and Predicted Results of Fatigue Life of SAE 1045 After Deleting the Unusual Data (Shear Fatigue Loading)

τ_a	N_{exp}	ER2 (%)	ER1 (%)
238.7	19500	−2.4	−2.6
238.7	18550	2.5	2.4
202.9	391900	−0.0	−0.1

TABLE 8.14
Experimental Data and Predicted Results of Fatigue Life of SAE 1045 After Deleting the Unusual Data (Multiaxial Fatigue Loading)

σ_a	τ_a	N_{exp}	ER2 (%)	ER1 (%)
318.3	216.5	5471	−36	−51
254.6	173.2	50660	−18	−22
254.6	173.2	55050	−24	−28
218	148.8	300000	−22	−16
218	148.8	280000	−16	−10
212	143.2	366400	−6	2
212	143.2	404900	−15	−7

8.5 CONCLUDING REMARKS

From the present study, the following conclusions can be made:

(1) When the numerical fitting R-square of the *S-N* equation is not near to 1 under the axial fatigue loading or under the shear fatigue loading or under the axial and shear fatigue loading, the predicted results by the similar *S-N* equation presented in this study are better than those by the *S-N* equation, including the multiaxial fatigue loading case.
(2) When the numerical fitting R-square of the *S-N* equation is very near to 1 under the axial and shear fatigue loading, the predicted results by the similar *S-N* equation almost are same as those by the *S-N* equation. At this time, it is not necessary for the similar *S-N* equation to replace the *S-N* equation in performing the fatigue life assessment.

By the way, it is pointed out that, by means of the similar *S-N* equation, the effects of the nonproportional loading and the mean stresses can be taken properly into account according to the approach in Ref [1] and are not discussed in this study.

Finally, a finding of this investigation is illustrated. In numerical fitting between $\left\{\frac{\bar{\sigma}_a}{\sigma_a}\right\}_i$ and $\left\{\frac{\sigma_a}{\sigma_0}\right\}_i$ $(i = 1,2,\ldots,M)$ or $\left(\frac{\bar{\tau}_a}{\tau_a}\right)_i$ and $\left\{\frac{\tau_a}{\tau_0}\right\}_i$, $(i = 1,2,\ldots,L)$, the *unusual data* can be easily found by observing $\left\{\frac{\sigma_a}{\sigma_0}\right\}_i$ $(i = 1,2,\ldots,M)$ or $\left\{\frac{\tau_a}{\tau_0}\right\}_i$, $(i = 1,2,\ldots,L)$. The *unusual data* here are the mathematical *outlier data*. The *unusual data* are deleted in the statistical processing of experimental data.

REFERENCES

1 Liu, B.W., Yan, X.Q. A new model of multiaxial fatigue life prediction with influence of different mean stresses. *International Journal of Damage Mechanics* 28(9), 1323–1343 (2009).
2 Liu, B.W., Yan, X.Q. A multiaxial fatigue limit prediction equation for metallic materials. To be published in *ASME Journal of Pressure Vessel Technology* (2020).
3 Susmel, L. *Multiaxial notch fatigue, from nominal to stress/strain quantities.* Cambridge, UK: Woodhead Publishing Limited, CRC Press; 2009.
4 Gough, H.J. Engineering steels under combined cyclic and static stresses. *Proceedings of the Institution of Mechanical Engineers* 160, 417–440 (1949).
5 Nishihara, T., Kawamoto, M. The strength of metals under combined alternating bending and torsion. *Memoirs of the College of Engineering, Kyoto Imperial University* 10, 177–201 (1941).
6 Kitaioka, S., Chen, J., Seika, M. The threshold of micro crack propagation under mixed mode. *Bulletin of the Japan Society of Mechanical Engineers* 29, 214–237 (1986).
7 Frith, P. H. Fatigue of wrought high-tensile alloy steel. *Proceedings of the Institution of Mechanical Engineers*, 462–499 (1956).
8 Nishihara, T., Kawamoto, M. The strength of metals under combined alternating bending and torsion with phase difference. *Memoirs of the College of Engineering, Kyoto Imperial University* 11, 85–112 (1945).

9 Achtelik, H., Jakubowska, I., Macha, E. Actual and estimated directions of fatigue fracture plane in ZI250 grey cast iron under combined alternating bending and torsion. *Studia Geotechnica et Mechanica* 5(2), 9–30 (1983).
10 Lempp, W. Festigkeitsverhalten von Stählen bei mehrachsiger Dauerschwingbeanspruchung durch Normalspannungen mit überlagerten phasengleichen und phasenverschobenen Schubspannungen. Dissertation, Universität Stuttgart, Germany; 1977.
11 Zenner, H., Heidenreich, R., Richter, I. Dauerschwingfestigkeit bei nichtsynchroner mehrachsiger Beanspruchung. *Zeitschrift für Werkstofftechnik* 16, 101–112 (1985).
12 Froeschl, J., Gerstmayr, G., Eichlseder, W., Leitner, H. Multiaxial fatigue of qt-steels: New fatigue strength criterion for anisotropic material behaviour. In: Fernando, U.S., editor. *Proceedings of 8th International Conference on Multiaxial Fatigue and Fracture*. Sheffield, UK: Sheffield Hallam University; 2007, Paper S3-1.
13 Matake, T. An explanation on fatigue limit under combined stress. *Bulletin of the JSME* 20(141), 257–263 (1977).
14 Altenbach, H., Zolochevsky, A. A generalised fatigue limit criterion and a unified theory of low-cycle fatigue damage. *Fatigue and Fracture of Engineering Materials and Structures* 19, 1207–1219 (1996).
15 Findley, W. N., Coleman, J. J., Hanley, B. C. Theory for combined bending and torsion fatigue with data for SAE 4340 steel. In: *Proceedings of International Conference on Fatigue of Metals*. London: Institution of Mechanical Engineers, London; 1956. p. 150–157.
16 Froustey, C. Fatigue multiaxiale en endurance de l'acier 30NCD16. PhD Thesis, Ecole Nationale Supérieure d'Arts et Métiers, Bordeaux, France; 1986.
17 Froustey, C., Lasserre, S. Multiaxial fatigue endurance of 30NCD16 steel. *International Journal of Fatigue* 11, 169–175 (1989).
18 Froustey, C., Lasserre, S., Dubar, L. Essais de fatigue multiaxiaux et par blocs. Validation d'un critère pour les matériaux métalliques. In: *Proceedings of METTECH 92*, Grenoble, France; 1992.
19 Fogué, M., Bahuaud, J. Fatigue multiaxiale à durée de vie illimitée. In: *Proceedings of Comptes Rendus 7ème Congrès Français de Mécanique*, Bordeaux, France; 1986. p. 30–31.
20 Delahay, T., Palin-Luc, T. Estimation of the fatigue strength distribution in high-cycle multiaxial fatigue taking into account the stress–strain gradient effect. *International Journal of Fatigue* 28, 474–484 (2005).
21 Palin-Luc, T., Lasserre, S. An energy based criterion for high cycle multiaxial fatigue. *European Journal of Mechanics—A/Solids* 17, 237–251 (1998).
22 Sonsino, C.M. Influence of load and deformation-controlled multiaxial tests on fatigue life to crack initiation. *International Journal of Fatigue* 23, 159–167 (2001).
23 Akrache, R., Lu, J. Three-dimensional calculations of high cycle fatigue life under out-of-phase multiaxial loading. *Fatigue and Fracture of Engineering Materials and Structures* 22, 527–534 (1999).
24 Gough, H.J., Pollard, H.V., Clenshaw, W.J. Some experiments on the resistance of metals to fatigue under combined stresses. Aeronautical Research Council, R and M 2522. HMSO, London; 1951.
25 Li, B.C., Jiang, C., Han, X., Li, Y. A new approach of fatigue life prediction for metallic materials under multiaxial loading. *International Journal of Fatigue* 78, 1–10 (2015).
26 Shamsaei, N., Fatemi, A., Socie, D.F. Multiaxial fatigue: an overview and same approximation models for life estimation. *Int J Fatigue* 33, 948–958 (2011).
27 Crossland B. Effect of large hydrostatic pressures on the torsional fatigue strength of an alloy steel. In: *Proceedings of the international conference on fatigue of metals.* London: The institution of Mechanical Engineers; 1956. p. 138–149.

28 Sines, G., Waisman, J.L., Dolan, T.J. *Metal fatigue*. London: McGraw-Hill; 1959.
29 Papadopoulos, I.V., Davoli, P., Gorla, C., Filippini, M., Bernasconi, A. A comparative study of multiaxial high-cycle fatigue criteria for metals. *Int J Fatigue* 19, 219–235 (1997).
30 Papadopoulos, I.V. Long life fatigue under multiaxial loading. *Int J Fatigue* 23, 839–849 (2001).
31 Gough, H.J., Pollard, H.V. Properties of some materials for cast crankshafts, with special reference to combined alternating stresses. *Proc Inst Automobile Eng* 31, 821–893 (1937).
32 Gough, H.J., Pollard, H.V. The strength of metals under combined alternating stress. *Proc Inst Mech Eng* 131, 3–18 (1935).
33 Verreman, Y., Guo, H. High-cycle fatigue mechanisms in 1045 steel under non-proportional axial-torsional loading. *Fatigue and Fracture of Engineering Materials and Structures* 30, 932–946 (2007). DOI: 10.1111/j.1460-2695.2007.01164.x.
34 Kardas, D., Kluger, K., Łagoda, T., Ogonowski, P. Fatigue life of AlCu4Mg1 aluminium alloy under constant amplitude bending with torsion. In: Sonsino, C.M., editor. *Proceedings of 7th International Conference on Biaxial and Multiaxial Fatigue and Fracture*. Berlin; 2004. p. 185–190.
35 Lee, S.B. A criterion for fully reversed out-of-phase torsion and bending. In: Miller, K.J., Brown, M.W., editors. *Multiaxial fatigue*. Philadelphia: ASTM STP 853; 1985. p. 553–568.
36 Chaudonneret, M. A simple and effi cient multiaxial fatigue damage model for engineering applications of macro-crack initiation. *Transactions of the ASME, Journal of Engineering Materials and Technology* 115, 373–379 (1993). DOI: 10.1115/1.2904232.
37 Kurath, P., Downing, S.D., Galliart, D.R. Summary of non-hardened notched shaft—round robin program. In: Leese, G.E., Socie, D.F., editors. *Multiaxial fatigue—analysis and experiments*. Warrendale, PA: SAE AE-14, Society of Automotive Engineers; 1989. p. 13–32.

Index

[illegible]inted in the United States
by Baker & Taylor [illegible]

Printed in the United States
by Baker & Taylor Publisher Services